T0231462

Design Criteria for Drill Rigs: Equipment and Drilling Techniques

By

C.P. Chugh, Ken Steele and V.M. Sharma

Taylor & Francis
Taylor & Francis Group

LONDON AND NEW YORK

Published by Taylor & Francis,
2 Park Square, Milton Park, Abingdon, Oxon, OX14 4RN
270 Madison Ave, New York NY 10016

Transferred to Digital Printing 2006

ISBN 90 5410 257 8

Publisher's Note

Printed and bound by CPI Antony Rowe, Eastbourne

PREFACE

Even though drilling is a field expertise it is necessary to have knowledge of the theoretical background in order to choose the correct path to be taken during concomitant exigencies while drilling operation is in progress.

Chapter one of the book is a reproduction of the Drilling Safety Guide of the International Drilling Federation for which permission was granted by the Federation. The first page of the Guide had to be slightly modified to maintain uniformity of style between the chapters.

The choice of method to be employed for the drilling operation is determined by the factors which affect the ultimate performance, the geology, the topography, accessibility of the site and the surrounding terrain, the intended depth of drilling, proximity to water, time frame restrictions due to weather access and myriad other factors which come into play in making such a judgement. Different drilling techniques have been discussed in chapter two.

The design of a drill machine is crucial in determining the depth rating, mast hookload capacity, drawworks capacity and pump capacity, aircompressor drilling programme diameter, horse power of the power unit, auxiliary equipment and details of operating componentry. The design aspects have been discused in detail in chapter three.

Geological formations vary considerably in different regions and so does the experience in use of prevailing technology. Over a period of time these experiences have contributed to various theories of rock mechanics, different classification system of rock mass, innovations in design and construction of support system and monitoring methods. These aspects have been discussed in chapter five.

Of the many facets of rock mechanics, the most important ones in the context of drilling are concerned with rock properties such as hardness and abrasiveness, rock structure and water flow. The theoretical aspects of rock drilling are of great value in understanding the process of rock drilling but of little help in predicting the output. Thus field experiences are very important and they have been discussed in chapter six and seven.

Since the commercial introduction of the down the hole (DTH) hammer in 1950's, the technique has developed into what is arguably the most widely used drilling system in use today. The hammer is used in mineral exploration drilling,

mining production, both surface and underground water well drilling, construction
for anchoring, piling, dewatering and oil and gas exploration. The potential uses
of this hammer have been amply illustrated in chapter eight and nine.

When considering deviation in a drill hole, it is probably more appropriate to
accept the generality that there is no such thing as a straight drill hole. Deviation
problems and their remedial corrective steps have been also discussed in chapter
11 to 13.

Other features of the book are lateral drilling system, horizontal and directional drilling of soils, modern survey support to drilling bore holes, geotechnical
investigation types and their economical aspects, design of foundations of dams
and work orders for drilling of exploration of bore holes.

It is hoped that this book will serve the needs of mining engineers, geologists
and geotechnical engineers.

December 1995   C.P. CHUGH
                                                          KEN STEELE
                                                          V.M. SHARMA

ACKNOWLEDGEMENTS

We are thankful to Superabrasive Engg., Pvt. Ltd., Rock Drill India, Pawan Automobiles, Atlas Copco, De Beers Industrial Division, Kores (India) Ltd., Mining Associates, Industrial Diamond Co. Ltd., London, Rawel Singh & Co., Shivalik Geotech Services, The Institution of Mining and Metallurgy, Baker Hughes INTEQ, Drillmark Consultants—West Australia, American Augers Inc., International Drilling Federation, AIMIL Sales & Agencies Pvt. Ltd., M/s. Geotech Consultants Pvt. Ltd., G.S. Jain & Associates Pvt. Ltd., Revathi C.P. Equipment Ltd., Chicago Pneumatic New York and George E. Failing Company for their assistance in providing textual material and photographs.

We are highly grateful to P.S. Misra, Anil Chowdhry, G.S. Jain, C.K. Jain for their token contributions to the text.

We shall be failing in our duty if we do not acknowledge the contributions of my many colleagues in different scientific institutions and technical organisations.

Ken Steele records separately his acknowledgements as follows for the services of his men and organisations in Australia "I wish to acknowledge with gratitude the assistance I have been given by the provision of material and illustrations which have been used in my chapters. The following companies and individuals have each assisted me in this way:

Ace Drilling, Canning Vale, Perth, Western Australia
Bulroc (UK) Ltd, Chesterfield, U.K
CBC Welnav, Tustin, USA
Downhole Surveys Pty Ltd, Kalgoorlie, Western Australia
Halliburton Energy Services, USA
International Drillquip Pty. Ltd. Perth Western Australia
Kennametallnt USA
Pontil Pty Ltd Dubbo NSW Australia
Reflex Instruments AB Sweden
Resource Review Perth Western Australia
SDS Digger Tools Perth Western Australia
Siesmic Supply International Pty. Ltd. Queensland Australia
Surtron Technologies Pty Ltd. Fremantle Western Australia

The pneumatic tool theoretical formulations were developed at the Colorado School of Mines USA. Originally published in the quarterly review, I regret that they are unascribed in my library notes".

C.P. Chugh would like to express his gratitute to his wife Pritma for her patient help in editing the manuscript and to Baby Philip for the pains he took in typing such difficult text material. He also records his appreciation of the assistance rendered by K.C. Pant and H.S. Bora in preparing the illustrations which were time consuming and complicated.

                                                        C P CHUGH
                                                        KEN STEELE
                                                        V.M. SHARMA

CONTENTS

CHAPTER 1

DRILLING SAFETY*

1. AN INTRODUCTION TO DRILLING SAFETY

The organization you work is interested in your safety, not only when you are working on or around a drill rig, but also when you are traveling to and from a drilling site, moving the drill rig and tools from location to location on a site or providing maintenance on a drill rig or drilling tools. This safety guide is for your benefit.

Every drill crew should have a designated safety supervisor. The safety supervisor should have the authority to enforce safety on the drilling site. A rig worker's first safety responsibility is to listen to the safety directions of the safety supervisor.

---

* This chapter is taken from *Drilling Safety Guide* published in 1985 by International Drilling Federation, 3008 Millwood Avenue, Columbia, with permission.

Fig. 1.1

2. GOVERNMENTAL REGULATIONS

All local, state and federal regulations or restrictions, currently in effect or effected in the future, take precedence over the recommendations and suggestions which follow. Government regulations will vary from country to country and from state to state.

3. THE SAFETY SUPERVISOR

The safety supervisor for the drill crew will in most cases be the drill rig operator.

• The safety supervisor should consider the "responsibility" for safety and the "authority" to enforce safety to be a matter of first importance.

• The safety supervisor should be the leader in using proper personal safety gear and set an example in following the rules being enforced on others.

• The safety supervisor should enforce the use of proper personal protective safety equipment and take appropriate corrective action when proper personal protective safety equipment is not being used.

• The safety supervisor should understand that proper maintenance of tools and equipment and general "housekeeping" on the drill rig will provide the environment to promote and enforce safety.

• Before drilling is started with a particular drill, the safety supervisor must be assured that the operator (who may be the safety supervisor) has had adequate training and is thoroughly familiar with the drill rig, its controls and its capabilities.

• The safety supervisor should inspect the drill rig at least daily for structural damage, loose bolts and nuts, proper tension in chain drives, loose or missing guards or protective covers, fluid leaks, damaged hoses and/or damaged pressure gauges and pressure relief valves.

• The safety supervisor should check and test all safety devices such as emergency shut-down switches at least daily and preferably at the start of a drilling shift. Drilling should not be permitted until all emergency shut-down and warning systems are working correctly. Do not wire around, bypass or remove an emergency device.

• The safety supervisor should check that all gauges, warning lights and control levers are functioning properly and listen for unusual sounds on each starting of an engine.

• The safety supervisor should assure that all new drill rig workers are informed of safe operating practices on and around the drill rig and should provide each new drill rig worker with a copy of the organization's drilling operations safety manual, and when appropriate the drill rig manufacturer's operations and maintenance manual. The safety supervisor should assure that each new employee reads and understands the safety manual.

• The safety supervisor should carefully instruct a new worker in drilling safety and observe the new worker's progress towards understanding safe operating practices.

• The safety supervisor should observe the mental, emotional and physical capability of each worker to perform the assigned work in a proper and safe manner. The safety supervisor should dismiss any worker from the drill site whose mental and physical capabilities might cause injury to the worker or coworkers.

Fig. 1.2

• The safety supervisor should assure that there is a first-aid kit on each drill rig and a fire extinguisher on each drill rig and on each additional vehicle and assure that they are properly maintained.

• The safety supervisor (and as many crew members as possible) should be well trained and capable of using first-aid kits, fire extinguishers and all other safety devices and equipment.

• The safety supervisor should maintain a list of addresses and telephone numbers of emergency assistance units (ambulance services, police, hospitals etc.) and inform other members of the drill crew of the existence and location of the list.

4. INDIVIDUAL PROTECTIVE EQUIPMENT

For most geotechnical, mineral and/or groundwater drilling projects, individual protective equipment should include a safety hat, safety shoes, safety glasses and close-fitting gloves and clothing. The clothing of the individual drill rig worker is not generally considered protective equipment; however, your clothing should be close fitting but comfortable, without loose ends, straps, drawstrings or belts or otherwise unfastened parts that might catch on some rotating or translating component of the drill rig. Rings and jewelry should not be worn during a work shift.

• Safety Head Gear. Safety hats (hard hats) should be worn by everyone working or visiting at or near a drilling site. All safety hats should meet the requirements of ANSI Z89.1. All safety hats should be kept clean and in good repair with the headband and crown straps properly adjusted for the individual drill rig worker or visitor.

• Safety Shoes or Boots. Safety shoes or boots should be worn by all drilling personnel and all visitors to the drill site that observe drilling operations within close proximity of the drill rig. All safety shoes or boots should meet the requirements of ANSI Z41.1

• Gloves. All drilling personnel should wear gloves for protection against cuts and abrasion which could occur while handling wire rope or cable and from contact with sharp edges and burrs on drill rods and other drilling or sampling tools. All gloves should be close fitting and not have large cuffs or loose ties which can catch on rotating or translating components of the drill rig.

• Safety Glasses. All drilling personnel should wear safety glasses. All safety glasses should meet the requirements of ANSI Z87.1.

• Other Protective Equipment. For some drilling operations, the environment or regulations may dictate that other protective equipment be used. The requirement for such equipment must be determined jointly by the management of the drilling organization and the safety supervisor. Such equipment might include face or ear protection or reflective clothing. Each drill rig worker should wear noise-reducing ear protectors when appropriate. When drilling is performed in chemically or radiologically contaminated ground, special protective equipment and clothing may and probably will be required. The design and composition of the protective equipment and clothing should be determined as a joint effort of management and the client who requests the drilling services.

Fig. 1.3

5. HOUSEKEEPING ON AND AROUND THE DRILL RIG

The first requirement for safe field operations is that the safety supervisor understands and fulfills the responsibility for maintenance and "housekeeping" on and around the drill rig.

• Suitable storage locations should be provided for all tools, materials and supplies so that tools, materials and supplies can be conveniently and safely handled without hitting or falling on a member of the drill crew or a visitor.

• Avoid storing or transporting tools, materials or supplies within or on the mast (derrick) of the drill rig.

• Pipe, drill rods, casing, augers and similar drilling tools should be orderly stacked on racks or sills to prevent spreading, rolling or sliding.

• Penetration or other driving hammers should be placed at a safe location on the ground or be secured to prevent movement when not in use.

• Work areas, platforms, walkways, scaffolding and other access ways should be kept free of materials, debris and obstructions and substances such as ice, grease or oil that could cause a surface to become slick or otherwise hazardous.

• Keep all controls, control linkages, warning and operation lights and lenses free of oil, grease and/or ice.

• Do not store gasoline in any portable container other than a nonsparking, red container with a flame arrester in the fill spout and having the word "gasoline" easily visible.

6. MAINTENANCE SAFETY

Good maintenance will make drilling operations safer. Also, maintenance should be performed safely.

• Wear safety glasses when performing maintenance on a drill rig or on drilling tools.

• Shut down the drill rig engine to make repairs or adjustments to a drill rig or to lubricate fittings (except repairs or adjustments that can only be made with the engine running). Take precautions to prevent accidental starting of an engine during maintenance by removing or tagging the ignition key.

• Always block the wheels or lower the leveling jacks or both and set hand brakes before working under a drill rig.

• When possible and appropriate, release all pressure on the hydraulic systems, the drilling fluid system and the air-pressure systems of the drill rig prior to performing maintenance. In other words, reduce the drill rig and operating systems

Fig. 1.4

to a "zero energy state" before performing maintenance. Use extreme caution when opening drain plugs and radiator caps and other pressurized plugs and caps.

• Do not touch an engine or the exhaust system of an engine following its operation until the engine and exhaust system have had adequate time to cool.

• Never weld or cut on or near a fuel tank.

• Do not use gasoline or other volatile or flammable liquids as a cleaning agent on or around a drill rig.

• Follow the manufacturer's recommendations for applying the proper quantity and quality of lubricants, hydraulic oils and/or coolants.

• Replace all caps, filler plugs, protective guards or panels and high-pressure hose clamps and chains or cables that have been removed for maintenance before returning the drill rig to service.

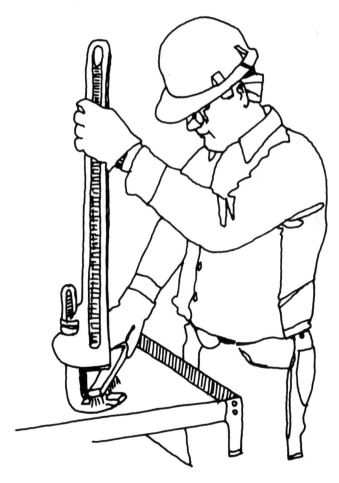

Fig. 1.5

7. SAFE USE OF HAND TOOLS

There are almost an infinite number of hand tools that can be used on or around a drill rig and in repair shops and more than an equal number of instructions for proper use. "Use the tool for its intended purpose" is the most important rule. The following are a few specific and some general suggestions which apply to safe use of several hand tools that are often used on and around drill rigs.

• When a tool becomes damaged, either repair it before using it again or get rid of it.

• When using a hammer, any kind of hammer, for any purpose, wear safety glasses and require all others around you to wear safety glasses.

• When using a chisel, any kind of chisel, for any purpose, wear safety glasses and require all others around you to wear safety glasses.

• Keep all tools cleaned and orderly stored when not in use.

• Use wrenches on nuts—don't use pliers on nuts.

• Use screwdrivers with blades that fit the screw slot.

• When using a wrench on a tight nut, first use some penetrating oil, use the largest wrench available that fits the nut, when possible pull on the wrench handle rather than pushing, and apply force to the wrench with both hands when possible and with both feet firmly placed. Don't push or pull with one or both feet on the drill rig or the side of a mud pit or some other blocking-off device. Always assume that you may lose your footing; check the place where you may fall for sharp objects.

• Keep all pipe wrenches clean and in good repair. The jaws of pipe wrenches should be wire brushed frequently to prevent an accumulation of dirt and grease which would otherwise build up and cause wrenches to slip.

• Never use pipe wrenches in place of a rod-holding device.

• Replace hook and heel jaws when they become visibly worn.

• When breaking tool joints on the ground or on a drilling platform, position your hands so that your fingers will not be smashed between the wrench handle and the ground or the platform, should the wrench slip or the joint suddenly let go.

8. CLEARING THE WORK AREA

Prior to drilling, adequate site clearing and leveling should be performed to accommodate the drill rig and supplies and provide a safe working area. Drilling should not be commenced when tree limbs, unstable ground or site obstructions cause unsafe tool handling conditions.

9. START-UP

• All drill rig personnel and visitors should be instructed to stand clear of the drill rig immediately prior to and during starting of an engine.

• Make sure all gear boxes are in neutral, all hoist levers are disengaged, all hydraulic levers are in the correct nonactuating position and the cathead rope is not on the cathead before starting a drill rig engine.

• Start all engines according to the manufacturer's manual.

10. SAFETY DURING DRILLING OPERATIONS

Safety requires the attention and cooperation of every worker and site visitor.

• Do not drive the drill rig from hole to hole with the mast (derrick) in the raised position.

• Before raising the mast (derrick) look up to check for overhead obstructions. (Refer to Section 11 on Overhead and Buried Utilities.)

• Before raising the mast (derrick), all drill rig personnel (with exception of the operator) and visitors should be cleared from the areas immediately to the rear and sides of the mast. All drill rig personnel and visitors should be informed that the mast is being raised prior to raising it.

• Before the mast (derrick) of a drill rig is raised and drilling is commenced, the drill rig must be first leveled and stabilized with leveling jacks and/or solid cribbing. The drill rig should be releveled if it settles after initial set-up. Lower the mast (derrick) only when the leveling jacks are down and do not raise the leveling jack pads until the mast (derrick) is lowered completely.

• Before starting drilling operations, secure and/or lock the mast (derrick) if required according to the drill manufacturer's recommendations.

• The operator of a drill rig should only operate a drill rig from the position of the controls. If the operator of the drill rig must leave the area of the controls, the operator should shift the transmission controlling the rotary drive into neutral and place the feed control lever in neutral. The operator should shut down the drill engine before leaving the vicinity of the drill.

• Throwing or dropping tools should not be permitted. All tools should be carefully passed by hand between personnel or a hoist line should be used.

• Do not consume alcoholic beverages or other depressants or chemical stimulants prior to starting work on a drill rig or while on the job.

• If it is necessary to drill within an enclosed area, make certain that exhaust fumes are conducted out of the area. Exhaust fumes can be toxic and some cannot be detected by smell.

• Clean mud and grease from your boots before mounting a drill platform and use hand holds and railings. Watch for slippery ground when dismounting from the platform.

• During freezing weather do not touch metal parts of the drill rig with exposed flesh. Freezing of moist skin to metal can occur almost instantaneously.

• All air and water-lines and pumps should be drained when not in use if freezing weather is expected.

• All unattended boreholes must be adequately covered or otherwise protected to prevent drill rig personnel, site visitors or animals from stepping or falling into the hole. All open boreholes should be covered, protected or backfilled adequately according to local or state regulations on completion of the drilling project.

• "Horsing around" within the vicinity of the drill rig and tool and supply storage areas should never be allowed, even when the drill rig is shut down.

• When using a ladder on a drill rig, face the ladder and grasp either the side-rails or the rungs with both hands while ascending or descending. Do not attempt to use one or both hands to carry a tool while on a ladder. Use a hoist line and a tool "bucket" or a safety hook to raise or lower hand tools.

An elevated derrick platform should be used with the following precautions:

• When working on a derrick platform, use a safety belt and a lifeline. The safety belt should be at least 4 in (100 mm) wide and should fit snugly but comfortably. The lifeline, when attached to the derrick, should be less than 6 ft (2 m) long. The safety belt and lifeline should be strong enough to withstand the dynamic force of a 250 lb (115 kg) weight (contained within the belt) falling 6 ft (2 m).

• When climbing to a derrick platform that is higher than 20 ft (6 m), a safety climbing device should be used.

• When a rig worker is on a derrick platform, the lifeline should be fastened to the derrick just above the derrick platform and to a structural member that is not attached to the platform or to other lines or cables supporting the platform.

• When a rig worker first arrives at a derrick platform, the platform should immediately be inspected for broken members, loose connections and loose tools or other loose materials.

• Tools should be securely attached to the platform with safety lines. Do not attach a tool to a line attached to your wrist or any other part of your body.

• When you are working on a derrick platform, do not guide drill rods or pipe into racks or other supports by taking hold of a moving hoist line or a traveling block.

• Loose tools and similar items should not be left on the derrick platform or on structural members of the derrick.

• A derrick platform over 4 ft (1.2 m) above ground surface should have toe boards and safety railing that are in good condition.

• Workers on the ground or the drilling floor should avoid being under rig workers on elevated platforms, whenever possible.

Be careful when lifting heavy objects:

• Before lifting any object without using a hoist, make sure that the load is within your personal lifting capacity. If it is too heavy, ask for assistance.

• Before lifting a relatively heavy object, approach the object by bending at the knees, keeping your back vertical and unarched while obtaining a firm footing. Grasp the object firmly with both hands and stand slowly and squarely while keeping your back vertical and unarched. In other words, perform the lifting with the muscles in your legs, not with the muscles in your lower back.

• If a heavy object must be moved some distance without the aid of machinery, keep your back straight and unarched. Change directions by moving your feet, not by twisting your body.

• Move heavy objects with the aid of hand carts whenever possible.

Drilling operations should be terminated during an electrical storm and the complete crew should move away from the drill rig.

11. OVERHEAD AND BURIED UTILITIES

The use of a drill rig on a site or project within the vicinity of electrical power lines and other utilities requires that special precautions be taken by both supervisors and members of the exploration crew. Electricity can shock, it can burn and it can cause death.

• Overhead and buried utilities should be located, noted and emphasized on all boring location plans and boring assignment sheets.

• When overhead electrical power lines exist at or near a drilling site or project, consider all wires to be alive and dangerous.

• Watch for sagging power lines before entering a site. Do not lift power lines to gain entrance. Call the utility and ask them to lift or raise the lines or deenergize (turn off) the power.

• Before raising the drill rig mast (derrick) on a site in the vicinity of power lines, walk completely around the drill rig. Determine what the minimum distance from any point on the drill rig to the nearest power line will be when the mast is raised and/or is being raised. Do not raise the mast or operate the drill rig if this distance is less than 20 ft (6 m), or if known, the minimum clearance stipulated by federal, state and local regulations.

• Keep in mind that both hoist lines and overhead power lines can be moved toward each other by the wind.

• In order to avoid contact with power lines, only move the drill rig with the mast (derrick) down.

• If there are any questions whatever concerning the safety of drilling on sites in the vicinity of overhead power lines, call the power company. The power

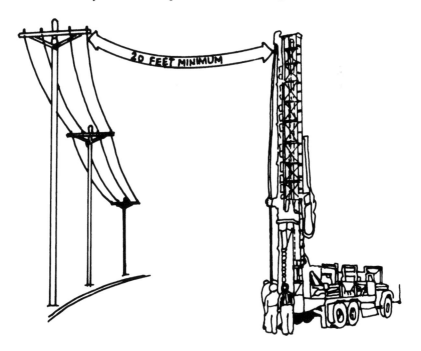

Fig. 1.6

Fig. 1.7

company will provide expert advice at the drilling site as a public service and at no cost.

Underground electricity is as dangerous as overhead electricity. Be aware and always suspect the existence of underground utilities such as electrical power, gas, petroleum, telephone, sewer and water. Ask for assistance:

● If a sign warning of underground utilities is located on a site boundary, do not assume that underground utilities are located on or near the boundary or property line under the sign: call the utility and check it out. The underground utilities may be a considerable distance away from the warning sign.

• Always contact the owners of utility lines or the nearest underground utility location service before drilling. Determine jointly with utility personnel the precise location of underground utility lines, mark and flag the locations and determine jointly with utility personnel what specific precautions must be taken to assure safety.

12. SAFE USE OF ELECTRICITY

Drilling projects sometimes require around-the-clock operations and, therefore, require temporary electrical lighting. In general, all wiring and fixtures used to provide electricity for drilling operations should be installed by qualified personnel in accordance with the National Electrical Code (NFPA70-1984) with consideration of the American Petroleum Institute's recommended practices for electrical installations for production facilities (API-RP-500B). Lights should be installed and positioned to assure that the work area and operating positions are well lit without

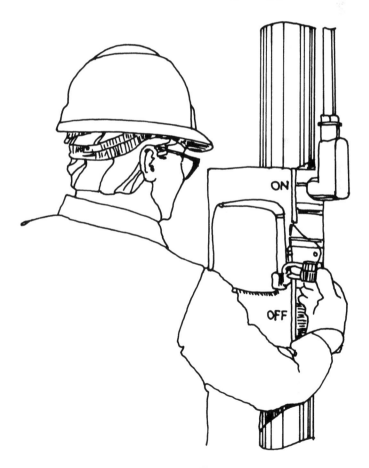

Fig. 1.8

shadows or blind spots. The following specific recommendations emphasize the safe use of electricity during land-based drilling operations:

• Before working on an electrical power or lighting system, lock-out the main panel box with your own lock and keep the key on your person at all times.

• All wiring should be installed using high-quality connections, fixtures and wire, insulated and protected with consideration of the drilling environment. Makeshift wiring and equipment should not be permitted.

• All lights positioned directly above working areas should be enclosed in cages or similar enclosures to prevent loose or detached lamps or vaportight enclosures from falling on workers.

• Lights should be installed to produce the least possible glare or "blind spots" on tools, ladders, walkways, platforms and the complete working area.

• Electrical cables should be guarded and located to prevent damage by drilling operations or by the movement of personnel, tools or supplies.

• All plug receptacles should be the three-prong, U-blade, grounded type and have adequate current-carrying capacity for the electrical tools that may be used.

• All electric tools should have three-prong, U-blade, ground-wire plugs and cords.

• Do not use electrical tools with lock-on devices.

• All electrical welders, generators, control panels and similar devices should be adequately grounded.

• Control panels, fuse boxes, transformers and similar equipment should have a secure, protective enclosure.

• Avoid attaching electrical lighting cables to the derrick or other: components of the drill rig. If this must be done, use only approved fasteners. Do not "string" wire through the derrick.

• Poles used to hold wiring and lights should not be used for any other purpose.

• Power should be turned off before changing fuses or light bulbs.

• When a drilling area is illuminated with electrical lighting, all workers should wear safety head gear that protects the worker's head, not only against falling or flying objects, but also against limited electrical shock and burn according to ANSI Z89.1 and Z89.2.

• Electrical equipment should only be operated by trained, designated personnel.

• If you are not qualified to work on electrical devices or on electric lines, do not go near them.

13. REACT TO CONTACT WITH ELECTRICITY

If a drill rig makes contact with electrical wires, it may or may not be insulated from the ground by the tires of the carrier. Under either circumstance the human body, if it simultaneously comes in contact with the drill rig and the ground, will provide a conductor of the electricity to the ground. Death or serious injury can be the result. If a drill rig or a drill rig carrier makes contact with overhead or underground electrical lines:

• Under most circumstances, the operator and other personnel on the seat of the vehicle should remain seated and not leave the vehicle. Do not move or touch any part, particularly a metallic part, of the vehicle or the drill rig.

• If it is determined that the drill rig should be vacated, then all personnel should jump clear and as far as possible from the drill. Do not step off—jump off, and do not hang onto the vehicle or any part of the drill when jumping clear.

• If you are on the ground, stay away from the vehicle and the drill rig, do not let others get near the vehicle and the drill rig, and seek assistance from local emergency personnel such as the police or a fire department.

• When an individual is injured and in contact with the drill rig or with power lines, attempt rescue with extreme caution. If a rescue is attempted, use a long, dry, unpainted piece of wood or a long, dry, clean rope. Keep as far away from the victim as possible and do not touch the victim until the victim is completely clear of the drill rig or electrical lines.

• When the victim is completely clear of the electrical source and is unconscious and a heart beat (pulse) cannot be detected, begin cardiopulmonary resuscitation (CPR) immediately.

14. SAFE USE OF WIRELINE HOISTS, WIRE ROPE AND HOISTING HARDWARE

The use of wireline hoists, wire rope and hoisting hardware should be as stipulated by the American Iron and Steel Institute *Wire Rope Users Manual.*

• All wire ropes and fittings should be visually inspected during use and thoroughly inspected at least once a week for: abrasion, broken wires, wear, reduction in rope diameter, reduction in wire diameter, fatigue, corrosion, damage from heat, improper reeving, jamming, crushing, bird-caging, kinking, core protrusion and damage to lifting hardware. Wire ropes should be replaced when inspection indicates excessive damage according to the *Wire Rope Users Manual.* All wire ropes which have not been used for a period of a month or more should be thoroughly inspected before being returned to service.

• End fittings and connections consist of spliced eyes and various manufactured devices. All manufactured end fittings and connections should be installed according to the manufacturer's instructions and loaded according to the manufacturer's specifications.

• If a ball-bearing type hoisting swivel is used to hoist drill rods, swivel bearings should be inspected and lubricated daily to assure that the swivel freely rotates under load.

• If a rod-slipping device is used to hoist drill rods, do not drill through or rotate drill rods through the slipping device, do not hoist more than 1 ft (0.3 m) of the drill rod column above the top of the mast (derrick), do not hoist a rod column with loose tool joints and do not make up, tighten or loosen tool joints while the rod column is being supported by a rod-slipping device. If drill rods should slip back into the borehole, do not attempt to brake the fall of the rods with your hands or by tensioning the slipping device.

• Most sheaves on exploration drill rigs are stationary with a single part line. The number of parts of line should not ever be increased without first consulting the manufacturer of the drill rig.

• Wire ropes must be properly matched with each sheave: if the rope is too large, the sheave will pinch the wire rope; if the rope is too small, it will groove the sheave. Once the sheave is grooved, it will severely pinch and damage larger sized wire ropes.

The following procedures and precautions must be understood and implemented for safe use of wire ropes and rigging hardware.

• Use tool-handling hoists only for vertical lifting of tools (except when angle hole drilling). Do not use tool-handling hoists to pull on objects away from the drill rig; however, drills may be moved using the main hoist if the wire rope is spooled through proper sheaves according to the manufacturer's recommendations.

• When stuck tools or similar loads cannot be raised with a hoist, disconnect the hoist line and connect the stuck tools directly to the feed mechanism of the drill. Do not use hydraulic leveling jacks for added pull to the hoist line or the feed mechanism of the drill.

• When attempting to pull out a mired down vehicle or drill rig carrier, only use a winch on the front or rear of the vehicle and stay as far as possible from the wire rope. Do not attempt to use tool hoists to pull out a mired down vehicle or drill rig carrier.

• Minimize shock loading of a wire rope; apply loads smoothly and steadily

• Avoid sudden loading in cold weather.

• Never use frozen ropes.

• Protect wire rope from sharp corners or edges.

• Replace faulty guides and rollers.

• Replace worn sheaves or worn sheave bearings.

• Replace damaged safety latches on safety hooks before using.

• Know the safe working load of the equipment and tackle being used. Never exceed this limit.

• Clutches and brakes of hoists should be periodically inspected and tested.

• Know and do not exceed the rated capacity of hooks, rings, links, swivels, shackles and other lifting aids.

• Always wear gloves when handling wire ropes.

• Do not guide wire rope on hoist drums with your hands.

• Following the installation of a new wire rope, first lift a light load to allow the wire rope to adjust.

• Never carry out any hoisting operations when the weather conditions are such that hazards to personnel, the public or property are created.

• Never leave a load suspended in the air when the hoist is unattended.

• Keep your hands away from hoists, wire rope, hoisting hooks, sheaves and pinch points as slack is being taken up and when the load is being hoisted.

• Never hoist the load over the head, body or feet of any personnel.

• Never use a hoist line to "ride" up the mast (derrick) of a drill rig.

• Replacement wire ropes should conform to the drill rig manufacturer's specifications.

15. SAFE USE OF CATHEAD AND ROPE HOISTS

The following safety procedures should be employed when using a cathead hoist.

• Keep the cathead clean and free of rust and oil and/or grease. The cathead should be cleaned with a wire brush if it becomes rusty.

• Check the cathead periodically, when the engine is not running, for rope-wear grooves. If a rope groove forms to a depth greater than 1/8 in (3 mm), the cathead should be replaced.

• Always use a clean, dry, sound rope. A wet or oily rope may "grab" the cathead and cause drill tools or other items to be rapidly hoisted to the top of the mast.

• Should the rope "grab" the cathead or otherwise become tangled in the drum, release the rope and sound an appropriate alarm for all personnel to rapidly back away and stay clear. The operator should also back away and stay clear. If the rope "grabs" the cathead, and tools are hoisted to the sheaves at the top of the mast, the rope will often break, releasing the tools. If the rope does not break, stay clear of the drill rig until the operator cautiously returns to turn off the drill rig engine and appropriate action is taken to release the tools. The operator should keep careful watch on the suspended tools and should quickly back away after turning off the engine.

• The rope should always be protected from contact with all chemicals. Chemicals can cause deterioration of the rope that may not be visibly detectable.

• Never wrap the rope from the cathead (or any other rope, wire rope or cable on the drill rig) around a hand, wrist, arm, foot, ankle, leg or any other part of your body.

• Always maintain a minimum of 18 inches of clearance between the operating hand and the cathead drum when driving samplers, casing or other tools with the cathead and rope method. Be aware that the rope advances toward the cathead with each hammer blow as the sampler or other drilling tool advances into the ground.

• Never operate a cathead (or perform any other task around a drill rig) with loose unbuttoned or otherwise unfastened clothing or when wearing gloves with large cuffs or loose straps or lacings.

• Do not use a rope that is any longer than necessary. A rope that is too long can form a ground loop or otherwise become entangled with the operator's legs.

• Do not use more rope wraps than are required to hoist a load.

• Do not leave a cathead unattended with the rope wrapped on the drum.

• Position all other hoist lines to prevent contact with the operating cathead rope.

• When using the cathead and rope for driving or back-driving, make sure that all threaded connections are tight and stay as far away as possible from the hammer impact point.

• The cathead operator must be able to operate the cathead standing on a level surface with good, firm footing conditions without distraction or disturbance.

Fig. 1.9

16. SAFE USE OF AUGERS

The following general procedures should be used when starting a boring with continuous-flight or hollow-stem augers:

   • Prepare to start an auger boring with the drill rig level, the clutch or hydraulic rotation control disengaged, the transmission in low gear, and the engine running at low RPM.

   • Apply an adequate amount of down pressure prior to rotation to seat the auger head below the ground surface.

   • Look at the auger head while slowly engaging the clutch or rotation control and starting rotation. Stay clear of the auger.

- Slowly rotate the auger and auger head while continuing to apply down pressure. Keep one hand on the clutch or the rotation control at all times until the auger has penetrated about one foot or more below ground surface.
- If the auger head slides out of alignment, disengage the clutch or hydraulic rotation control and repeat the hole starting process.
- An auger guide can facilitate the starting of a straight hole through hard ground or a pavement.

The operator and tool handler should establish a system of responsibility for the series of various activities required for auger drilling, such as connecting and disconnecting auger sections, and inserting and removing the auger fork. The operator must assure that the tool handler is well away from the auger column and that the auger fork is removed before starting rotation.

- Only use the manufacturer's recommended method of securing the auger to the power coupling. Do not touch the coupling or the auger with your hands, a wrench or any other tools during rotation.
- Whenever possible, use tool hoists to handle auger sections.
- Never place hands or fingers under the bottom of an auger section when hoisting the auger over the top of the auger section in the ground or other hard surfaces such as the drill rig platform.
- Never allow feet to get under the auger section that is being hoisted.
- When rotating augers, stay clear of the rotating auger and other rotating components of the drill rig. Never reach behind or around a rotating auger for any reason whatever.
- Use a long-handled shovel to move auger cuttings away from the auger. Never use your hands or feet to move cuttings away from the auger.
- Do not attempt to remove earth from rotating augers. Augers should be cleaned only when the drill rig is in neutral and the augers are stopped from rotating.

17. SAFETY DURING ROTARY AND CORE DRILLING

Rotary drilling tools should be safety checked prior to drilling:

- Water swivels and hoisting plugs should be lubricated and checked for "frozen" bearings before use.
- Drill rod chuck jaws should be checked periodically and replaced when necessary.
- The capacities of hoists and sheaves should be checked against the anticipated weight to the drill rod string plus other expected hoisting loads.

Special precautions that should be taken for safe rotary or core drilling involve chucking, joint break, hoisting and lowering of drill rods:

- Only the operator of the drill rig should brake or set a manual chuck so that rotation of the chuck will not occur prior to removing the wrench from the chuck.
- Drill rods should not be braked during lowering into the hole with drill rod chuck jaws.
- Drill rods should not be held or lowered into the hole with pipe wrenches.

• If a string of drill rods are accidentally or inadvertently released into the hole, do not attempt to grab the falling rods with your hands or a wrench.

• In the event of a plugged bit or other circulation blockage, the high pressure in the piping and hose between the pump and the obstruction should be relieved or bled down before breaking the first tool joint.

• When drill rods are hoisted from the hole, they should be cleaned for safe handling with a rubber or other suitable rod wiper. Do not use your hands to clean drilling fluids from drill rods.

• If work must progress over a portable drilling fluid (mud) pit, do not attempt to stand on narrow sides or cross members. The mud pit should be equipped with rough surfaced, fitted cover panels of adequate strength to hold drill rig personnel.

• Drill rods should not be lifted and leaned unsecured against the mast. Either provide some method of securing the upper ends of the drill rod sections for safe vertical storage or lay the rods down.

18. SAFETY DURING TRAVEL

The individual who transports a drill rig on and off a drilling site should:

• Be properly licensed and should only operate the vehicle according to federal, state and local regulations.

• Know the traveling height (overhead clearance), width, length, and weight of the drill rig with carrier and know highway and bridge load, width, and overhead limits, making sure these limits are not exceeded.

• Never move a drill rig unless the vehicle brakes are in sound working order.

• Allow for mast overhang when cornering or approaching other vehicles or structures.

• Be aware that the canopies of service stations and motels are often too low for a drill rig mast to clear with the mast in the travel position.

• Watch for low-hanging electrical lines, particularly at the entrances to drilling sites or restaurants, motels or other commercial sites.

• Never travel on a street, road or highway with the mast (derrick) of the drill rig in the raised or partially raised position.

• Remove all ignition keys when a drill rig is left unattended.

19. LOADING AND UNLOADING

When loading or unloading a drill rig on a trailer or a truck:

• Use ramps of adequate design that are solid and substantial enough to bear the weight of the drill rig with carrier, including tooling.

• Load and unload on level ground.

• Use the assistance of someone on the ground as a guide.

• Check the brakes on the drill rig carrier before approaching loading ramps.

• Distribute the weight of the drill rig, carrier and tools on the trailer so that the center of weight is approximately on the centerline of the trailer and so that some of the trailer load is transferred to the hitch of the pulling vehicle. Refer to the trailer manufacturer's weight distribution recommendations.

Fig. 1.10

   • The drill rig and tools should be secured to the hauling vehicle with ties, chains and/or load binders of adequate capacity.

20. OFF-ROAD MOVEMENT

The following safety suggestions relate to off-road movement:
   • Before moving a drill rig, first walk the route of travel, inspecting for depressions, stumps, gulleys, ruts and similar obstacles.
   • Always check the brakes of a drill rig carrier before traveling, particularly on rough, uneven or hilly ground.
   • Check the complete drive train of a carrier at least weekly for loose or damaged bolts, nuts, studs, shafts and mountings.
   • Discharge all passengers before moving a drill rig on rough or hilly terrain.

• Engage the front axle (for 4 × 4, 6 × 6, etc. vehicles or carriers) when traveling off highway on hilly terrain.

• Use caution when traveling sidehill. Conservatively evaluate sidehill capability of drill rigs because the arbitrary addition of drilling tools may raise the center of mass. When possible, travel directly uphill or downhill. Increase tire pressures before traveling in hilly terrain (do not exceed rated tire pressure).

• Attempt to cross obstacles such as small logs and small erosion channels or ditches squarely, not at an angle.

• Use the assistance of someone on the ground as a guide when lateral or overhead clearance is close.

• After the drill has been moved to a new drilling site, set all brakes and/or locks. When grades are steep, block the wheels.

• Never travel off-road with the mast (derrick) of the drill rig in the raised or partially raised position.

21. TIRES, BATTERIES AND FUEL

Tires on the drill rig must be checked daily for safety and during extended travel for loss of air and must be maintained and/or repaired in a safe manner. If tires are deflated to reduce ground pressure for movement on soft ground, the tires should be reinflated to normal pressures before movement on firm or hilly ground or on streets, roads and highways. Underinflated tires are not as stable on firm ground as properly inflated tires. Air pressures should be maintained for travel on streets, roads and highways according to the manufacturer's recommendations. During air-pressure checks, inspect for:

• Missing or loose wheel lugs.
• Objects wedged between duals or embedded in the tire casing.
• Damaged or poorly fitting rims or rim flanges.
• Abnormal or uneven wear and cuts, breaks or tears in the casing.

The repair of truck and off-highway tires should only be made with required special tools and following the recommendations of a tire manufacturer's repair manual.

Batteries contain strong acid. Use extreme caution when servicing batteries.

• Batteries should only be serviced in a ventilated area while wearing safety glasses.

• When a battery is removed from a vehicle or service unit, disconnect the battery ground clamp first.

• When installing a battery, connect the battery ground clamp last.

• When charging a battery with a battery charger, turn off the power source to the battery before either connecting or disconnecting charger leads to the battery posts. Cell caps should be loosened prior to charging to permit the escape of gas.

• Spilled battery acid can burn your skin and damage your eyes.

Spilled battery acid should be immediately flushed off of your skin with lots of water. Should battery acid get into someone's eyes, flush immediately with large amounts of water and see a medical physician at once.

• To avoid battery explosions, keep the cells filled with electrolyte, use a flashlight (not an open flame) to check electrolyte levels and avoid creating sparks around the battery by shorting across a battery terminal. Keep lighted smoking materials and flames away from batteries.

Special precautions must be taken for handling fuel and refueling the drill rig or carrier.

• Only use the type and quality of fuel recommended by the engine manufacturer.

• Refuel in a well-ventilated area.

• Do not fill fuel tanks while the engine is running. Turn off all electrical switches.

• Do not spill fuel on hot surfaces. Clean any spillage before starting an engine.

• Wipe up spilled fuel with cotton rags or cloths—do not use wool or metallic cloth.

• Keep open lights, lighted smoking materials and flames or sparking equipment well away from the fueling area.

• Turn off heaters in carrier cabs when refueling the carrier or the drill rig.

• Do not fill portable fuel containers completely full to allow expansion of the fuel during temperature changes.

• Keep the fuel nozzle in contact with the tank being filled to prevent static sparks from igniting the fuel.

• Do not transport portable fuel containers in the vehicle or carrier cab with personnel.

• Fuel containers and hoses should remain in contact with a metal surface during travel to prevent the buildup of static charge.

22. FIRST AID

At least one member of the drill crew, and if only one, preferably the drilling and safety supervisor, should be trained to perform first aid. First aid is taught on a person-to-person basis, not by providing or reading a manual. Manuals should only provide continuing reminders and be used for reference. It is suggested that courses provided or sponsored by the American Red Cross or a similar organization would best satisfy the requirements of first-aid training for drill crews.

For drilling operations it is particularly important that the individual responsible for first aid should be able to recognize the symptoms and be able to provide first aid for electrical shock, heart attack, stroke, broken bones, eye injury, snake bite and cuts or abrasions to the skin. Again, first aid for these situations is best taught to drill crew members by instructors qualified by an agency such as the American Red Cross.

A first-aid kit should be available and well maintained on each drill site.

23. DRILL RIG UTILIZATION

Do not attempt to exceed manufacturers' ratings of speed, force, torque, pressure, flow, etc. Only use the drill rig and tools for the purposes for which they are intended and designed.

24. DRILL RIG ALTERATIONS

Alterations to a drill rig or drilling tools should only be made by qualified personnel and only after consultation with the manufacturer.

* * *

MISCELLANEOUS INFORMATION ON DRILLING SAFETY

Schramm Inc, has published some safety recommendations relevant to use of its rotadrills. As these directions and instructions are applicable to all the rotary and rotary-cum-DTH drills manufactured by various suppliers throughout the world, they are presented below.

PRE-DRILLING PERIODIC CHECKLIST

Never climb mast without Safety Belt engaged in Tube.
Make sure machinery guards are in place.
Pipe handling sling cable and connections are in good condition.
Pipe handling sling clamshell will safely retain drill pipe to be handled Check closed clearance and material condition.
Check feed chains for equal tension.
Check rotation gearbox trunnion rollers.
Fan belts are tensioned properly. See the engine manufacturer's Maintenance Manual for adjustment procedure.
Winch cable is in good condition.
Winch cable support material at top of mast is in good operating order.
Outrigger check valves are sealing properly (unit maintains position).
Check engine oil level.
Lube level in rotation gearbox.
Fluid level in compressor air/oil tank.
Heat exchanger coolant level and solution temperature protection level.
Check condition of rotary seals in housing on top of rotation gearbox, and the finish of sealing sleeve on which the seals run. Coat parts with grease on re-assembly.
Grease points outlined in Lubrication Section of Maintenance.

DRILL SETUP CHECKLIST

Avoid setups near high-voltage power transmission lines. Make sure a safe distance exists between lines and closest part of drill.
Always call your local underground locating service for utilities in drill area.
Always position drill support equipment and drill pipe supply to take best advantage of the drill rig pipe handling equipment.
Elevate drill on outriggers, no higher than necessary for work zone and level condition.
When raising mast, check for any unknown loose objects in mast.

Always position drill rig where there is adequate clearance for operation.
Check the condition of the material to be drilled to prevent cave-ins or slides.
Store all cold weather starting aids and flammable substances off the rig and away
from sparks, heat or open flames.
Check lubrication of Drill Rig.
Check for fluid leaks.
Inspect components. Make certain that air pressure is correct and their are no leaks.
Check Safety devices to make sure they are working properly.
Inspect drill rig for any physical damage. Do not operate damaged drill rig.
Replace all loose or missing hardware.
Never operate without protective guards and panels in place.
Use caution when starting the drill rig with other persons in the area.

WARNING

Use extreme caution when working in the vicinity of electric power lines.
Never raise the mast where it could touch overhead obstructions. Equipment
may become electrically charged when working in the area of high-frequency
transmitters.

OPERATING CHECKLIST

Items to be checked prior to and during start-up:
1) Engine oil check daily, if diesel engine equipped.
2) Compressor oil checked daily.
3) Bleed water out of compressor air/oil tank and the air delivery control filters
   daily before starting. Recheck oil level.
4) Start diesel with throttle in idle position. If rig is electric-motor driven, close
   compressor inlet valve, start motor; open compressor inlet when electric motor
   is up to speed.
5) After start-up, check for safety circuit green light "ON" after a few seconds
   running time.
6) Avoid sudden changes in engine rpm to obtain best fan belt life. High-capacity
   fan should be brought up to speed gradually.
7) Occasionally check downfeed relief valve by fully extending traversing cylinders to end of stroke. See specifications for proper pressure setting.
8) Occasionally check holdback relief valve by fully retracting traversing cylinders
   to end of stroke (rotation gearbox up). See specifications for proper pressure
   setting.
9) Occasionally lubricate feed chains with SAE 20, 30 or SAE 40 motor oil. Even
   in dusty conditions, lubrication will improve chain life. Use the heaviest oil that
   will penetrate the chains.

ABNORMAL CONDITIONS

TOWING

These towing instructions are for an emergency only. Always haul the machine if long distances must be traveled.

Use a tow bar if the machine is to be moved more than a few feet. If a tow bar is not available, attach a machine of equal size to the rear of the towed machine to provide braking when going downhill.

Be sure the tow bar is strong enough and is in good condition.

Attach the tow bar only to the tow hooks on the frame. Do not tow faster the 3 mph (4.8 kph).

Shielding must be provided on the towing vehicle to protect the operator if the tow bar breaks.

Do not have tension on the tow bar when inspecting it. Avoid shock loading on the tow bar.

Do not allow riders on a machine that is being towed.

Always block both tracks before disengaging the final drives. When the final drives are disengaged, the brakes are also disengaged. The machine can move.

Use precautions to maintain control of the machine when using another machine to tow it.

Avoid towing machine with mast up on grades exceeding 5%.

Block the tracks when parking if the final drives are disengaged. Brakes are also disengaged.

Refer to Operator's section of Manual for track final drive disengaging procedure.

SAFETY

Read the warning and service information provided on the machine. Follow servicing instructions carefully. Do not start or operate the machine while it is being serviced. Attach a warning tag to the controls.

There are certain hazards which must be recognized as potential causes of personal injury. Be aware of these hazards and follow the recommendations which are listed in this manual.

CRUSHING OR CUTTING

Never ride on rotation gearbox to climb the mast.

Stand clear of mast bottom when rotation gearbox is being moved up or down.

Never attempt adjustments while the machine is moving or the engine is running.

Make sure there are no loose drive subs, tools or other items in the mast before raising it.

Never ride the mast up or down.

Support equipment when working beneath it. Do not depend on hydraulic cylinders to hold it up.

Stand clear of any hydraulic cylinder or motor-powered device when being operated. Even the smallest hydraulic device can generate tremendous power or force.

The fan blades will throw or cut any object or tool that falls or is pushed into them. Check positioning and condition of fan guard.

Drive shaft and universal joints, when rotating, can catch loose clothing, rags, or hair. Check positioning and condition of machinery guards.

Wear gloves when handling wire rope or cable. Do not use kinked or frayed cable as they may not meet original design strength.

Wear protective glasses when hammering on metal drifts, punches, or chisels. Chips can fly from the object or hammer.

Raise or lower mast only when the cab door on the mast side is closed. Never extend arms or legs out of cab when raising or lowering mast.

Never climb the mast. Service should be performed in the lowered position.

Keep hands and loose clothing away from chains and sprockets at all times.

FLUIDS

Cooling system conditioners contain alkali. Avoid prolonged contact.

Never check coolant levels with a hot cooling system.

Battery electrolyte is an acid. Avoid contact with eyes, skin or clothing.

Keep all lubricants stored in properly marked containers.

When draining engine crankcase or hydraulic system, avoid contact with hot oil.

Do not touch rotation or pump gearboxes when running or just after shutdown. Internal lubes will heat gearboxes to skin-burn temperatures.

Avoid unknown compressed air leaks. Find unknown leak source at lowest possible pressure levels. High-pressure air is explosive in nature.

Make sure air-line safety chains are in good working order on all lines that could whip, in the event of fitting failures.

Always check tank air-pressure gauge before performing air system maintenance. Tank pressure must be zero before safe maintenance can be performed.

Never add compressor oil with air pressure in tank. Check pressure gauge on sump position only.

Never adjust or tamper with factory settings on compressed-air system valves or regulators.

Air Compressor sump pressure must never exceed safety valve setting.

FIRE OR EXPLOSION

Diesel fuel and all lubricants are flammable. Do not weld on pipes or tubes that contain oil. Clean them thoroughly with non-flammable solvent before welding.

Make sure that any rig equipped fire extinguishing system is in good working order. Check bottle pressure periodically.

Do not smoke when refueling.

To avoid fires, clean up oil spills and built-up debris. Keep machinery and machine deck as clean as practically possible.

Make sure that any fluid system leak is fixed before continuing operation.

The vapor (hydrogen gas) from a charging battery is explosive. Do not smoke when checking batteries or working around batteries.

Check condition of fluid heating system elements for wear spots, bare electrical system wires or connections.

Tighten loose hose clamps or fasteners. Replace missing ones.

Tighten loose fluid fittings or connections.

Check for misalignment of tubes, hoses, or items containing fluid. Reroute lines if necessary to prevent interference.

Look for loose, frayed, damaged, or disconnected wires.

Inspect electrical junction boxes for water tight condition. Repair seals as necessary.

Replace or tighten loose or missing mufflers or exhaust system parts.

PERSONAL SAFETY

Always use a Safety Belt when climbing the mast.

Wear a hard hat, protective shoes, and protective glasses when doing lubrication and maintenance work as well as operating the machine.

Limit air pressure to 30 psi (205 kPa) when cleaning with air.

Never do air system maintenance work unless compressor sump is at zero pressure.

Never point an air nozzle at anyone.

Know the weight limits of cable, chains, and slings before using them. Do not use frayed or kinked cables.

Check tong assembly for worn or broken parts.

Use a "DO NOT OPERATE" or similar WARNING tag on the vehicle starter switch or controls when working on a machine.

Use steps and grab irons when servicing the machine.

Store rags that have oil or other flammable material on them in a safety-type container. Keep the container away from open fires, welding, or flame cutting areas.

Operate the engine only in a well-ventilated area. In a closed area, vent exhaust fumes to the outside.

Always be alert.

Never operate the machine under the influence of alcohol, drugs or medication.

Never leave the machine while in operation.

Keep hands, rags, tools away from moving parts.

Keep all unnecessary persons away from drill area. Warn all persons in the area when you are about to begin drilling.

Do not lubricate or service machine while running.

Keep work area clear of all objects such as tools and cuttings.

Keep work areas, ladders and hand rails clean of grease and oil.

Do not use equipment for anything other than the intended purpose.

Read all warning and instruction labels.

Use proper lighting for night time operation.

Periodically check belts, chains and hoses for indication of wear, looseness, cracking or fraying.

Remove all loose objects stored in mast and on deck before raising mast.

Never allow unqualified personnel to operate the rig.

Attempt to drill on a level work area.

Never raise or lower the rig mast unless the mast side cab door is closed.

If rig is electric-motor driven and equipped with a trailing cable reel, never remove collector ring enclosure unless power is off.

Electric model drills are equipped with starter control cabinet interlocks that protect personnel from high voltage contact; however, high voltage input contacts still exist in the cabinet. Only qualified personnel should attempt entry in the high voltage compartment of the motor starter cabinet.

CAUTION

Use caution when removing caps, drain plugs, fittings, or pressure taps.

Turn the disconnect switch off and remove the key before servicing the electrical system.

CHAPTER 2

DRILLING TECHNIQUES AND DEVELOPMENTS

Soils
Classifying Rocks
Rotary Drilling
Dual-Tube Drilling Systems
Auger Boring
Overburden Drilling
Jet Grouting
Coprod
Boltec Rock Bolter
Equipment for Precision Drilling
Tunnel Boring Machines

Drilling techniques apply to many different fields pertinent to soils and rocks. The main applications include oil-well drilling, water-well drilling, pile driving, underpiling, sheet piling, geotechnical surveying—prospecting for ore deposits or conducting geological surveys prior to construction of dams, power-houses and tunneling, trench pipe laying, grout injection and soil mechanics. The drilling involved is carried out with rotary drilling units, diamond drills, DTH hammers and hydraulic top hammers. Jet grouting and a number of other important tasks, such as back-anchoring and steerable drilling under road banks, constitute common procedure.

SOILS

Soil has been given various scientific definitions by agricultural engineers, geologists and others. However, the highway engineer usually considers soil as being any earth material that he encounters in his work except for embedded rock and shale. Thus soil may vary from clay to glacial debris and rock on mountain slopes.

Boulder: Any particle whose largest dimension is 12 in/305 mm or more.

Cobble: Any particle whose largest dimension is greater than 3 in/76 mm but less than 12 in/305 mm.

Gravel: A soil which contains 50% or more which is greater than 0.1875 in/5 mm but less than 3 in/76 mm. The portion of the soil which comprises less than 50% of the total is used to describe the type of gravel, such as sandy gravel, silty gravel and so forth.

Coarse gravel: Particle size between 0.75 in/19 and 3 in/76 mm.

Fine gravel: Particle size between 0.1875 in/5 and 0.75 in/19 mm.

CLASSIFYING ROCKS

To better understand our earth you must study the materials of which it is made, the most important being the rock crust. Because there are so many variations in the earth's materials, it has been necessary to assign names to many groups, classes and varieties of these materials. It will be to your advantage to become familiar with these names and their meanings.

Rocks are naturally formed mixtures of minerals (minerals are substances of definite chemical composition) and are divided into three general groups: (1) Igneous (meaning fire), or rocks that have cooled from a molten state. Some common examples are granite, rhyolite and basalt. (2) Sedimentary (settled), or rocks that are made from the solutions or pieces of other rocks. Some common examples are shale, sandstone, conglomerate and limestone. (3) Metamorphic (changed), or rocks that have been modified or changed by heat and/or pressure. Common examples are gneiss, schist, slate and marble.

Hardness

Hardness in rocks, minerals and gems is rated relatively by the capability a substance/mineral has to resist abrasion when a pointed fragment of another mineral is drawn across it.

The mineral scale of hardness established by F. Mohs is given in Table 2.1. Of the 10 minerals selected by him, each will scratch any with a number lower as per scale and will not scratch one higher as per seriation of this table. Variation in hardness is not denoted in equal degrees as per numbers of this scale. The difference between 9 and 10 is immensely greater than between 9 and 8. The Knoop scale units were added for comparison. They are determined by the ability of a mineral

Table 2.1

Minerals	Mohs	Knoop Scale Units
Talc	1	12
Gypsum	2	32
Calcite	3	135
Fluorite	4	163
Apatite	5	395
Feldspar	6	560
Quartzite	7	710–790
Topaz	8	1250
Corundum	9	1700–2200
Diamond	10	8000–8500

to withstand indentation by a special wedge-shaped diamond point under a specific
load applied in a Knoop testing apparatus.

Application data and geological information are presented for some miner-
als/rocks in alphabetical order in Table 2.2. The logical sequence of decisions in
choosing a drilling method/type of bit/spc of diamond tools is based on hard-
ness/Moh's scale of the mineral/rock to be drilled.

In addition to hardness as per Moh's scale, physical and mechanical properties
of rocks guide the choice of drilling method(s) to be adopted. These constitute
abrasiveness, strength, toughness, brittleness, density, plasticity, porosity, specific
gravity, jointing, permeability, friability and stability.

In addition, structure, texture and mineralogical composition differ in rocks of
different origin and also determine the choice of drilling method to be adopted.

ROTARY DRILLING

In this drilling method a hole is drilled by a rotating bit to which a downward
force (drill collars) is applied. The bit is fastened to and rotated by the drill stem,
which also provides a passage for the circulating fluid. A rotary drilling rig is
basically hoisting equipment necessary for well drilling. It includes a rotary table,
draw works, kelly, swivel, hook, blocks, line engines, mud pumps and piping (steel
mud pits if used), utilities unit, dog-house, toolhouse, mud-house etc., and electric
generators, motors and wiring if used.

The various rotary drilling systems are illustrated in Fig. 2.1a, b, c, d.

The airlift method operates on the principle of introducing air bubbles into the
drilling fluid returning up the centre of the drill pipe.

The introduction of air into the returning fluid makes it less dense than the
fluid between the hole wall and the outside of the drill pipe. The weight of the
fluid is thus accelerated by displacement down the drill hole, through the bit, and
with cuttings up the inside of the drill pipe.

Air is introduced into conductor pipes located on the inside of the flanged drill
pipe, either through the head, or by means of an air swivel under the head.

The means whereby the system is operated is best described by detailing the
actions taken in adding pipe during drilling of the hole:

First, 30ft/9 m are drilled using a centrifugal pump and airlift lines attached,
which mate and seal where the pipe is added (see Fig. 2.2a).

The drill bit has an outlet from one of these airlift lines. The other line is
blocked off at the head.

If 100 psi/7 kg/cm^2 air is available at approximately 70 m below the static
water level, the groundwater pressure will equal the air compressor pressure (14 psi
loss for every 10 m drilled below the water table). In other words at 70 m using
100 psi/7 kg/cm^2 air, groundwater pressure would prevent air in the airline pipe
entering the conductor pipe (assuming a water table at or near ground level).

Therefore, having drilled 20 m, one introduces a conductor pipe to the string
(see Fig. 2.2b) but continues to drill using the initial start-up air line.

Table 2.2

Name and Description	Origin	Usual colour	Average Hardness (MOH Scale)	Weight (pounds)		Specific Gravity
				Cu/Ft	Cu/Yd	
ANDESITE Fine-grained igneous rock. Easier to drill than Basalt	Igneous	Dark Grey	7.2	173	4660	2.4/2.9
ASBESTOS (see serpentine)						
BASALT Fine-grained compact rock. Hard, tough but not too abrasive	Igneous	Dark Grey, Blackish	7.0	188	5076	2.8/3.0
BAUXITE Hydrated oxide of aluminium. Soft, earthy mineral suitable for mechanical excavation or rotary drilling.	Mineral	Reddish, Grey	2.0	159	4293	2.4/2.6
BORNITE Form of copper ore	Mineral	Purple, Blue	3.0	165	4455	4.9/5.4
CALICHE Type of limestone. Generally easy drilling.	Sediment	White, Light Grey		90	2430	2.9/3.0
CHALCOCITE Form of copper ore	Mineral	Grey, Black	2.7	165	4455	5.5/5.8
CHALCOPYRITE Form of copper ore	Mineral	Coppery Yellow	3.7	160	4320	4.1/4.3
CHALK Pure marine limestone, usually rotary drilled	Sediment	White, Light Grey	1.0	137	3699	2.4/2.6
CHERT Form of silica found in many limestones. Flint sometimes called variety of chert. Tough to drill.	Sediment	Light Grey	6.5	160	4320	2.5
CLAYS Soft, earthy	Sediment	White to Dark Grey	1.0	100	2700	2.4/2.6
COALS						
Anthracite (Hard Coal)	Sediment	Black	3.0	94	2646	1.4/1.8
Bituminous (Soft Coal)	Sediment	Black	2.5	84	2218	1.2/1.5
Lignite (Brown Coal)	Sediment	Brown	2.0	78	2106	1.1/1.4

(Contd.)

Table 2.2   Continued

Name and Description	Origin	Usual colour	Average Hardness (MOH Scale)	Weight (pounds)		Specific Gravity
				Cu/Ft	Cu/Yd	
CUPRITE Copper ore	Mineral	Red, Brown	3.5	165	4455	5.8/6.1
DIABASE, DOLERITE Medium grained. Very strong due to inter lock of grain	Igneous	Grey	7.8	175	4725	2.8
DIORITE Coarse grained. For drilling estimate, con- sider as granite	Igneous	Grey	6.5	185	4995	2.8
DOLOMITE Harder than limestone. Mineral which forms Magnesium limestone and the rock itself.	Sediment	Pink, White	3.7	180	4860	2.7
FELSITE Dyke rock akin to granite	Igneous	Any Light Colour	6.5	170	4590	2.65
FLINT Concretionary form of silica. Hardness equal to quartz. Brittle	Sediment	Grey, Black	7.0	160	4320	5.5/7.0
GABBRO Coarse-grained rock. Softer than granite, not as abrasive	Igneous	Dark Grey	5.4	180	4860	2.8/3.0
GALENA Common ore of lead.	Mineral	Grey black	2.6	465	12555	7.5
GANNISTER or QUARTZITE Pale grey sandstone. Very abrasive	Meta-morphic	Variable	7.0	160	4320	2.65
GNEISS Coarse rock resembling granite when drilling	Meta-morphic	Variable	5.2	180	4860	2.6/2.9
GRANITE Coarse-grained rock containing quartz, feldspar and mica, uniformly distributed.	Igneous	Light Grey Pink	4.2	170	4590	2.6/2.9
GYPSUM Soft white mineral found as crystals in clay, and as compact beds (alabaster)	Mineral	White, Grey	1.8	175	4725	2.3/2.4

HALITE (SALT) Native salt	Sediment	White	2.5	145	3915	2.1/2.6
HEMATITE Form of iron ore	Mineral	Red	6.0	306	8263	4.9/5.3
KAOLIN (see Clays)						
LIMESTONE Rock variable in colour texture and hardness	Sediment	Variable	3.3	163	4400	2.4/2.9
LIMONITE Form of iron ore	Mineral	Grey	5.3	189	6399	3.6/4.0
MAGNETITE Form of iron ore	Mineral	Grey or Black	4.2	315	8505	4.9/5.2
MALACHITE Form of copper ore	Mineral	Vivid Green, Blue	3.7			3.9/4.0
MARBLE Crystalline metamor-phosed limestone	Meta-morphic	White	3.0	160	4320	2.1/2.9
MARL A calcareous clay	Sediment	Grey, Brown	3.0	140	3780	2.2/2.4
MICA Soft, flaky, elastic minerals	Igneous	Variable	2.3	180	4860	2.8/3.1
MICA SCHIST Laminated form of mica	Meta-morphic	Variable	3.7	170	4590	2.5/2.9
MUDSTONE Indurated clay which doesn't split into laminations	Sediment	Brown	2.0	110	2970	2.4
OBSIDIAN A volcanic glass	Igneous	Black	5.5	180	4860	1.3/5.0
PEGMATITE Coarse-grained vein granite	Igneous	Light Grey Pink	6 to 7			3.1/3.2
PORPHYRY Granite rock with large crystals of quartz in fine crystalline matrix.	Igneous	Variable	5.5	159	4293	2.8
PUMICE Hardened volcanic glass	Igneous	Grey	6.0	40	1080	1.5
QUARTZ Very abrasive. Harder than glass, very com-mon mineral	Igneous	White, Pink	7.0	185	4455	2.65
QUARTZITE Rock formed of sand grains cemented with silica	Meta-morphic	Variable	7.0	160	4320	2.0/2.8

(Contd.)

Table 2.2   Continued

Name and Description	Origin	Usual colour	Average Hardness (MOH Scale)	Weight (pounds)		Specific Gravity
				Cu/Ft	Cu/Yd	
SALT (see Halite)						
SANDSTONE	Sediment	Brownish	3.8	145	3915	2.0/2.8
Rock formed of cemented sand grains. If silica drilling difficult. If iron or lime, drilling good, with bit wear. If clay, drilling easy						
SCHISTS	Meta-morphic	Variable	5.0			2.8
Foliated rock contain-ing flaky layers of minerals. Will split into sheets. Drilling difficult parallel to planes of foliation.						
SERPENTINE	Meta-morphic	Green	4.0	171	4671	2.5/2.6
Magnesium mineral and rock. Very soft						
SHALE	Sediment	Dark Grey	2.0	160	4320	2.4/2.8
Laminated clay						
SILICA (see Quartz, Flint and Chert)						
SLATE	Meta-morphic	Grey	3.1	168	4535	2.5/2.8
Formed from laminated clay under strong pressure						
SPHALERITE	Mineral	Brown	3.7	253	6831	4.0
Principal ore of zinc						
SYENITE	Igneous	Variable	5.8	165	4455	
Coarse rock						
TACONITE	Mineral	Grey		150	4050	2.6/3.0
A form of iron ore				200	5400	
TALC	Sediment	White, Grey	1.0	168	4535	2.5/2.8
Soft mineral, soapy feel, Found in folia-ted granular or fibrous masses						
TRAP ROCK	Igneous	Brown	7.0	180	4870	2.6/3.0
Basalts, dolerites etc						

At 30 m one introduces a second conductor pipe into the string (see Fig. 2.2c), at the same time blocking off the initial airline supply and opening the second airline supply conductor pipe (Fig. 2.2b). The second airline supply flows down the string

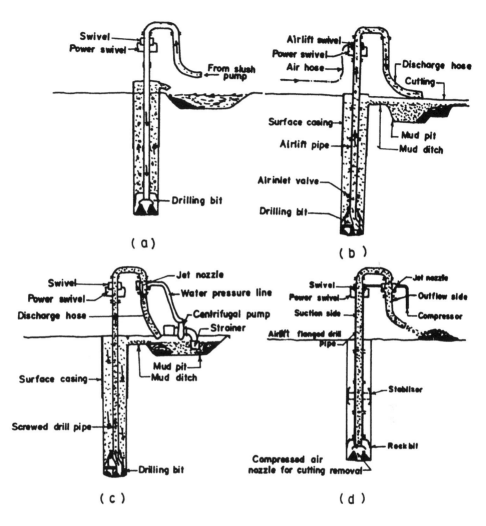

Fig. 2.1: (a–d)

(a) Direct circulation drilling; (b) airlift drilling system; (c) jet suction drilling system; (d) compressed air jet suction dry drilling system.

and into the conductor pipe which was introduced at 20 m depth but is now at 30 m depth.

At 40 m total depth introduce another conductor pipe to the string and close off airline supplying to conductor pipe (Fig. 2.2b), opening airline to conductor pipe (Fig. 2.2c).

Proceed drilling the hole in this fashion to its entire depth. The principle is one of displacement of the aerated column (see Fig. 2.3). The area of air injection

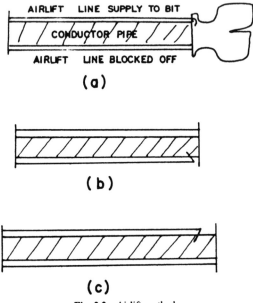

Fig. 2.2: Airlift method.

providing a less dense column of fluid must be displaced by the weight of the fluid in the drill hole through the drill bit. This will work regardless of depth subject only to friction losses.

The advantages of this method against the centrifugal pump method are, firstly, considerably greater depths can be achieved, with air pressure being increased to increase the depth of submergence of the aerated column, thus counteracting friction losses.

Secondly, since the fluid return does not pass through a pump, a stone catcher is not required. This is of considerable advantage when drilling large gravels or river-bed type aquifers.

The disadvantage of the system is that the hole must be drilled to 10 metres plus before an effective submergence of an aerated column is obtained. This can be done by using either a bucket bit or a centrifugal pump.

Rotary Rigs

These rigs are truck-mounted all-hydraulic rotary drill rigs intended primarily for well drilling but also used for blast-hole drilling in quarries and open pit mines. They are designed to meet the requirements of modern drilling technology for an efficient rotary drill rig. Accurate control of pulldown and holdback (positive and negative feed), a high maximum torque and a wide spindle feed range make the drills just as suitable for drilling with downhole drills as for drilling with roller or drag bits.

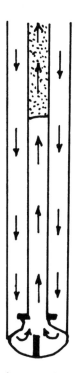

Fig. 2.3: Displacement of aerated column.

DUAL-TUBE DRILLING SYSTEMS

Several terms are synonymous with the dual-tube drilling system, including 'double-wall pipe drilling system' and 'rotary continuous sample system'. In either case the concept is the same, as illustrated in Fig. 2.4. An alternate view is illustrated in Fig. 2.5. The drill pipe is constructed with two concentric tubes, one within the other. When drilling with a double-wall drill pipe the circulation medium, usually air, is forced down the annulus between the inner and outer pipes to the drill bit and then directed to the centre of the pipe carrying the cuttings, chips or core-like samples to the surface continuously at high velocities. Depending on the formation, either drag, open-face, tricone or hammer bits can be used. The double-wall drill pipe is usually flush jointed, permitting the borehole to be cut with a minimum of clearance. With this type of reverse circulation drilling, the samples are forced to the surface through the centre tube and sample contamination by caving formations or particles eroded from the wall of the hole is eliminated. The danger of losing samples into voids and fractures is minimalised.

There are several important advantages to the dual-tube system.
A) Continuous samples: The system delivers a representative sample at high velocities continuously to the surface for collection. A larger volume of sample material is recovered per inch diameter of hole than with core drilling. Sample

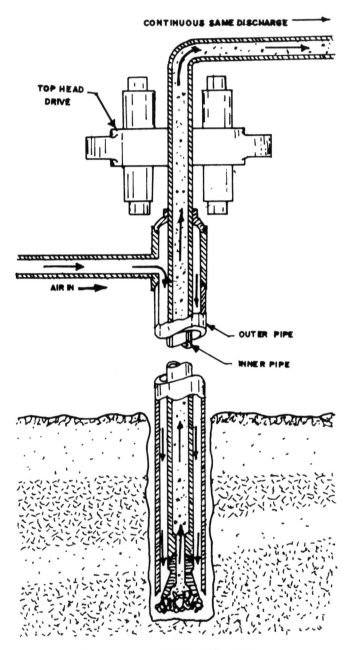

Fig. 2.4: Typical dual-tube drilling system.

recovery will equal core drilling and in many situations, surpass recovery
percentages.

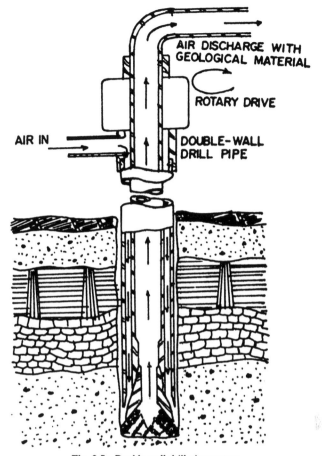

Fig. 2.5: Double-wall drill pipe system.

B) Rate of Penetration: With top-head drive rotaries, hourly and daily production rates are higher than with most conventional core drills, even as much as 18 times greater in some situations.

C) Operating Costs: Operating costs are less than with conventional techniques. Bit costs are usually less than when using other techniques. Operating costs in some cases have been reduced by 75%.

D) Hole Deviation: Because of the flush wall and packed assembly, deviation is less than with conventional drilling techniques.

E) Surface Casing: Surface casing can be eliminated because of the configuration of the dual-wall pipe. The outer pipe supports the hole while circulation is maintained internally.

F) Lost Circulation: The pipe configuration, maintaining circulation inside, can provide circulation to the surface even while drilling in vugs, fractures, voids, joints and low-pressure zones.

G) Surface Equipment: Wherever possible air or air with water injection is used to provide a cleaner sample and faster return to the surface. Some projects require water or mud drilling; in either event smaller volumes than those used in conventional drilling are required to supply adequate bailing velocities and sample return.

The dual-tube system is used in the following specific mineral exploration and sampling applications:

A) Coal sampling, primarily for surface mining deposits
B) Placer deposits including gold, magnetite, cassiterite, rutile
C) Phosphate sampling
D) Laterite sampling (nickel)
E) Uranium exploration
F) Mercury sampling
G) Lithium prospecting
H) Copper exploration
 I) Other sulphide sampling projects
J) Vein sampling such as gold
K) Bauxite exploration
L) Diatomaceous earth sampling
M) Lignite prospecting.

The number of dual-tube projects are increasing as more drillers become aware of the potential of the system for providing the information required faster and more economically than with conventional drilling or core drilling.

Sampling procedures vary from application to application. Proper collection of the sample, whether in the form of chips or cores, is the singlemost important phase of the system. The sample is returned to the surface at high velocity and passed into a pneumatic separator. Then the sample is split, tubed, bagged and boxed depending on the requirements.

The dual-tube drilling system for continuous sampling is becoming a rapidly accepted exploratory tool. The technology is expanding and more people are realising the advantages of the system over other techniques that have been in use for many years.

Section drilling methods in dual-tube systems are illustrated in Fig. 2.6.

AUGER BORING

Back in the Iron Age, after people learned how to make iron and steel, it was not very long before the first steel auger was invented to do sample drilling in the earth in search of minerals. This technology has evolved continuously since then and there are many different types of augers and auger drilling equipment. The astronauts of APOLLO XI in 1969 and APOLLO XVII in 1972 used augers to penetrate the crust of the moon. These samples of soil are displayed for the general public at the National Air & Space Museum in Washington, D.C. For centuries, the auger was a solid flight auger, meaning that the construction is a solid centre stem with a flat steel wrapped around this stem to bring the cuttings to the surface.

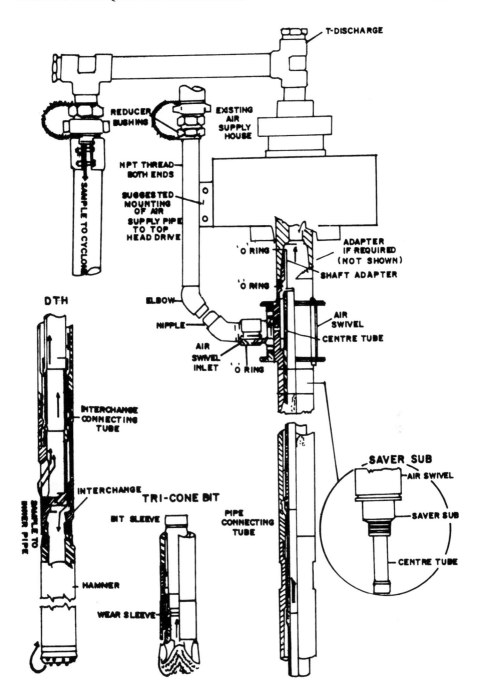

Fig. 2.6: Section drilling methods in dual-tube systems.

The current KPA (Environmental Protection Agency) requirements in the USA have forced manufacturers to improve auger boring so that a means would be provided to allow sampling to the bottom of the hole into undisturbed soil for investigation. The new augers (see Figs. 2.7a, b) are known as hollow-stem augers and are classified by size based on the inside diameter of the auger. For example, a 7 in/178 mm and an outside diameter measured to the outside of the flight of 11-1/2 in/292 mm and the bit cut a 12-in diameter hole in the ground. The original standard test hole would require a 2 in/51 mm size to be 4 in/102 mm, 6 in/152 mm, 8 in/202 mm and larger, requiring larger hollow-stem augers to do this job. The larger augers have forced drill rig manufacturers to design very powerful portable auger machines, such as the Failing Model F-10, with a maximum torque of 30,000 ft/lb/4050 kg/m available to turn these augers. Please refer to Fig. 2.8a and notice the construction of the tubulars to drill a well. The inner pipe is connected to the centre bit which plugs the opening of the hollow-stem auger. Drilling is done by the main auger bit. An undisturbed sample can be taken, as illustrated in Fig. 2.8b, utilising a sampling device working through the inside of the hollow-stem auger, after removing the centre bit.

Let us now take a look at Fig. 2.8c. After the hole has been drilled to its total depth, the casing can be inserted inside the hollow-stem auger, after removal of the inner pipe and the centre bit. Pulling the main auger up a little ways, at 5 ft/1.5 m intervals, the use of a tremie pipe will allow the driller to fill the annular space between the casing and the borehole with material of his choice, such as grout, gravel pack or cement. Fig. 2.8d shows a completed well.

This type of drilling has been adapted more and more to drill domestic water wells to depths of 100–200 ft—30–60 m. This type of drilling requires no mud pump or water for circulation. Therefore, problems of lost circulation are totally eliminated. There is also no wallcake, which eliminates the need for dissolving such after drilling the hole before setting the casing. Also, no air compressor is required for air drilling, which would necessitate a large compressor with a big engine. The simplicity of drilling thus avoids many problems encountered during normal drilling, while the casing can be set inside a hole that is already cased and the gravel pack is very easy as described in the various procedures below.

The machine to do this is portable and very easily maintained since the top drive is driven mechanically and not by hydraulics. The few hydraulic components necessary on the machine all operate on a low-pressure hydraulic open loop circuit. The mast can be adapted to utilise drilling rods of 10 ft/3 m, 15 ft/4-1/2 m or 20 ft/6 m drill pipe. It is a very effective and simple way to drill water wells without mud pumps or compressors and therefore requires a minimum investment.

This machine has proven to be very versatile. Should a water well be required in hard zones, the hollow-stem auger drilling method can be used to drill through the overburden to the hard formation. The driller can then simply use an auxiliary air compressor unit and a swivel and drill pipe combination for drilling by the down-the-hole hammer method through the inside of the auger or he can switch over to mud rotary drilling by hooking up to an auxiliary mud pump unit. Also,

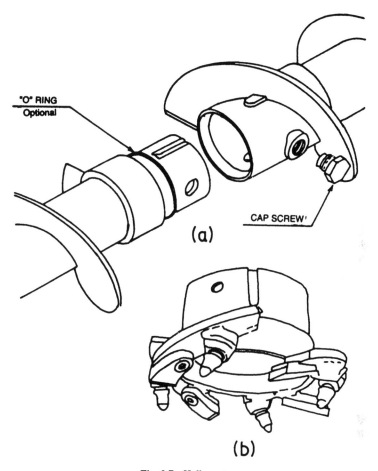

Fig. 2.7: Hollow-stem augers.

the units are equipped with a high-speed top drive to allow wireline coring, again using the hollow-stem auger as a surface casing while drilling through the inside with the core rods and barrel.

The new design hollow-stem auger drill, such as the Failing Model F-10, is extremely versatile and can do a large number of drilling operations. The hoisting capacity, also known as the pullback capacity, is 25,000 1b/11,340 kg, which can easily be upgraded to 30,000 1b/13,610 kg. Operation and maintenance have been reduced to a minimum in this newly designed drilling rig.

OVERBURDEN DRILLING

As much as 90% of the land surface of the earth is covered with loose, unconsolidated material, such as soil, clay, silt, sand, gravel and boulders, which varies in depth from a few centimetres to hundreds of metres.

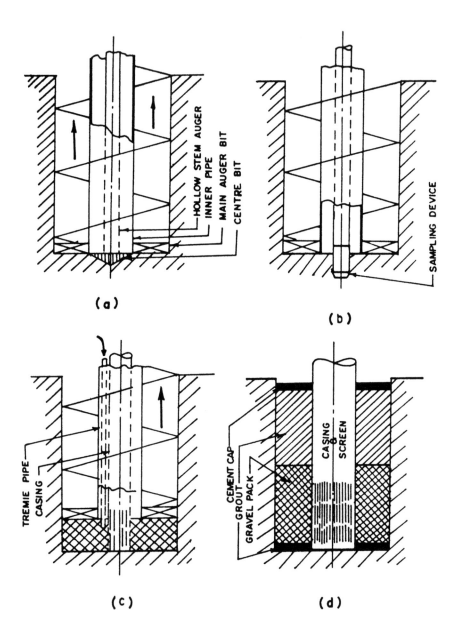

Fig. 2.8: (a–d)

(a) Construction of tabulars; (b) Undisturbed sampling; (c) Tremie pipe and (d) Completed well.

Drilling through this so-called overburden is often problematic due to the tendency of the earth to cave in behind the drill bit. This makes it difficult to retrieve the drill string after the hole has been drilled. In practice, the borehole is often lost before a casing tube can be inserted to support it.

Other problems are caused by cavities or porous ground, which interfere with circulation of the flushing medium and prevent drill cuttings from being flushed out of the hole.

In places where overburden strata are missed, or when their drillability is not known, it is difficult for the driller to decide what tools to use in order to get the best overall results without risking the loss of equipment in the hole.

The best solution for dealing with such problems is to use ODEX equipment (see Figs. 2.9 and 2.10).

ODEX equipment enables simultaneous drilling and casing of deep holes in all types of formation, even those with large boulders. Casing diameters from 89 mm to 222 mm can be used. This method is based on a pilot bit and an eccentric reamer, which together drill a hole slightly larger than the external diameter of the casing tube. This enables the casing tube to follow the drill bit down the hole.

When using ODEX together with DTH hammers, part of the impact energy is diverted to the casing tube via a shoulder on a guide device, which in turn impacts a special casing shoe at the lower end of the casing. ODEX equipment for the top hammers operates in a traditional manner, with impact and rotation transmitted through extension rods. To drive the casing down the hole, the shank adopter is used to transfer part of the impact energy from the rock drill to a driving cap above the casing tube.

In both DTH and top-hammer drilling the casing is driven down into the hole without rotation. When the casing enters the bedrock, drilling is stopped briefly and reverse rotation applied carefully, which causes the reamer to turn in, thus reducing the overall diameter of the drill-bit assembly. When this has been accomplished, the entire drill string can be pulled up through the inside of the casing tubes, leaving the latter embedded in the bedrock. Drilling can then be continued into the bedrock by moving a conventional drill string.

The flushing medium that carries the cuttings to the surface when using ODEX for pneumatic or hydraulic top hammers is air, which is directed through the shank adopter.

To improve flushing, the ODEX guide device has backward-directed flushing holes. In difficult conditions a foaming additive may be added to the compressed air to further improve flushing performance.

Commercially available steel tubes in standard dimensions are used for the casing. They are welded together and left in the ground after the hole has been completed. In certain situations the casings may be reused. In this case it generally pays to use threaded casing tubes.

Optimum utilisation of any product is naturally dependent on correct training in handling and in maintenance.

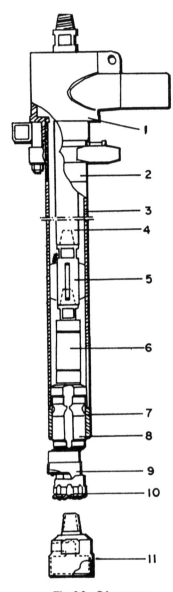

Fig. 2.9: Odex system.

1. Discharge/diverter head; 2. Adapter sleeve; 3. Casing tube; 4. Drill pipe; 5. Guide sleeve; 6. DTH hammer; 7. Casing shoe; 8. Guide device; 9. Reamer; 10. Pilot bit; 11. Grouting device.

JET GROUTING

This technology has led to reconsideration of the ground consolidation concept. Atlas Copco and CCP International, which are in the van of this field, are introducing on the market a complete range of machines with high performance and productivity.

	With Top hammer	With DTH hammer				
	ODEX 76	ODEX 90	ODEX 115	ODEX 140	ODEX 165	ODEX 190
Water-well drilling			●	●	●	●
Installation of tie back anchors	●	●	●			
Pile installation			●	●	●	●
Soil investigation	●	●				
Grouting	●		●			
Exploration	●	●	●	●	●	●
Underwater drilling	O		O			
Road embankment drilling	●	●	●	●	●	●

● = Ideal combination   O = Possible application

Fig. 2.10: Common applications of ODEX equipment.

A general diagram of a CCP system is illustrated in Fig. 2.11.

Jet grouting applications are shown in Figs. 2.12, 2.13, 2.14 and 2.15

An operating diagram of the CCP system is shown in Fig. 2.16. It is a nomograph reflecting relationships among grout density, pressure, injector diameter, number of injectors, requisite horse power in addition to requirement of water in litres per minute. The dotted line indicates one such relationship: 3 indicators of 2 mm diameter involving 400–450 hp with 400 bar/5800 psi at grout density of 0.8 will require 140 litres of water.

COPROD

Conventional extension drill rods are the product of technical compromise since they have to perform several conflicting functions. For drilling to take place, they must be able to:

1) Transmit impact energy from the rock drill piston to the drill bit.
2) Transmit rotational torque to the drill bit.
3) Transmit feed free to the drill bit.
4) Convey a flushing medium to the drill bit.
5) Couple easily and reliably together to enable deep holes to be drilled (they must also be easy to uncouple from each other).
6) Transmit retraction force to the drill bit.
7) Give a long service life.

It is well known that any change in the cross-sectional area of the drill string interferes seriously with transmission of impact energy to the drill bit. The main offenders in this respect are the threads and the flushing hole in the drill steel.

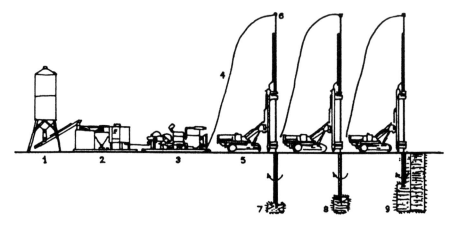

Fig. 2.11: Jet Granting—Colla CCP International Pumps and mixing plants, Atlas Copco Mustang Jetting drilling rigs.

1. Cement silo.

2. Mixing plant:
Weight measuring (300/500 kg); Manual and programmable automatic operations; Remote control; 2,700 l water tank; 3,800 l grout agitator; Up to 16 m³/h total capacity;

3. Triplex T-10 high-pressure pump
FIAT or Mercedes 250 up to 650 hp engines; Fuller 8/15 ratio mechanical transmission; Pistons: 3 3/8″ (3″) bore, 10″ stroke; Max pressure: 450 bar (3 3/8″); 700 bar (3″); 1,000 bar (2 1/2″)*; Max delivery: 320 l/min at 400 bar; 800/1,000 l water tank;

4. High-pressure hose

5. Mustang Jetting drilling rig
Max torque: 600 up to 1,000 kgm; Max pullback/pushdown: 5,000 up to 8,000 kg; Max shaft speed: 450 up to 630 rpm; Crawler carrier with 360° swivel assembly for best approaching drilling point; Hydraulic chuck with 92 mm max bore; Sliding upper rod guide for easy transport; Max bore depth: about 15 m with 18 m drill string; Hydraulic and electronic control system for timed lifting; Fast switching from 'jetting' to ordinary drilling rig;

6. High-pressure swivel head

7. Drilling step

8. Lifting back and injection step

9. Process cycle finishing and repeating

With the trend in development of hydraulic rock drills moving towards larger and more powerful machines, it became clear that conventional threaded extension rods would not be able to transmit the extra impact power efficiently. For this reason Atlas Copco developed the COPROD system (see Figs. 2.17a, b, c) which enables transmission of impact and feed to be separated from transmission of rotation and flushing, by means of separate components. This makes it possible to optimise the respective components of the system for the functions they are to perform.

Coprod Drilling Principle

Coprod combines the speed of top-hammer drilling with hole straightness and dimensions of tube drilling.

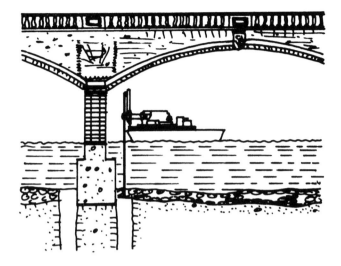

Fig. 2.12: Underpinning.

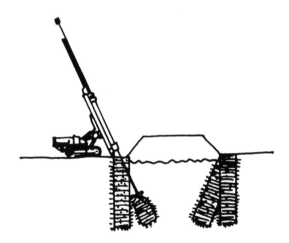

Fig. 2.13: Railway embankment injection.

The working principle is simple and effective: Coprod features both an impact rod (non-rotating) and a drill tube. This means that transmission of impact energy is completely separated from rotation in the drill string.

Impact from the rock drill portion is transmitted over an anvil (adapter) and through the impact rods just as in conventional drilling. The difference is that the rods simply rest one on top of the other, pressed together firmly by the thrust free of the rock drill. With such an arrangement, there are virtually no impact energy losses when the shock waves are transmitted from one component of the drill string to another.

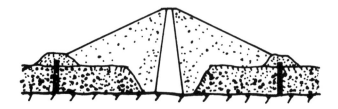

Fig. 2.14: Waterproofing cutoff walls.

Fig. 2.15: Tunnelling protection by sloping diaphragms.

The impact rods in the Coprod string therefore meet demand Nos. 1 and 3 listed above. The impact rods are enclosed in the wall tubes, which are joined together by means of conventional conical threads. The rotation chuck transmits the rotation motor torque to the drill bit via a tube driver, the tubes and the Coprod head. The bit is turned by means of a spline connection, similar to that in a DTH Hammer.

When the tubes are pulled up, the drill bit drops down in the chuck and is arrested by a stop ring, as in a DTH hammer.

This means that the drill tubes in the Coprod system meet demand Nos. 2 and 6. The impact rods are centred inside the tubes by exchangeable polymer guides at the ends of each tube. The guides have four axial slots along the rods. The impact rods have forged heels, which prevent them from falling out of their respective tubes.

The flushing medium is conveyed from the rock drill to the drill bit via the annular space between the impact rods and the tubes. The axial slots in the polymer guides permit the flushing medium to flow between the joints in the drill string. In this way, the combination of impact rods and tubes in the Coprod drill string meet demand Nos. 4 and 5.

Compared with conventional extension rods, transmission of impact and rotation energy to the drill bit via separate components gives much greater efficiency. This is because the absence of threads on the impact rods, which are simply 'stacked' one on top of the other, enables a much higher impact power to be fully utilised in an economical way.

Hole straightness is achieved by the combination of a rigid tube supporting an inner impact rod and is equal to that typical of DTH drilling.

This together with separate transmission of impact and rotation forces, increases the service life of the Coprod drill string to such an extent that it is terminated by wear and not by fracture.

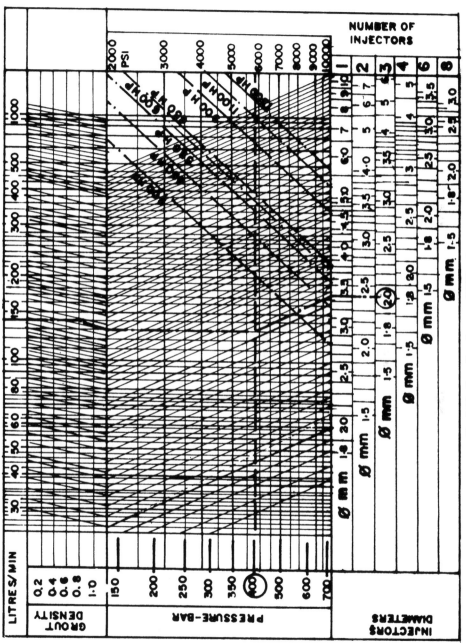

Fig. 2.16: Operating diagram of CCP system.

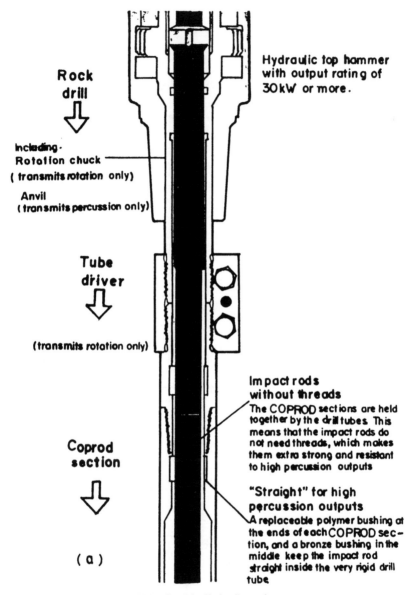

Rock drill ⇩

Including·
Rotation chuck
(transmits rotation only)

Anvil
(transmits percussion only)

**Hydraulic top hammer
with output rating of
30 kW or more.**

Tube driver ⇩

(transmits rotation only)

**Impact rods
without threads**
The COPROD sections are held
together by the drill tubes. This
means that the impact rods do
not need threads, which makes
them extra strong and resistant
to high percussion outputs

Coprod section ⇩

**"Straight" for high
percussion outputs**
A replaceable polymer bushing at
the ends of each COPROD section, and a bronze bushing in the
middle keep the impact rod
straight inside the very rigid drill
tube

(a)

Fig. 2.17(a): Coprod—Hydraulic top hammer.

This meets demand No. 7.

Coprod enables the latest in powerful impact mechanisms (30 kW and more) to be fully utilised with good drilling economy.

At the same time the Coprod user benefits from: 1) higher penetration rates, 2) very straight holes and 3) greatly improved service life.

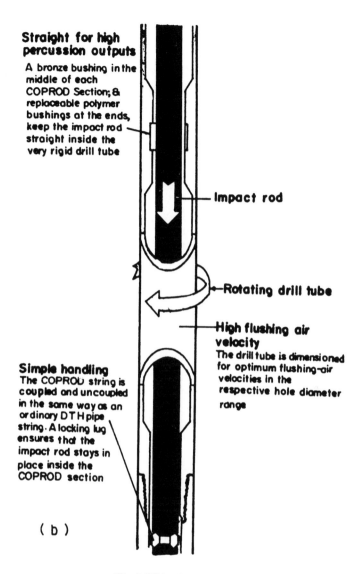

Straight for high percussion outputs

A bronze bushing in the middle of each COPROD Section; & replaceable polymer bushings at the ends, keep the impact rod straight inside the very rigid drill tube

— Impact rod

— Rotating drill tube

High flushing air velocity
The drill tube is dimensioned for optimum flushing-air velocities in the respective hole diameter range

Simple handling
The COPROD string is coupled and uncoupled in the same way as an ordinary DTH pipe string. A locking lug ensures that the impact rod stays in place inside the COPROD section

(b)

Fig. 2.17(b): Coprod—impact rod.

BOLTEC ROCK BOLTER

Atlas Copco's new mechanised bolting rig, Boltec, increased productivity by some 20% during field tests compared to the mine's existing bolters. Boltec can be used to instal any kind of bolt commonly used.

It has completely new single feed drilling/bolting unit (see Fig. 2.18) which is extremely compact and lightweight.

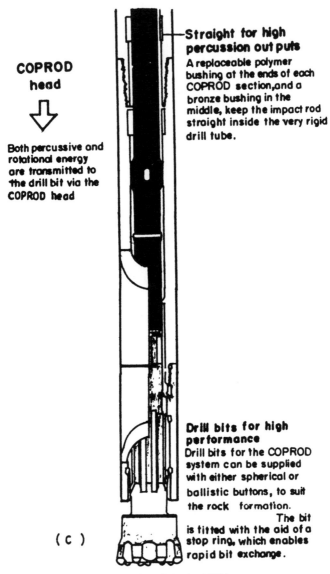

COPROD head

⬇

Both percussive and rotational energy are transmitted to the drill bit via the COPROD head

Straight for high percussion outputs
A replaceable polymer bushing at the ends of each COPROD section, and a bronze bushing in the middle, keep the impact rod straight inside the very rigid drill tube.

Drill bits for high performance
Drill bits for the COPROD system can be supplied with either spherical or ballistic buttons, to suit the rock formation. The bit is fitted with the aid of a stop ring, which enables rapid bit exchange.

(c)

Fig. 2.17(c): Coprod—drill bit.

By means of a simple switch-over arrangement the operator changes from drilling to bolting without moving the feed from the hole position. In addition, the operator has full visibility over the bolting area.

The new bolter gives better accessibility, requires less space and is easier and faster to manoeuvre. There are fewer moving parts and the rig is also extremely stable, which means greater precision in positioning.

THE HEAD

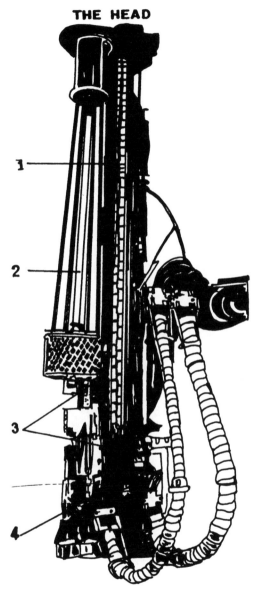

Fig. 2.18: Boltec rod bolter: 1. Light aluminium feed; 2. Bolt magazine; 3. Switch positive of work tools; 4. Hydraulic rock drill.

It has an out-of-the-way bolt magazine and the support swings away automatically for the rock drill and the bolt chuck, enabling the full length of the feed to be utilised. This means bolts can be set at their full length even in difficult-to-access areas and when the rock face is rough.

EQUIPMENT FOR PRECISION DRILLING

Atlas Copco has developed a system for drilling long holes very straight, with a deviation of less than 1%. In normal drilling deviation is often as high as 5%. Interest in straight-hole drilling is growing in the mining and civil engineering industries. Applications include extremely long holes in mines, as well as sewage and communications holes.

The straight-drilling component incorporates four projecting guide devices placed evenly around the periphery. The guide device (see Fig. 2.19) uses the walls of the hole to prevent the bit from changing direction.

Other parts include standard Atlas Copco DTH components. Three important factors contribute to the drilling of very straight holes:
— A perfect fit between guides and drill hole
— Rigidity in all parts of the guiding device
— Placing guides close to the hammer.

Straight-hole drilling is economical. Apart from the practical need for straight holes, the equipment also fulfils a number of other important requirements, which help to make operations very economical.
— Increased hole depth
— Reduction of number of holes and charges
— Less secondary blasting.

Fig. 2.19: Equipment for precision drilling.

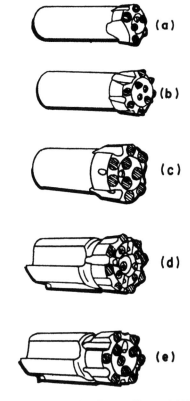

Fig. 2.20: Button bits for tunnelling and drifting.

Button Bits

Button bits of various types are used with extension drilling equipment; these are shown in Fig. 2.20

Other types of button bits are used in tunnelling and drifting operations; these are shown in Fig. 2.21.

Down-the-hole bits are manufactured in many dimensions. Each dimension is available with one or several of the figurations illustrated in Fig. 2.22

TUNNEL BORING MACHINES

The future belongs to tunnellers. There is simply no doubt about it. The world's infrastructure systems are expanding rapidly and that puts tunnelling contractors who can guarantee high-speed, high-quality tunnels in first position.

In a comparison between tunnels, bridges, ring roads and other ways of dealing with natural obstacles, tunnels invariably take precedence.

For one thing, they meet today's and tomorrow's environmental demands. They do not disturb the landscape. They provide an effective way of reducing

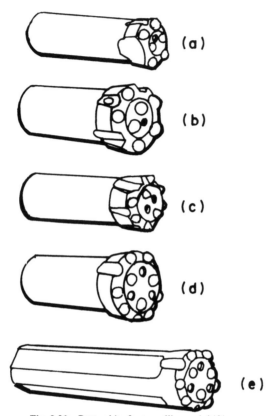

Fig. 2.21: Button bits for tunnelling and drifting.

traffic noise and exhaust fumes. And because they do not take up valuable space on the surface, they help to keep overall costs to a minimum.

Furthermore, in the long term tunnels are cheaper than other alternatives simply because they require less maintenance. Thus being so, the demand for high-speed, high-quality tunnelling is constantly increasing and contractors are expected to offer the best available technology.

Development of equipment needed for tunnelling in hard rock has progressed rapidly over the last few years. The new rigs with booms have made the drill and blast method as easy and as productive as possible. These new-generation drilling rigs are characterised by high speed and high precision. The Robot Boomer machine of Atlas Copco combines the operator's personal skill with the accuracy and reliability of automation. It is designed to give optimum results, whatever the task, and has the added advantage of a computer which continuously stores all the data from drilling and positioning and provides a printout at any time to verify that quantity demands are being met.

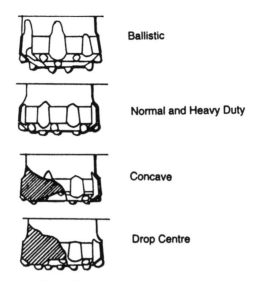

Ballistic

Normal and Heavy Duty

Concave

Drop Centre

Fig. 2.22: Down-the-hole hammer bits.

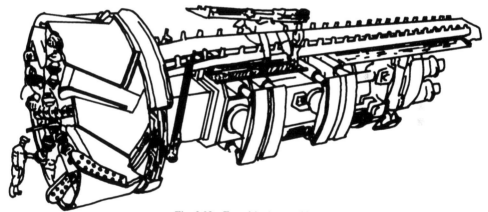

Fig. 2.23: Tunnel boring machine.

The new generation of Boomer rigs can be used manually, backed by sophisticated instruments, semi-automatically for optimum speed and accuracy, or completely automatically when accuracy is the only priority. In other words, freedom of choice is the name of the game.

Another kind of freedom is the contractor's freedom to select the method of rock excavation and type of equipment he prefers. TBM technology has now made full-face tunnelling a very competitive alternative in hard and mixed rock, long distance tunnels, or for boring pilot tunnels as a complement to drill and blast. In addition, the TBM method has proved to be the perfect solution in projects where noise and vibration are critical factors.

Figure 2.23 illustrates the most powerful tunnel-boring machine ever built to tackle a tough job at a power-station project in northern Sweden.

A newly developed Atlas Copco Jarva Mark 27 TBM will drive a water intake tunnel 10-8 kilometres long near Tarnaby.

The Atlas Copco machine has a cutter head with a diameter of 6.5 m for full-face boring and is the strongest TBM ever developed for this tunnel size. The cutter head power is an incredible 4600 hp which can be further expanded to 6100 hp.

CHAPTER 3

DESIGN CRITERIA FOR DRILL RIGS

HOW TO DETERMINE YOUR RIG'S DEPTH LIMIT

The depth a rig is capable of attaining is of major concern to the contractor when deciding on a drilling programme. So the question 'What is the depth limit of your equipment?' is a viable one. The answer, however, requires a great deal of calculation and discussion, and only after all rig components have been carefully selected can a satisfactory one be found.

A rig equipped for drilling with air or mud in direct circulation may have a 'base' rating of 8000 ft/2400 m with 4 in/101.6 mm drill pipe. But 8000 ft. for this rig is vague because the rating is termed basic and is determined using a specific size of drill pipe. The size of drill pipe to be used with a rig, however, is only one of many factors determining the depth rating.

All the following factors are important when determining depth and most manufacturers offer optional components to match the drilling programme spelled out by the customer:

— Mast hookload capacity
— Drawworks capacity (single-line pull, brake capacity, spooling capacity)
— Mud pump capacity (volume in gal/min, pressure in psi)
— Air compressor capacity (volume in cfm, pressure in psi)
— Drilling programme (diameter of hole to be drilled, strata to be drilled)
— Horsepower of power units
— Auxiliary equipment (stand-by pumps, stand-by compressors, well control units, mud system)

The list can go on and on. But perhaps most important in determining depth rating is the discussion between the customer and the rig manufacturer. The complete drilling programme must be discussed so the sales engineer can select those components best suited to the job. There are no 'one rig does it all' packages that drill all holes economically at maximum efficiency based on equipment investment and utilisation.

Exact Mast Rating

The exact rating of the mast should always be determined because gross load, gross nominal rating, or just mast rating does not give the necessary information. What is needed are the various hook-load ratings based on corresponding string-up of the travelling block. Trying to modify the crown block to get more lines strung up will void the crown and mast warranty because loads in the crown are balanced and distributed to the main mast legs. Unequal distribution will result in twisting of the mast.

The gross load rating represents the maximum load imposed on the mast by the crown block. (see Fig 3.1 to calculate the maximum static hook-load capacity.)

With an imaginary line, cut the mast in half between the crown block and the travelling block. It is now obvious how many lines are pulling down on the crown (symbolised by double arrows) and how many lines are pulling up on the travelling block (symbolised by single arrows). Therefore, the maximum gross load divided by the number of lines pulling down will dictate the maximum allowable single-line pull.

In this case, it is 312,500 lb divided by 10, resulting in 31,250 lb. The load in the wireline, expressed in pounds of tension, is constant in the line from drum to deadline anchor. Thus the usable maximum static hook-load for Fig. 3.1 is the number of lines pulling up on the block multiplied by the single-line pull. Eight times 31,250 lb/14,200 kg equals 250,000 lb/113,500 kg hook-load capacity.

This basic arithmetic will work for all non-AP masts up to approximately 350,000 lb/149,820 kg gross load. Fig 3.2 depicts other combinations.

The total dry weight of the entire drill string should not exceed 75% of the hook-load. The maximum usable dry drill steel weight should not exceed 75% of the maximum static hook-load. This allows a reserve pulling capacity of 25% in a dry hole and approximately 25% + 12% = 37% reserve pulling capacity in a

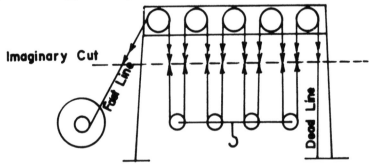

Fig. 3.1: Dividing the mast in half between the crown block and the travelling block shows that 10 lines are pulling down on the crown, while 8 are pulling up on the travelling block. Assuming this mast has a gross load capacity of 312,500 lb, the maximum single-line pull would be 31,250 lb and maximum static hook load 250,000 lb.

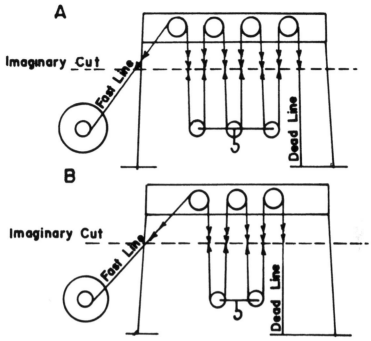

Fig. 3.2: Configuration A, with a gross load of 312,500 lb, has a hook-load capacity of 234,375 lb and a maximum single-line pull of 39,062 lb. Configuration B, with the same gross load, has a hook-load of 208,333 lb and single-line capacity of 52,083 lb.

mud-filled hole at the total depth of the well to be drilled. This percentage may vary slightly, based on different mud weights.

With all this information it is possible to calculate the depth rating based on mast capacity. It is obvious that a smaller drill pipe will allow more depth; however, the mud pump requires as large a drill pipe as possible. Hydraulics are as vital to engineering a hole into the ground as is the mast capacity. A happy medium is needed to satisfy both mast and mud pump.

Sometimes a separate, partial string of lightweight drill pipe will allow the operator to get an extra 500 or 1000 ft/150 or 300 m depth capacity.

Drawworks Capacity

The clutch is the main factor in determining the maximum single-line pull of the drawworks. The brakes may be single or double, air cooled or water cooled and are matched to the single-line pull. The sprocket or sheave ratio will provide the block and hook speeds and the combination of load and speed will calculate back to horsepower.

For example, to move 60 ft/18 m of drill pipe with a total hook load of 100,000 lb/45,400 kg in 20 seconds will require 545 hp. A smaller drawworks will move this load in less time.

Stringing up more lines to the travelling block will increase the block hoisting capacity while the block speed decreases. Also, by spooling more line onto the hoisting drum, there is a percentage loss of single-line pull each time the line jumps to the next layer. So a full drum can apply only 66% of the designed bare drum single-line pull (Fig. 3.3.).

Heavy loads lowered entirely on the brakes could build up a live load horsepower far beyond the drawworks capacity. For example, the load of 100,000 lb, after a free fall of just 8 ft/2.4 m will develop a speed of 22 ft/6.7 m/s and an energy capacity in excess of 3,000 hp, based on 1/2 s of stopping time. Runaway loads, as well as any shock loads, must be avoided at all times.

Correct Mud Pump

Correct selection of a mud pump is only possible when the full drilling programme is known: hole diameter, hole depth, diameter and style of drill pipe, and all related equipment.

In direct circulation an annulus return velocity of the drilling fluid is rated as follows:

— 50 ft/15 m/min: minimum recommendation
— 100 ft/30 m/min: fair
— 150 ft/45 m/min: good
— 200 ft/61 m/min: exceptional.

The selected return velocity in combination with the area of the annulus will yield the required volume in gallons per minute. This volume is calculated into

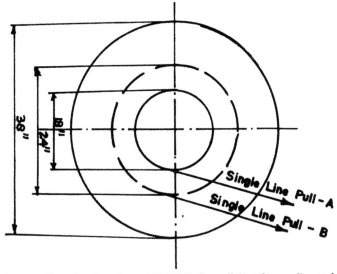

Fig. 3.3: Assume a line pull at bare drum of 30,000 lb (line pull A). After spooling on four layers of 1-in line, the diameter of the spool will be 24 in and the line pull (B) will be 20,000 lb, which is only 66% of the initial line pull. Calculations are based on identical torque loading of the clutch.

friction loss curves of drill pipe and drill collars, resulting in the total pressure due to the restrictions of swivel openings, tool joints, drill collars, and bit pressure drops. Once these factors are calculated, an adequate pump may be selected.

In general, for a high-volume and low-pressure pump, the duplex dual-acting piston pumps are best suited. This choice would enable drilling a large-diameter well with a large drill pipe at a limited depth due to the weight of the larger drill pipe. A low-volume, high-pressure triplex single-acting pump may be used when drilling small-diameter, deeper wells with a small drill pipe.

In both cases a series of different liner and piston sizes is available to accommodate the best hydraulic system suited to the well to be drilled. Different well designs may dictate a change in linear size or the use of stand-by pump units.

Pump performance depends greatly on the horsepower available to drive the pump, which must be at least 10% more than the hydraulic horsepower of the pump due to mechanical and hydraulic efficiencies. When all the data is available, the optimum choice of components to drill a well at maximum efficiency may be calculated. A change in drill pipe size may lower fuel consumption by the pump engine, due to lower pump pressure, enough to pay the difference in price of a more expensive drill-string in less than a year.

For example, if 400 gal of mud are pumped through a $4\frac{1}{2}$ in/114 mm (internal flush) drill pipe and also a $3\frac{1}{2}$ in/89 mm drill pipe, each 5000 ft/1024 m long, the friction loss in the smaller pipe will be 1400 psi but only 250 psi/172 kg sq. cm in the $4\frac{1}{2}$ in drill pipe. The difference is 1150 psi/80 kg/sq. cm at 400 gpm/ 1820 litres and a hydraulic horsepower of 201. Add 10% for efficiency and a diesel of 225 bhp is needed to overcome the difference in pressure.

This results in an approximate savings of 21 gal/hr. At a cost of $ 1.25/gal, it would take just 6 hr to pay the difference in cost of the $3\frac{1}{2}$ in and $4\frac{1}{2}$ in/89 mm and 114 mm drill pipe, based on a 30 ft/9 m pipe.

The mud pump is protected from overpressuring by a small, relatively inexpensive relief valve that shears a pliable steel pin when maximum pressure is exceeded. The size of the pin must match the pressure rating of the liner used. Using an alloy, high-strength steel pin, such as an Allen Wrench, or an improper size pin will void the pump warranty. Of course, as the operating pressure increases inside the pump, the life of expendables, such as piston cups and liners, will decrease accordingly.

Air Compressors

Selecting an air compressor is just as difficult as selecting the mud pump. A minimum annulus return velocity of 3000 ft/900 m/min is needed for proper hole clearing. A return of 5000 ft/1500 m/min is considered good and will prevent the bit from crushing the same chips over and over.

The area of the annulus, combined with 3000 ft/900 m will calculate into the required cubic feet per minute volume of the compressor. This volume does not depend on whether the well is drilled with a rock bit or hammer tools.

Pressure rating of the compressor depends on the type of tool being used. Different brands will operate at maximum efficiency at various pressures. In addition, the pressure is also affected by the amount of water that must be blown from the hole, since 1 psi/0.07 kg/sq. cm equals 2.31 ft /0.711 m water. And before circulation begins, the air pressure must exceed the hydrostatic water pressure at the air exits in the bit.

Compressors, either piston-type or screw-type, are protected by one or more pressure relief valves. Tampering with these valves or replacing them with inadequate substitutes with regard to pressure and flow characteristics is extremely dangerous.

The proper compressor can only be selected when the type of formation and the required hole size are known.

Drilling Programme

Study and past experience have taught every contractor the best means of designing a well. This may be by direct circulation, reverse circulation, foam drilling, air drilling, or other programmes.

Relaying this information to the sales engineer is important since selection of the proper components for the rig must be done jointly with him. All aspects, from beginning to well completion, are important—surface casing, final casing, screen etc.—and must be reviewed.

Horsepower Ratings

Horsepower is defined as a unit of measure equal to the work accomplished by lifting 550 lb/250 kg of weight 1 ft/0.305 m in 1 s. This may be applied to calculate hook horsepower. The rotary table uses another form of horsepower, a combination of rotary torque and rotary speed. The mud pump uses yet another form of horsepower, expressed in hydraulic horsepower. It can be calculated from the fluid requirements in gallons per minute and pressure in pounds per square inch. All are actual, final horsepower ratings.

When this power is transmitted from one component to another, it gets smaller. To obtain a final hydraulic horsepower of 100, the engine may produce much as 125 hp, depending on the power train through which it flows. The loss of horsepower is due to friction which turns into heat and is dissipated in the air. Selecting the proper power units is important: too much power will result in damage to other components, while too little power results in poor rig efficiency.

For example, all the following components may be used at the same time: rotary table, mud pump, supercharging pump, rod oiler system, mast lights, and others. Note that horsepower does not increase by shifting gears, adding a transmission, or changing the travelling block. In fact, horsepower will decrease by each of these actions due to the efficiency factor.

Auxiliary Equipment

A change in drilling programme can result in the need for auxiliary components such as one or more stand-by pumps or compressor units. As pointed out before, a drilling rig's depth rating is linked directly to the capacity of the mud or air circuit. Additional mud or air could enlarge the capacity; therefore, each component must be checked out thoroughly to avoid hooking a high-pressure mud pump to a low-pressure surface piping or a low-pressure swivel hose.

In addition, all rigs must be equipped with instruments that enable driller to observe the mast loads, pump circuit pressure, control air pressure and drilling air pressure.

A string of reverse circulation equipment will increase the capacity of a rig to drill large-diameter wells because the annulus return velocity is no longer a problem. This return velocity is now inside the relatively small diameter drill pipe. Holes with diameters of 30 and 36 in/760 and 914 mm are not uncommon and can be successfully drilled by the airlift reverse circulation method.

Many other instruments and equipment are available for the drilling contractor to enable him to accomplish his specific drilling programme.

DRILLING TECHNOLOGY (WATER-WELL DRILLING)

One of the most important factors in purchasing equipment for a successful drilling programme is the discussion between the customer and the sales engineer. This discussion must be completely up front to avoid any unpleasant surprises later.

Virtually any component needed on a drilling for a specific job can be supplied by drilling equipment manufacturers. However, the customer must be specific in discussing the needs of his drilling programme in order to avoid buying more components than are needed. The purchase must forecast the type of drilling to be done with this unit and special items such as hole diameter, drill-pipe size, casing size and depth of hole. Deviation from this forecast could result in the customer buying the wrong size drill rig.

If a customer simply requests a machine to drill 5000 feet, all types of holes, all types of formations, with diameters up to 30–40 in/760–1016 mm, it would probably cost around 10 million dollars a piece and take 24 trucks to move. For each hole drilled, of course, perhaps only two trucks of this load might be used. This would not be a wise return on investment and maximum equipment usage; so it is strongly recommended that extensive research be done to specifically determine the equipment that should be purchased. This includes a good understanding of the various rig components necessary for an effective drilling programme.

Mud Pump

A mud pump is a piece of equipment used to circulate the fluid that carries the cuttings up from the bottom of the hole. The cuttings will settle out in the pit and the pump will use the same fluid over again. Most mud pumps are double-acting duplex piston pumps and their pressure and flow capacity will vary with the size of the liners and stroke of the pistons. The mud pump suction must be inspected very carefully since a small leak would not emit fluid rather but suck in air. This would lower the performance of the mud pump.

The suction line must be fitted with a suction screen to keep foreign objects out of the fluid since any such could damage not only the mud pump, but the swivel and other components as well.

The valve and valve seats must be inspected for wear when priming the mud pump. A leaky suction valve will make it virtually impossible to prime the mud pump unless the bottom end of the hose is also equipped with a foot valve. The foot valve, in turn, could also contribute to the problem if it does not operate properly. The foot valve and strainer must be of adequate size to supply fluid to the pump. Surging of the hose while in operation should be regular. Any irregular movement of the hose indicates a stuck open suction valve.

The mud pump discharge line should always be equipped with a pressure relief valve, installed between the mud pump discharge flange and the first cut-off valve. This relief valve protects the mud pump from overpressuring.

The air chamber, also called the surge chamber, must be installed on top of this high-pressure line to absorb the peak pressure and cushion the flow. The basic operation of this chamber is to compress air trapped inside it. A chamber hung upside down from the bottom of the line will definitely not trap air. The top of the chamber is usually fitted with a plug. If this plug leaks water, this means that there is no air trapped inside and, consequently, the advantage of the surge chamber is nil.

The operator should adjust the size of the piston to the calculated friction loss of the drill string. Also the uphole return velocity must be calculated to assure an adequate flow. The mud pump drive must be checked periodically for proper clutch operation.

Centrifugal Pump

A centrifugal pump may be used for drilling. However, the pump characteristics are such that a large volume is delivered at a fairly low pressure. So, typically, a centrifugal pump could be used for water-well drilling if the hole diameter is large, the opening through the drill pipe large and the hole depth not too deep. A common problem with a centrifugal pump is that it loses, its prime. This can be caused by a leaking suction hose or an air leak in the input shaft of the impeller. The seal around this drive shaft must be lubricated properly and adjusted so that a few drops of fluid leak out from the pump.

Overtightening of this packing gland will result in burning of the packing and may very well destroy the shaft along with the seal in a very short period of time. This, of course, would allow air to be sucked into the fluid end, causing the pump to lose its prime.

When using a centrifugal pump, a foot valve must be installed on the bottom end of the suction hose and must be in a continuous rise from the mud pit to the pump to allow air to be drawn out of the hose while priming the centrifugal pump. Any amount of air trapped will enter the centrifugal pump and simply expand inside, causing the pump to lose its prime.

Grout Pump

At various stages in the drilling operation, well conditions may dictate that cement has to be pumped into the bottom of the well. The cement can be pumped with the mud pump but the clean-up is quite time consuming. So a grout pump with a hydraulic drive may be installed to do the cementing or grouting of the well. Several pumps differing in capacity in gallons per minute and psi pressures are available. These pumps are reversible in operation so that cleaning does not require much time nor much fresh water.

Air compressor

Different-sized air compressors may be installed on drilling equipment. Two types of compressors are available: piston compressors and screw compressors. The function of the compressor is to compress air to a high pressure, which will then be pumped through a reservoir through the drill pipe to the bottom of the well to either be used as a driving force for the down-the-hole hammer or simply to return the cuttings while drilling with air.

A compressor is a very delicate piece of machinery and requires constant attention from the driller. Air-intake filters must be checked for contamination as well as the oil filter. Relief valves must not be tampered with although they must be inspected from time to time for proper operation. The screw-type air compressor is equipped with a safety shut-down device to protect the compressor. Again, this is not to be tampered with. Lubrication of the compressor varies with the type of compressor. A flooded screw compressor may use a Dexron-type automatic transmission fluid or a synthetic lubricant. These two fluids will not mix; therefore the driller must know the type of fluid in the compressor before adding more.

A number of gauges are furnished on the compressor to indicate the air-discharge temperature and injection oil temperature. These must be observed closely for proper operation. The driller must familiarise himself with the operation of the compressor and understand all valves, gauges and safety features. Compressed air can be very dangerous and any components containing pressurised air must not be disassembled while any amount of air pressure remains inside. The compressor drive must be checked periodically for proper clutch adjustment and operation. The common problem with compressors is poor performance due to clogged intake filter or improper lubrication.

Water Injection

Water injection is required when drilling with an air compressor in a dry and hard formation. The cuttings will come out of the hole as dust. To protect the driller against breathing this dust, water is injected into the air stream, which traps the dust particles which then appear as small balls of mud.

The water volume should be adjustable since the conditions of the hole determine as to how much water is to be injected. For example, a well may actually produce a little water so injection of water is not necessary; contrarily, a totally dry well requires all the water that can be pumped in. The water injection pump can also be used as a wash-down pump to clean the equipment.

Foam Pump

The foam injection pump will inject a small amount of chemical into the water discharge of the water injection pump, which will then flow into the main flow line and mix with the compressed air to a thick foam when it changes from compressed air to atmospheric pressure after exiting from the bit at the bottom of the hole. This stiff foam will serve as the main carrier of cuttings from out the hole.

Foam drilling requires a much lower uphole velocity compared to air drilling so that a much larger diameter hole can be drilled with the same size compressor. For example, a 900 cu ft/25 cu m compressor would have to struggle to drill a 10-inch hole whereas that same compressor could easily drill a 24-inch hole using stiff foam. The equipment must be thoroughly flushed to clean the inside of the pipe and tubulars to avoid excessive rusting.

Gearboxes

Drilling equipment would not be complete without incorporating a number of gearboxes. Gearboxes of different types may be used to reduce rpm, for example, from a hydraulic motor to a low-output speed for driving a rotary table. Gearboxes are also used to turn around corners. These boxes are known as bevel gearboxes and are commonly used in drawworks assemblies. A rotary table is actually a hollow spindle gearbox.

Oil contamination and low oil levels are the common causes of gearbox failure. The proper adjustment with measured backlash between the gears must be checked by a mechanic to assure a long life. The wear pattern of the gears will indicate any problem of assembly. Overfilling of the boxes with oil will result in a very high operating temperature, which will eventually burn the seals and cause the oil to leak out. The oil used in a gearbox should be the transmission gear type, such as SAE-90 weight or an SAE-140 weight.

Chain Drives

Fully enclosed, oil-splash lubricated chain drives are used throughout the design of drilling equipment to transmit the horsepower from one drive shaft to another. Chains are commonly used in PTO subdrive or transfer cases, mud pump drives and even some top drives. The lubricating oil for all chain drives, including the power end of the Failing manufactured mud pumps, should be a non-detergent motor oil, such as a 30W or 40W.

The oil must be low enough in viscosity to penetrate between the chain pin and the chain roller to provide lubrication. For example, a heavy weight SAE-90 gear oil will not penetrate this tight space between the roller and pin, which would cause the chain to wear from the inside out. Caution must be taken when assembling chain drives not to subject common ball-bearings to a thrust loading by preloading the bearings.

Oil levels and operating temperatures must be checked periodically. Some chain drives may not be fully enclosed nor oil-bath lubricated. Chains like this are used as pulldown chains in the mast or as drawworks chains. The chains must be tensioned properly to assure a long life. A loose chain will jump the sprocket and cause great stress, followed by a shock load which is severe to the life of the chain. Idlers may be used to provide this proper tensioning. A dry-type lubricant, such as a Molykote, may be used to lubricate chains of this type.

V-Belt Drives

V-belt drives are used for several purposes, such as driving the fan on the engine radiator, driving small hydraulic pumps or large mud pumps and compressors. The tension on all V-belt drives is critical since a loose belt will cause slippage. Slippage, in turn, will create excessive heat on the belt, which will cause the belt material to burn or become brittle and, consequently, the belt will fail prematurely.

Belts may be dressed with a proper belt dressing. Caution must be taken not to use ordinary oil as it would lubricate the pulleys and cause the belts to slip. If a V-belt drive is equipped with a tensioning pulley, the pulley centre line must be exactly parallel to the centre line of the main drive pulley to avoid climbing of the belt. Climbing of the belt means that it rides up to the high side of the idler, causing misalignment of the belt, excessive stretch, and thus a shorter life.

Drawworks

A drill rig will be equipped with some type of wireline hoist to handle tubulars in the mast or to handle heavy equipment around the drill site. On a Kelly-driven machine, a large drawworks assembly will provide drums which allow wireline to be spooled up, which in turn, will operate travelling blocks in the mast to extract drill pipe from the borehole or lower casing back into the borehole. This drum is the main hoisting drum. Drums are equipped with some type of brake, band or disc, and care must be taken that no lubrication whatsoever touch this braking surface as it would void operation.

A clutch assembly will allow the driller to choose when to engage or disengage this drum assembly. The clutch may be air controlled or mechanically controlled. A mechanical lever works best for the driller. An air clutch is more easily controlled from the driller's station by the use of an air valve. This air valve may be the on-off or the incremented type. The incremented type allows the driller to apply a small amount of air pressure in the clutch to operate it. This is good for picking up the slack of the wirelines. However, slipping the clutch at low pressure for a long period of time overheats it and can destroy the friction material within a matter of ten minutes. Therefore, an incremented clutch control should not be used in the middle range but should be fully engaged when hoisting heavy loads.

Most air clutches are self-adjusting but nonetheless depend entirely on the amount of air pressure. Just because the gauge at the driller's station reads 125 psi/8.75 sq. cm does not mean that the clutch pressure in the fully engaged position is operating at 125 psi/8.75 sq. cm. An air leak, for example, could bleed off most of the pressure, causing the clutch to slip. To detect an air leak, follow this simple procedure: build up air pressure in the air-control circuit of the rig and then shut the machine down. At this time engage the clutch line and listen to the air flow into the clutch. It should become quiet after the clutch builds up pressure since there is no more air flowing. A high-pitched noise will indicate an air leak, which could be in the hose, fittings or clutch bladder. Air leaks must be fixed immediately.

A drawworks can be equipped with drums for different purposes, such as the main hoisting drum, drums of equal capacity which are used on rigs featuring the dual travelling block operation plus a third drum, commonly known as a bailing drum or a sand reel. This third drum will usually not use a travelling block and it will have a great spool capacity at a fairly small single-line pull. This drum is used to operate a bailer, which is lowered into the borehole to clean material out of it. The drum can also be used to pick up heavy items around the drill site. Caution must be taken not to operate the drum at a severe fleet angle to the crown block sheave since this would cause the wireline to jump the sheave and get caught in the crown block structure.

Hydraulic Winch

One or more hydraulic winches may be installed on the machine with the same basic function as a drawworks. The winches are a lot smaller in capacity of single-line pull and wireline strength and are usually not operated in conjunction with travelling blocks.

A hydraulic winch will give a very fine control to position equipment at a specific point. However, operation is very slow. A common problem with the hydraulic winch is that the wireline is spooled onto the drum backwards, causing the internal brake not to work properly.

Cathead

A cathead may be installed on the machine to operate the catline, a manila rope wound on the cathead in 3–4 wraps. When the operator pulls the wraps tight on the spool, the cathead will exert tension on the line which enables the operator to pick up heavy equipment or do certain operations, such as sample driving.

The cathead spool must always be free of grease and rust. A good cathead spool will be shiny without aid of paint or primer. The cathead will be equipped with a rope splitter, also known as a divider, which will ensure that the rope spooled onto the drum will advance from one end to the other without stacking.

Rotary Table

As mentioned before, a rotary table is basically a hollow-spindle bevel gearbox. It provides twisting or turning power to the drill pipe by use of a Kelly drive bushing and Kelly combination.

Common problem with a rotary table is that soil and water from the drilling operations get into the oil, causing the gears and bearings to wear. The rpm of the table can be controlled to suit the drilling programme selected. For example, a drag-bit operation would use a low table speed. A tricone bit selection would require the table to run fairly fast, whereas a coring operation would require the table to run at high speeds, up to 500–600 rpm. The table may be equipped on the bottom end with a dust or dirt deflector to overcome the contamination problem.

A table must be inspected quite frequently for proper oil level and for proper clutch adjustment. A table may be hydraulically retracted or manually pivoted to allow a greater opening for the casing setting.

Top Drive

A top drive applies rotary power to the top of the drill pipe and advances down or up as the drill pipe is moved in or out of the hole. A swivel is mounted on top to inject the flushing media inside the drill pipe.

A hydraulic cable pulldown system, illustrated in Fig. 3.4, absorbs shocks associated with drilling. An air-driven top-head rotary drive is illustrated in Fig. 3.5. It can be conveniently used for drilling through the overburden/rock formations.

Top drives are generally not as powerful as rotary tables and are usually driven by hydraulic motor(s). A high-torque machine, such as that used with hollow-stem auger drilling, is equipped with a mechanically driven top drive eliminating the hydraulic motor(s). A top drive should be equipped with a floating sub to allow making connections without applying too much downthrust pressure onto the threads. The drive mechanism can be either a chain drive or a gear drive and may be a low-speed or high-speed drive or a combination.

Lubrication is based on the function of the head and the level must be properly maintained with the proper lubricant. A swivel may be installed below the top drive as well as one above the top drive to allow for simultaneous dual-wall air reverse circulation drilling.

Several systems are used for pipe handling, including a single joint loader or a carousel or a hook and sling assembly. A number of different sizes are used to support the drill pipe and to make and break the connections.

Crown Block

A crown block assembly, illustrated in Fig. 3.6, is mounted on top of the mast and incorporates a number of wireline sheaves mounted on bearings to allow stringing up of travelling blocks as required. It may also include bearing-mounted sprocket assemblies to support the continuous chain for holdback and pulldown assemblies, especially when used with a top drive. The wireline sheaves must be grooved to fit the proper wireline and bearings must be lubricated periodically to avoid seizure.

Weight Indicator

A weight indicator is a mechanical or hydraulic load cell-operated device that tells the driller the amount of pull being exerted by the hook on the travelling block. Different scales are available to match the different string-up combinations

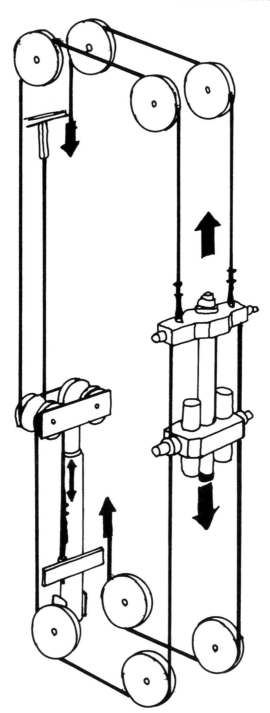

Fig. 3.4: Hydraulic cable pulldown system.

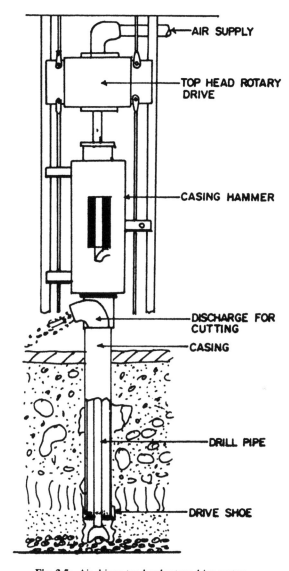

Fig. 3.5: Air-driven top head rotary drive system.

and dials may be furnished in metric or SAE scales. A weight indicator must be calibrated and checked for proper operation frequently.

Pulldown and Holdback

A pulldown and/or holdback circuit may be installed in the mast. A pulldown assembly will add additional pulldown to the weight on the bit for fast penetration through hard formations. A problem may arise if the pulldown is used too long;

Fig. 3.6: Crown block assembly.

the hole may veer off the vertical centre line of the well. This can cause problems when setting the casing.

It is best to drill with the weight of the drill string and actually holding back some of this weight to keep the drill string in tension. This almost automatically ensures a vertical hole. The amount for pulldown or holdback will vary, based on the desired weight on the bit, as specified by the bit manufacturer.

Hydraulic Circuit

The hydraulic circuit on a drilling rig can vary from very simple to extremely complicated. It is up to the driller to understand the circuit fully so that he can diagnose problems and undertake repair before an entire circuit becomes contaminated and self-destructs.

Cleanliness of the hydraulic circuit is of utmost importance and care must be taken when adding hydraulic fluid that it is clean and is handled properly with clean equipment in a clean environment. The hydraulic fluid may have to be cooled if heat sources are present in the design, such as pumping fluid over a relief valve on a continued basis or when operating a closed loop hydrostatic circuit. The driller must understand what adjustments he can make to the system, such as flow control and pressure control. The most common problem of hydraulic circuits is contamination of the fluid, which causes internal destruction of the pump and/or motor.

Various Other Components

There are a number of items that remain to be discussed, such as the driller's control panel, the different platforms that fold up or swing out, and mechanical jacks or hydraulic jacks. A thorough understanding of the control panel is extremely important since a wrong move can very easily injure someone or cause extensive material damage.

SAFETY PROCEDURES

Safety begins and ends with the operator. Thousands of dollars are spent by manufacturers and drilling companies to inform the operator of various hazardous situations. This is conveyed to the operator by expensive engraved plates or printed stickers mounted on the machine. None can save lives unless close attention is paid to them and drill-rig operators practice what the signs advise.

Safety starts in the morning when dressing for work. Safety shoes must be worn when working around heavy machinery. Loose clothing is very hazardous around all types of machinery that have exposed rotating parts. A very important safety point is that the operator must have a complete understanding of the job to be done and knowledge of the equipment used to do the job. He must also know the capability of his helper because, in an emergency, he must be able to rely on him to save his life. Naturally, the driller is in charge on the drill site not only of the drilling rig, but also of the entire drilling operation, including safety procedures.

Upon receipt of a new piece of equipment, the unit must be checked in the yard by the driller and the maintenance worker before being put to work. All oil levels must be checked, grease fittings must be greased and all paint on control rods, shafting cables and hydraulic valve linkage must be scraped off to avoid sticky controls. A visual inspection must assure that all shipping material has been removed from the machine and that instrumentation is intact. The machine must be thoroughly cleaned and any kind of rust immediately removed. Do not assume that the equipment is fully lubricated upon delivery from the factory since many things can happen, such as people draining the cases for the valuable oils.

If the unit is mounted on a truck, check the truck engine, transmission and gearboxes as well as axles for proper lubrication. Make sure that the brakes of the truck are in good condition and that the air-control circuit has no leaks. Check to make sure that all controls are operating correctly. If the equipment furnished includes a deck-mounted engine, the engine must be checked for proper oil, cooling fluid and transmission lubrication. All drive-lines must be checked to assure that all nuts and bolts are tightened before starting up the equipment.

All the aforesaid items must be considered part of the safety programme since malfunction of any one could cause a great hazard.

Equipment Start-up

The machine should be started up and test-run while still in the yard, i.e., before being taken to the drill site. All controls must be thoroughly checked and the driller must familiarise himself with this particular piece of equipment. Even though he may have drilled for many years on the same type and model, the controls may have changed due to new rules and regulations by various government agencies. For example, the throttle control may have to be moved in a different direction to give more throttle.

If the above visual inspection turns out okay, the machine may be started but all personnel not necessary to the start-up should be excluded. Only necessary personnel, the driller and the helper, should be present and, of course, safety clothing, including hard hats and safety shoes, must be worn.

After start-up of the equipment, the driller needs to check each control of the unit. After levelling the rig using either hydraulic jacks, mechanical jacks or both, he may raise the mast. The mast may have been banded to the front mast support for shipment, which must be removed before attempting to raise the mast. No loose items should be in the mast so, prior to raising it, the driller must inspect the mast assembly to ensure that components skipped inside it have been removed.

To raise a mast, first move the mast raising control lever into the downward mode and watch the hydraulic pressure gauge build up pressure to ensure that there is hydraulic fluid on top of the piston; thus when the mast breaks over centre, it will be cushioned into the final upright position. While raising the mast, the driller must pay careful attention to the wirelines that run from the various drums to the crown block. When these lines align correctly, he must take time to release the brakes slightly and give them some slack. The brakes should never be released prior to raising the mast to let the wirelines feed off by themselves. A heavy travelling block and Kelly will more than likely be attached to the other end of one of the wirelines on raising the mast after loosing the brakes of this particular line, the Kelly is apt to fall, causing damage to the equipment or personal injury.

Once the mast is raised and locked in the vertical position, the driller should perform more inspections and checks. The wireline sizes need to be checked for proper diameter and correct spooling on the drums. For example, hydraulic winches are designed in such a way that an internal brake will hold the load if properly spooled onto this winch. If the wireline is spooled backwards on this winch, the brake does not operate. Wirelines must also be inspected for proper anchoring to the drums and the deadline and proper installation of the wireline clips. The main body of the wireline clips should be on the long end of the line and the U-bolt should be wrapped around the end of the wireline. An easy way to remember this procedure is 'Never put a saddle on a dead horse'. The 'saddle' would be the body of the wireline clip and the 'dead horse' would be the end of the wireline.

Operating Equipment

Most operating equipment is pulled from the warehouse and shipped directly with the unit without further inspection. So, upon receipt of all this equipment, the driller must make sure that all the threaded connections of the Kelly, drill pipe, drill collars, drill bits and subs will match or make sure that crossover subs are available.

When installing the swivel hose or any other mud hose, make sure that the end of the hose is connected with a chain to the rigid piping of the mast or drill frame so that in case of hose failure, the loose end of the hose is attached to this chain. Particular attention must be given to the pipe handling provisions of whatever type they may be. This can vary from a hook-and-sling assembly to an automatic pipe handler or pipe tongs, slips and clamps. All these items must be checked for proper fit.

Preventive Maintenance

Preventive maintenance will save the drilling contractor thousands of dollars even though the machine is shut down and non-productive for a short while. Many times a small item in disrepair will be found and can be quickly repaired during preventive maintenance. If not repaired at this time, it could deteriorate and cause major damage. For example, the bolts that mount the drive-line to a gearbox can become loose; if tightened properly during preventive maintenance, they will cause no further damage. Should the bolt come out of the gearbox flange, the loose end of the drive-line could literally beat components, such as the compressor, engine, radiator and others, to a pulp before the driller could take action to kill the power to this drive-line.

Checking on levels must be a daily routine as it indicates to the driller what cases are running cool, clean and without leakage. All lubricated grease fittings must be checked even though they are hard to reach. Undergreasing is certainly bad for the equipment but in some places, overgreasing is just as bad. When a band brake pin grease fitting is lubricated too much, the excessive grease will end up on the friction surface of the band brake, causing this brake to malfunction. This can be very hazardous. A bearing that is not receiving enough grease can run hot and, consequently, fail much earlier than its expected design life.

Cleaning the drilling rig and all related equipment is part of preventive maintenance. Cleaning includes removal of rust and applying a new coat of paint. Leaks of any kind can be an indicator of an internal condition of the equipment asking for attention. Leaks of any kind are not good for the equipment. For example, a small leak in the control air pressure will drop the operating air pressure and allow clutches to slip. A low-volume air compressor working around the clock will wear out much faster than if it were allowed to build up to a proper pressure and run idle for a while.

A leak in the main mud pump discharge line can cause extensive damage to the connecting threads of the coupling and, of course, leaks of lubricating oil and hydraulic oil will cause the drill frame to be extremely slippery.

V-belt drives and open-chain drives need to be checked for proper adjustment during preventive maintenance. Unusual noises coming from different components of the drill rig always indicate an unusual condition, such as stuck valves in the fluid end of the mud pump. A knocking noise could be caused by loose piston rods. Smoke and unusual odours around the drilling rig can indicate slipping clutches or hot brakes and hot gearboxes. This must be checked out immediately and the situation corrected.

SIZING COMPONENTRY

Sizing componentry can be confusing in an international marketplace where drill rig manufacturers offer a large variety of sizes of masts, drawworks, drill pipes, mud pumps and compressors that can be installed on the same basic model drill rig. The components must be chosen carefully by the drilling contractor. However, the different measurement units used by drill-rig manufacturers around the world do pose difficulties for many contractors in selecting and sizing componentry.

Several international formulae have been developed to make these tasks much easier. The formulae described below can be used to calculate the volume of a hole, uphole velocity, size of a required air compressor, volume of well-sealing material, and size of a mud pump and mast.

Hole Volume

Figure 3.7 should be referred to for calculating the volume of a hole. For drillers who are used to the American system, which uses dimensions of a hole in inches and depths in feet, the cubic footage of the hole volume must be calculated. This can be readily done using the formula detailed in this Figure. The international system is also explained in this Figure for those contractors mixing the hole diameter expressed in inches and the hole depth expressed in metres.

Uphole Velocity

Actual uphole velocities for direct circulation mud drilling can be calculated without the use of a calculator, by using the simplified formulae detailed in Fig. 3.8. The phrase $(D^2 - d^2)$ will appear in many figures given in this chapter. It is very easily calculated, as illustrated in Fig. 3.8. Since the size of the borehole is not very precise, it is common to round off bit sizes, for example $7\frac{7}{8}$ to 8 in and the drill pipe size of, for example, $2\frac{7}{8}$ to 3 in diameter. This simplifies the calculation of this phrase, which can then be done without the use of a calculator.

Mud Drilling

The size of the mud pump can be selected once the size of the drill pipe and hole are established. Figure 3.9 is based on the American SAE system and the international system. In the latter system, simply plug in the mud pump requirements in gallons per minute and multiply by the minimum uphole velocity 0.25 m/s. The calculations used are based on the US gallon which equals 3.782 litres. Actual uphole velocity can be calculated based on the two equations at the bottom of this Figure.

Air Drilling

Figure 3.10 enables the driller to calculate the size of the air compressor required for an air-drilling programme. The equations are expressed in the American SAE system and the international system. To calculate the minimum required size air compressor, simply take the phrase $(D^2 - d^2)$ times 1/2 and this will equal the minimum volume required of the air compressor expressed in cubic metres per minute. No further conversion is required. Once the minimum size air compressor is known, the actual uphole velocity can be calculated based on the existing air compressor of a given size. The uphole velocity is based on a minimum of 3000 fpm, which is equal to 15 m/s. To determine the pressure required on the air compressor, the operator must first select whether the type of drilling will be done with a down-the-hole hammer or simply a hard formation tricone button bit.

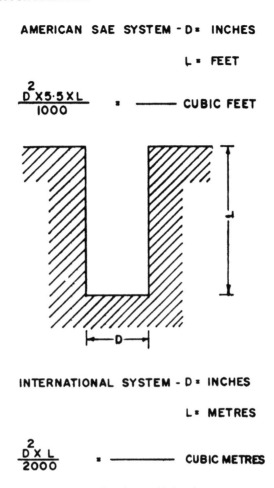

AMERICAN SAE SYSTEM - D = INCHES

L = FEET

$$\frac{D^2 \times 5.5 \times L}{1000} = \text{———— CUBIC FEET}$$

INTERNATIONAL SYSTEM - D = INCHES

L = METRES

$$\frac{D^2 \times L}{2000} = \text{———— CUBIC METRES}$$

Fig. 3.7: Calculation of hole volume.

If the air compressor must drive a hammer mechanism, this hammer assembly will dictate the pressure of the compressor. For example, a hammer that operates at 180 psi will require a compressor of 220 psi minimum.

The pressure of an air compressor is required to start circulation in the hole after making a connection. The air pressure will have to be greater than the hydrostatic water pressure in the bottom of the hole. For easy calculation, one psi equals 2.3 feet of water column. One bar of pressure equals 10 metres of hydrostatic water column in the metric system.

Well-sealing Volume

Once the hole has been drilled and a casing has been inserted into the borehole, it is time to calculate the annular volume of gravel or sealing material to finalise the well.

$$(\overset{2}{D} - \overset{2}{d})$$

D — INCHES — HOLE SIZE

BIT DIAMETER

d — INCHES — DRILL PIPE

DRILL COLLAR

CASING

D X D = ―――――

-d X d = ‗‗‗‗‗

▬▬▬ SQUARE INCHES

ROUND OFF HOLE SIZE AND PIPE SIZE TO SIMPLIFY CALCULATION.

EXAMPLE : $7\frac{7}{8}''$ HOLE - $2\frac{7}{8}''$ PIPE

$(7\frac{7}{8}^2 - 2\frac{7}{8}^2) = 53 \cdot 75$

ROUNDED OFF : $(\overset{2}{8} - \overset{2}{3}) = 55$

Fig. 3.8: Calculation of uphole velocity.

Calculation of this volume can be very difficult, especially in the international system, whereby the hole and casing diameters are expressed in inches and the length of the borehole to be filled is expressed in metres. Figure 3.11 includes the formulae for the American SAE system and the international system. Once again, the equation should be entered in the proper denominations with no further conversion.

Mud Pump Sizing

Refer to Fig. 3.12 to determine the proper size of a mud pump. This Figure illustrates the friction loss caused inside three different sizes of drill pipes pumping the same amount, 500 gpm, through length of drill pipe, 3000 ft (approximately 900 m). Horsepower required to produce pressure at 500 gpm is displayed at the bottom of the chart. More horsepower is required for a $3\frac{1}{2}$ in/89 mm drill pipe.

To further explain the importance of proper drill-pipe selection, the fuel consumed to provide power to the mud pump to overcome the friction equals 24.2 gallons per hour in the $3\frac{1}{2}$ in/89 mn drill pipe. It is thus very obvious that the larger

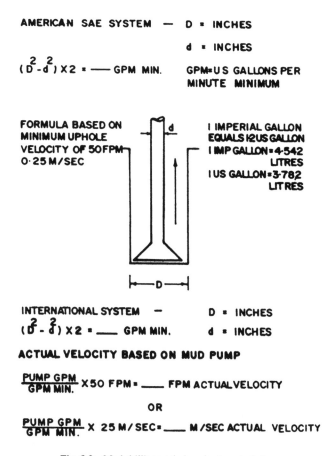

Fig. 3.9: Mud drilling: uphole velocity calculation.

drill pipe should be used when setting the drill pipe even though more costly. It will pay off in time in terms of fuel and mud pipe efficiency.

Mast Sizing

The difference in mast rating must be understood in order to determine mast capacity. For example, mast rating may be spelled out as mast rating, gross load rating, hook-load rating and static hook-load rating. The only rating that is useful to the contractor is the last (static hook-load rating). This is the maximum load that may be pulled with the mast, based on a specific type of string-up. For example, if the hook-load rating is based on operation with a two-sheave travelling block deadlined to the crown, then a two-sheave travelling block must be used when pulling to the maximum hook-load.

Figure 3.13 shows different reeving diagrams illustrating the difference between hook-load and mast load. This Figure is readily understood if an imaginary line is

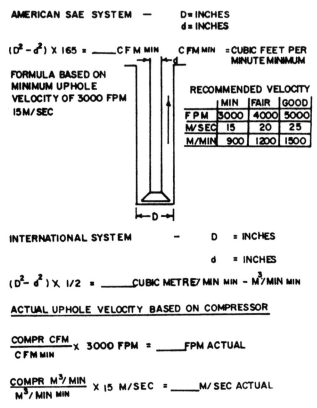

Fig. 3.10: Air drilling: uphole velocity calculation.

drawn cutting all wireline in half and applying a load in each piece to support the hook-load. The hook-load in D would be carried by four lines and would be equal to four times the maximum single-line pull of the drawworks. At the same time, there are five lines pulling down from the crown block of the mast in D, meaning that the mast loading is equal to five times the maximum single-line pull of the drawworks.

'A' (Fig. 3.13) demonstrates that when cutting the line, it is obvious that only one line will pull up on the hook, which will equal the maximum single-line pull on the drawworks; yet there are two lines pulling down on the top of the mast, loading this mast twice the maximum single-line pull of the drawworks. Example A allows the hook-load to be only 50% of the mast load, whereas this has improved to 66% in B, increases to 75% in C, and rises to 80% in illustration D. In summary, the efficiency of the mast load increases by adding a number of sheaves.

After the contractor decides on the weight of the drill string, including the swivel, Kelly, drill pipe, subs, drill collars and bit, and calculates the total weight of the casing, he can then decide on the hook-load capacity of the mast. The calculated dry weight of the drill string should never exceed 75% of the maximum

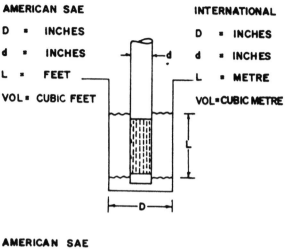

AMERICAN SAE

$$\frac{(D^2 - d^2) \times 55 \times L}{1000} = \underline{\quad\quad} \text{VOLUME CUBIC FEET}$$

INTERNATIONAL

$$\frac{(D^2 - d^2) \times L}{2000} = \underline{\quad\quad} \text{VOLUME CUBIC METRE}$$

Fig. 3.11: Calculation of annular space volume.

static hook-load capacity of the mast. A casing load may be higher than the mast capacity if this casing is to be set in a mud or water-filled hole and a casing float sub is used to float the casing. To say the least, this would still be a very risky operation.

Once the hook-load capacity of a mast is known and the number of sheaves in the travelling block has been selected, the single-line pull will appear by dividing the hook-load by the number of wirelines pulling on the travelling block. This single-line pull will determine the size of the drawworks. The drawworks, of course, must be able to spool up the necessary wireline to give the appropriate hoisting travel on the travelling block in order to make a connection or to run the top drive up and down.

The brakes must be of adequate size to hold the maximum load, whether it be drill-string load or casing load. The horsepower of a drawworks will determine the maximum speed at which a hook-load may be moved in the mast. A small horsepower drawworks will move the load slowly whereas a larger drawworks will move the load more rapidly. There are many engineering formulae available to calculate hoisting speeds and hoisting weights back to horsepower; this is referred to as hook horsepower.

DRILLPIPE COMPARISON	Bombay July.19, 1982 John L' Espoir		
PIPE SIZE	3½" IF	4" IF	4½" IF
Now. weight in lbs / ft.	13·5	14·0	16·6
Including tool 'joints	14·0	15·5	18·2
Torsion yield psi, Grade E	18 520 psi	23,250 psi	30,750 psi
Tensile in pounds	271·570	285,360	330,560
Dry Weight 3000 ft	42,000 lbs.	46,500 lbs	54,600 lbs
Buoyancy 15 %	6,300 lbs.	6,975 lbs	8,190 lbs.
Wet Weight	35.700 lbs.	39,525 lbs	46,410 lbs.
Friction Loss at 500 GPM for 3000 ft.	425 psi/1000 ft. 1275 psi	170 psi/1000ft. 510 psi	85 psi/1000 ft. 255 psi
Hydraulic Horsepower	372 HP	148 HP	74 HP
Mechanical Horsepower (1·2)	446 HP	177 HP	88 HP.
Fuel GPH	24·2	9·6	4·8

ADVANTAGES:

Lower pump pressure

Longer piston-liner life.

Less fuel consumption

Longer engine life

DISADVANTAGES:

Restriction to depth capacity based on mast and drawworks

Larger pipe is more costly

Fuel consumption based on standard Detroit Diesel -
approximate 0·38 lbs/BHP-HR
1 gallon - 7 pounds

Fig. 3.12: Advantages/disadvantages of drill pipe size for mud pump.

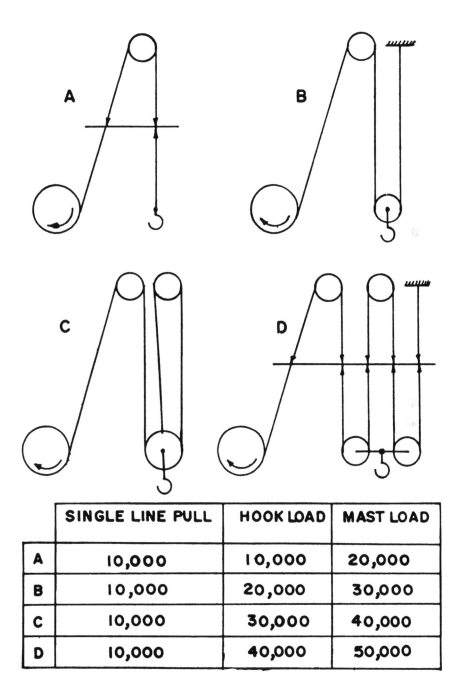

	SINGLE LINE PULL	HOOK LOAD	MAST LOAD
A	10,000	10,000	20,000
B	10,000	20,000	30,000
C	10,000	30,000	40,000
D	10,000	40,000	50,000

Fig. 3.13: Differences between hook-load and mast load.

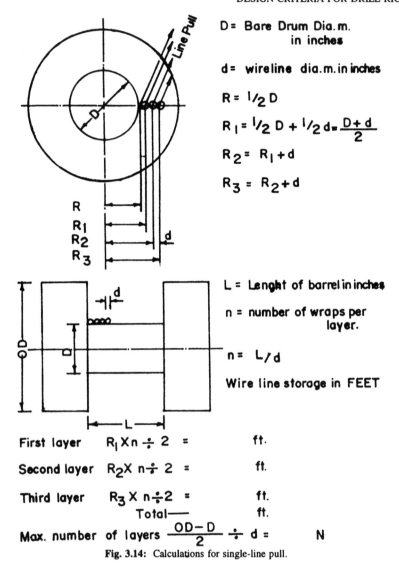

$$D = \text{Bare Drum Dia. m. in inches}$$

$$d = \text{wireline dia.m. in inches}$$

$$R = \tfrac{1}{2}D$$

$$R_1 = \tfrac{1}{2}D + \tfrac{1}{2}d = \frac{D+d}{2}$$

$$R_2 = R_1 + d$$

$$R_3 = R_2 + d$$

$$L = \text{Lenght of barrel in inches}$$

$$n = \text{number of wraps per layer.}$$

$$n = L/d$$

Wire line storage in FEET

First layer $R_1 \times n \div 2$ =   ft.

Second layer $R_2 \times n \div 2$ =   ft.

Third layer $R_3 \times n \div 2$ =   ft.

                  Total——   ft.

Max. number of layers $\dfrac{OD-D}{2} \div d =$ N

Fig. 3.14: Calculations for single-line pull.

When calculating a drill string based on the mud pump pressure loss characteristics, it becomes obvious that as large a diameter of pit as possible should be used. When calculating the drill string from the standpoint of the mast and drawworks, it appears that as small a drill pipe as possible should be used So the actual size selection of the drill pipe could be very difficult. A large drill pipe weight means a large drilling rig, which has problems of its own, such as transportation and physical size, since it is extra wide and extra tall.

Calculations for single-line pull may be worked out for Fig. 3.14.

The estimated performance for drilling 24 in/610 mm hole 100 ft/30 m deep with a T-650-W drill is given in Fig. 3.15.

DRILL SET-UP TIME	.3 hr
MUD TANK SET-UP TIME	5.0 hr

$$\text{DRILLING TIME (hr)} = \frac{\text{Hole Area} \times \text{Depth}}{\text{Machine Capacity (Hole Area} \times \text{Depth/Time)}}$$

MACHINE CAPACITY (Est.) = 6″ Hole 100′ Deep in 2.5 hr

$$\text{MACHINE CAPACITY} = \frac{\text{Hole Area} \times \text{Depth}}{\text{Time}} = (\pi 3^2)100'/2.5 \; hrs$$

100′ REMAINS CONSTANT = C

MACHINE CAPACITY = 28.3 C/2.5·11.3 C/hr

DRILLING TIME 6″ HOLE $\dfrac{28.3\phi}{11.3\phi}$		= 2.5 hrs
CHANGE BIT TO REAMER		.5
REAMING TIME 12″ HOLE $\dfrac{113 - 28.3}{11.3} = \dfrac{85}{11.3}$		= 7.5 hr
CHANGE REAMER		1.0
REAMING TIME 18″ HOLE $\dfrac{254 - 113}{11.3} = \dfrac{141}{11.3}$		= 12.5 hr
CHANGE REAMER		1.0
REAMING TIME 24″ HOLE $\dfrac{452 - 254}{11.3} = \dfrac{198}{11.3}$		= 17.5 hr
TEAR DOWN AND CLEAN UP		2.0
TOTAL TIME TO COMPLETION		49.8 hr

NOTE: This is an estimate of the time that could be required to drill a large diameter hole, using a standard T-650-W, equipped with a mud pump. Actual drilling time is estimated to be 40.0 hours, while 9.8 hours are required for rigging up the equipment. Compare this with the estimated performance of other types of drills.

Fig. 3.15: Esimated Performance for Drilling 24″ Hole 100′ Deep With T-650-W

Return Velocity of Drilling Fluid

The return velocity of drilling fluid V is calculated from the formula:

$$V = \frac{24.5 \times gpm}{D^2 - d^2}$$

where D is hole size/bit diameter in inches and d is drill pipe/drill collar/diameter in inches.

In Figures 3.16 to 3.23 return velocities with various size holes (bit sizes), and different pump deliveries are illustrated using $6\frac{5}{8}$ in/175 mm, $5\frac{1}{2}$ in/140 mm, 5 in/152 mm, $4\frac{1}{2}$ in/114 mm, 4 in 101 mm, $3\frac{1}{2}$ in/89 mm, $2\frac{7}{8}$ in/72 mm and $2\frac{3}{8}$ in/60 mm drill pipes respectively.

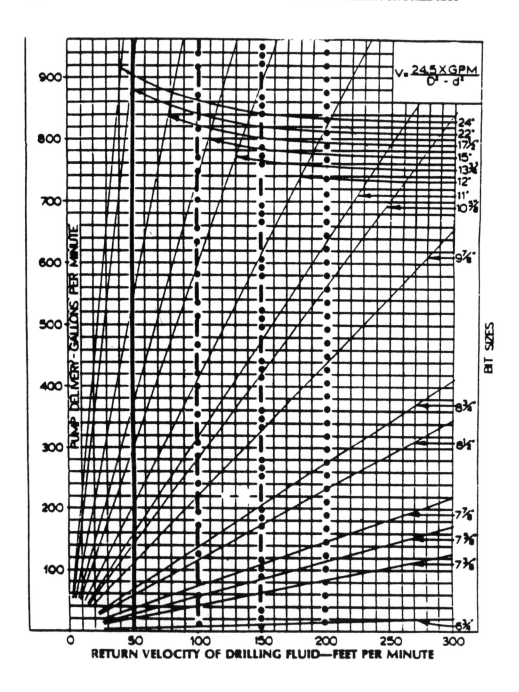

Fig. 3.16: Return velocity of drilling fluid using $6\frac{5}{8}$ inch drill pipe.

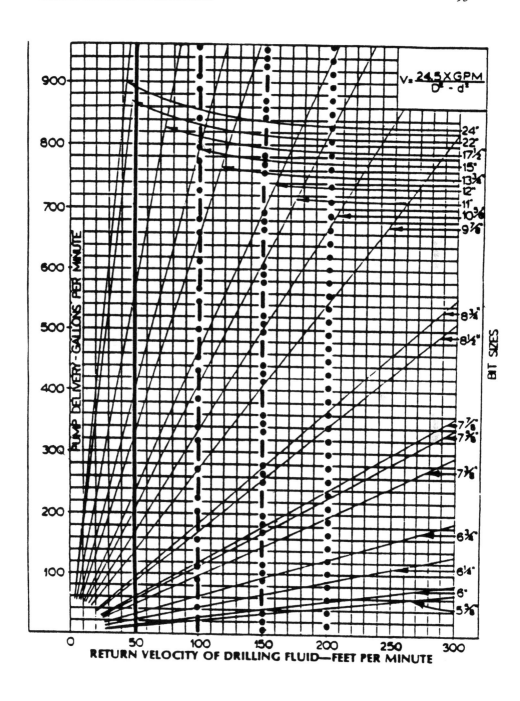

Fig. 3.17: Return velocity of drilling fluid using $5\frac{1}{2}$ inch drill pipe.

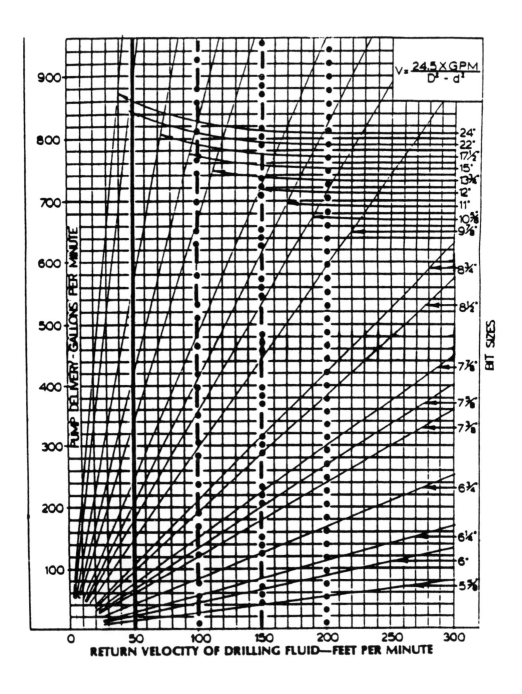

Fig. 3.18: Return velocity of drilling fluid using 5 inch drill pipe.

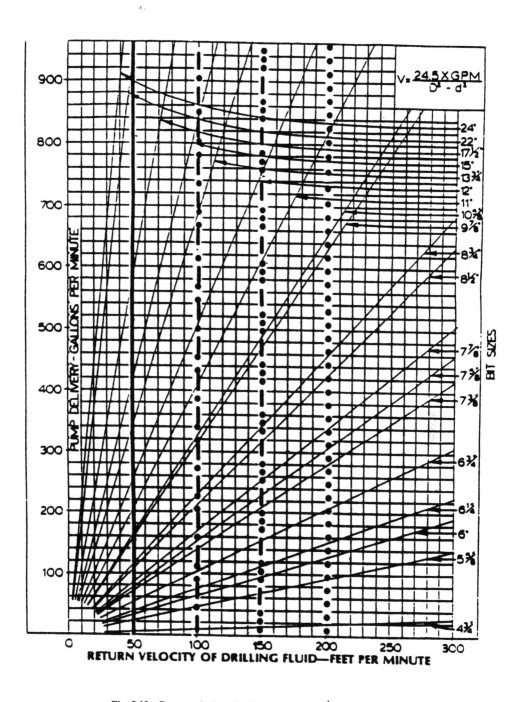

Fig. 3.19: Return velocity of drilling fluid using $4\frac{1}{2}$ inch drill pipe.

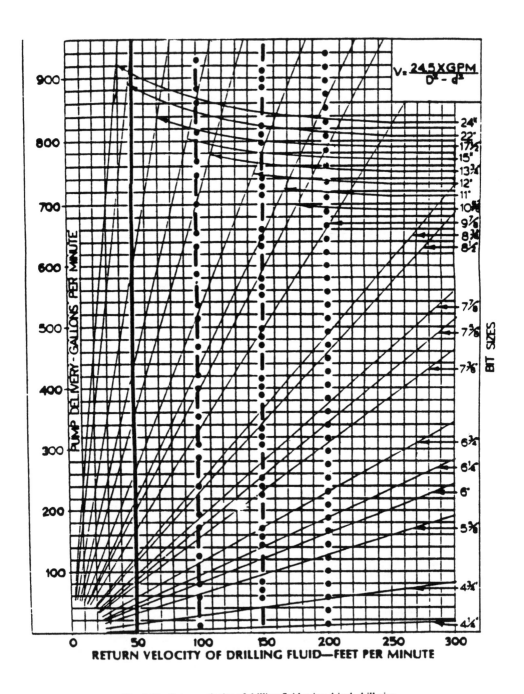

Fig. 3.20: Return velocity of drilling fluid using 4 inch drill pipe.

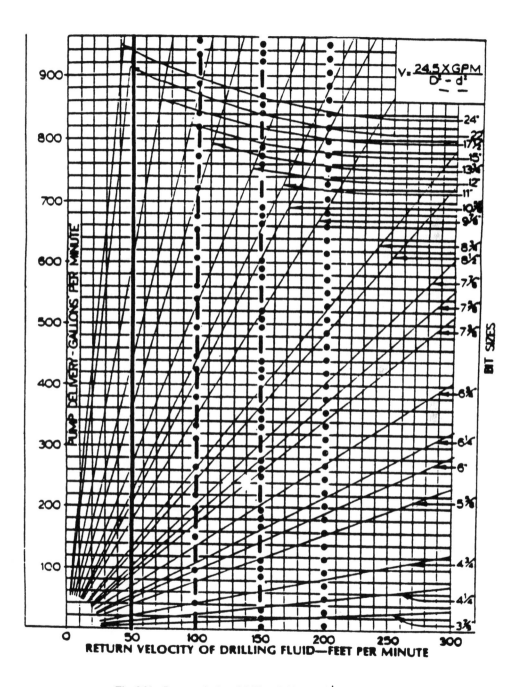

Fig. 3.21: Return velocity of drilling fluid using $3\frac{1}{2}$ inch drill pipe.

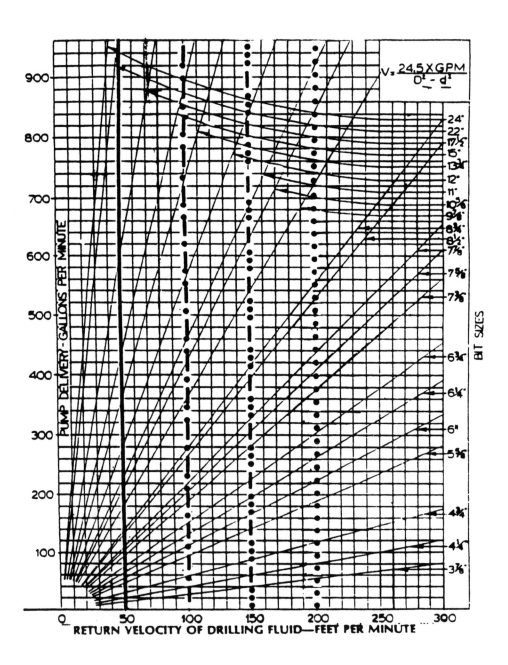

Fig. 3.22: Return velocity of drilling fluid using $2\frac{7}{8}$ inch drill pipe.

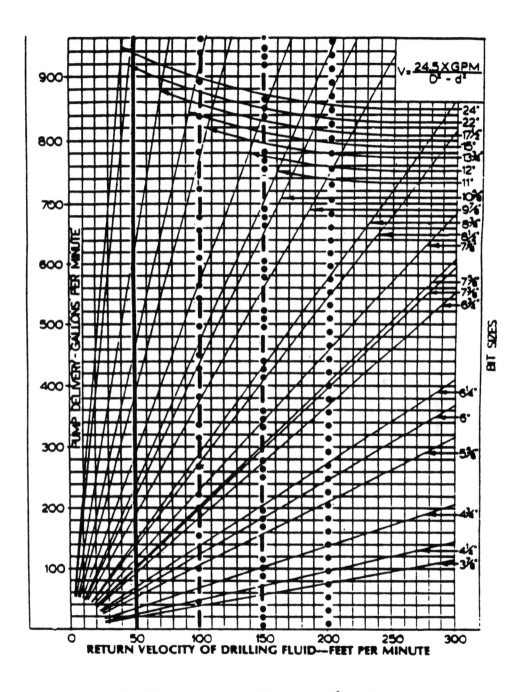

Fig. 3.23: Return velocity of drilling fluid using $2\frac{3}{8}$ inch drill pipe.

Practical conversion data from horsepower SAE to horsepower metric and uphole velocities is given in Fig. 3.24.

Friction loss in psi per 1000 ft/300 m in internal flush and full-hole drill pipes is presented in Table 3.1.

Friction loss in psi per 100 ft/30 m in various size bores through drill collar/Kelly is shown in Table 3.2.

ANGLE DRILLING MODEL F-10

The nomograph in Fig. 3.25 enables determination of actual drill footage requirement and burden change for various inclinations of holes. Inclined holes are necessary in certain topography.

Pipe length L, vertical depth V, and horizontal deviation H for 5°, 10°, 15°, 20° and 30° inclinations are illustrated in Tables 3.3, 3.4 and 3.5 respectively.

Slant-Hole Drilling

Slant-hole drilling is accepted as a viable alternative to many previously used conventional drilling practices. Many slant-hole wells have been drilled in recent years and increased use is anticipated.

Table 3.1. FRICTION LOSS IN PSI PER 1000 FEET.

| | DRILL-PIPE (in inches) | | | | | | |
| | IF—INTERNAL FLUSH | | | | | FH—FULL HOLE | |
PUMP GPM	$2\frac{3}{8}$	$2\frac{7}{8}$	$3\frac{1}{2}$	4	$4\frac{1}{2}$	$3\frac{1}{2}$	$4\frac{1}{2}$
50	40	20					
75	100	40					
100	160	70	20			20	
125	240	110	25			30	
150	350	140	40	10		50	
175	475	200	60	15		65	
200	600	250	75	20		80	
250	950	400	110	40	20	120	20
300		570	160	60	25	175	30
350		760	220	85	40	235	50
400		970	280	110	50	300	60
450			350	140	70	370	75
500			425	170	85	450	95
550			510	200	100	535	110
600			600	240	120	630	135
650			700	280	140	740	155
700			800	325	165	840	180
750			920	370	185	960	195
800				420	210		235
850				460	230		265
900				515	260		295
950				570	285		325
1000				630	320		360

HORSE POWER SAE	HORSE POWER METRIC
Hook HP $= \dfrac{W \times S}{33,000}$ W = lbs S = ft/min Hydr. HP $= \dfrac{GPM \times PSI}{1,714}$ GPM = U.S. Gallons/minute Rotary HP $= \dfrac{T \times RPM}{5,252}$ T = ft.lbs. Rotary HP $= \dfrac{T \times RPM}{33,000}$ T = inch.lbs.	Hook HP $= \dfrac{W_m \times S_m}{4,562}$ W_m = Kg. S_m = m/min Hydr. $HP_m = \dfrac{LPH \times Kg/cm^2}{457.5}$ LPM = litres per minute. Rotary $HP_m \dfrac{T_m \times RPM}{726}$ T_m = Kgm.
SP GR $= \dfrac{\text{lbs/cu.ft.}}{62.4}$	SP GR $= \dfrac{Kg}{cub \cdot dm.} = \dfrac{Kg}{litre}$

UPHOLE VELOCITY

50 ft/min——15.3 m/min		150 ft/min——45.7 m/min	
100 ft/min——30.5 m/min		200 ft/min——61.0 m/min	

MULTIPLIERS   (Inverted)

lbs to Kg —————————— multiply X 0.4536 ——————————		(2.204)
ft. to m —————————— multiply X 0.3048 ——————————		(3.28)
Inch to mm———————— multiply X 25.40 ——————————		(0.039)
GPM to LPM———————— multiply X 3.786————————		(0.264)
Gallon to cubic feet—— multiply X 0.1336 ————————		(7.48)
PSI to Kg/cm^2———————— multiply X 0.0705 ————————		(14.184)
ft.lbs. to Kgm ———————— multiply X 0.1383————————		(7.231)
Inch lbs. to Kgm ———————multiply X 0.0115 —————————		(86.957)
CFM to cubic metre/min.——multiply X 0.0283————————		(35.31)

1 PSI = 2.31 feet of water   1 Kg/cm^2 = 10 metres of water

Fig. 3.24: Data for converting horsepower SAE into horsepower metric.

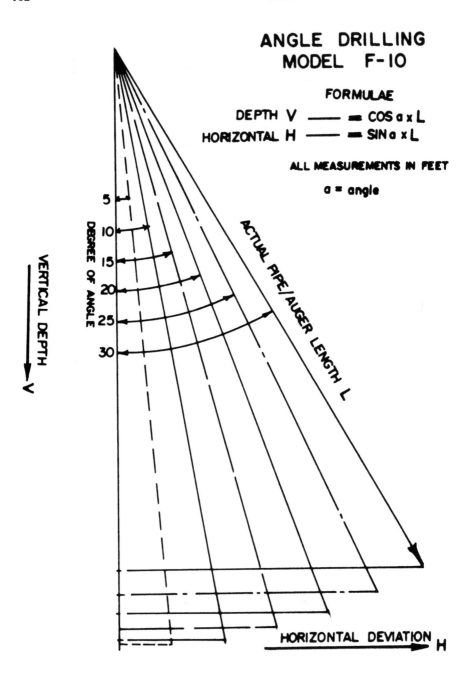

Fig. 3.25: Nomograph for determination of actual drill footage

Table 3.2. FRICTION LOSS IN PSI PER 100 FEET.

PUMP GPM	BORE THROUGH DRILL COLLAR/KELLY (in inches)						
	$1\frac{1}{2}$	$1\frac{3}{4}$	2	$2\frac{1}{4}$	$2\frac{1}{2}$	$2\frac{3}{4}$	3
50	10	10					
75	25	15					
100	45	20	10				
125	70	30	15				
150	90	40	20				
175	120	55	25	10			
200	150	75	35	20	10		
250	225	105	55	30	15	10	
300	310	150	80	45	25	15	10
350	425	200	110	60	35	20	15
400	550	255	135	75	45	25	20
450		320	170	95	55	35	25
500		395	205	115	70	40	30
550		475	250	140	80	50	35
600		560	290	165	95	60	40
650			335	190	110	70	45
700			390	220	130	80	55
750			440	250	150	90	60
800			500	275	165	105	70
850				310	190	115	80
900				350	210	130	85
950				380	230	140	95
1000				430	260	155	105

Slant-hole drilling (Fig. 3.26) is a method for achieving horizontal displacement at shallow depths to access targets that can not be reached by conventional directional drilling. Slant-hole drilling techniques have also been used for multiwell programmes where well heads are spaced closely together on a drill site or pad and the well directionally drilled to hit reservoir targets.

In multiwell drilling programmes the emphasis is on improved economics and cost savings by minimising land use, surface gather systems and maintenance during the producing life of the reservoir.

MODEL FM-45

(Mudmaster $5'' \times 6\frac{1}{43}''$ Slush Pump)

The Failing model FM-45 (Fig. 3.27, Table 3.6) is a duplex reciprocating type power slush pump designed for maximum volume without sacrifice of working pressure and compactness.

Its power end is a rugged, one-piece unit with integral supports for jackshaft and eccentric shaft bearings. The cover is removable for easy access to the power-end mechanism. The main sprocket and eccentrics are cast in one piece and keyed

Table 3.3. (all measurements in feet).

PIPE LENGTH L	5° ANGLE		10° ANGLE	
	V	H	V	H
5	4.98	.43	4.9	.8
10	9.96	.87	9.8	1.7
15	14.94	1.30	14.7	2.6
20	19.92	1.74	19.7	3.4
25	24.90	2.17	24.6	4.3
30	29.88	2.61	29.5	5.2
35	34.86	3.05	34.4	6.0
40	39.84	3.48	39.4	6.9
45	44.82	3.92	44.3	7.8
50	49.80	4.35	49.2	8.6
60	59.77	5.22	59.0	10.4
70	69.73	6.10	68.9	12.1
80	79.69	6.97	78.7	13.8
90	89.65	7.84	88.6	15.6
100	99.61	8.71	98.4	17.3
120	119.54	10.54	118.1	20.8
140	139.46	12.20	137.8	24.3
160	159.39	13.94	157.5	27.7
180	179.31	15.68	177.2	31.2
200	199.23	17.43	196.9	34.7

Table 3.4. (all measurements in feet).

PIPE LENGTH L	15° ANGLE		20° ANGLE	
	V	H	V	H
5	4.8	1.3	4.7	1.7
10	9.6	2.5	9.4	3.4
15	14.5	3.8	14.1	5.1
20	19.3	5.1	18.8	6.8
25	24.1	6.4	23.5	8.5
30	28.9	7.7	28.2	10.2
35	33.8	9.0	32.9	11.9
40	38.6	10.3	37.5	13.6
45	43.4	11.6	42.2	15.4
50	48.3	12.9	47.0	17.1
60	57.9	15.5	56.4	20.5
70	67.6	18.1	65.7	23.9
80	77.2	20.7	75.1	27.3
90	86.9	23.3	84.5	30.7
100	96.6	25.8	93.9	34.2
120	115.9	31.0	112.7	41.0
140	135.2	36.2	131.5	47.8
160	154.5	41.4	150.3	54.7
180	173.8	46.5	169.1	61.5
200	193.1	51.7	187.9	68.4

Table 3.5.   (all measurements in feet).

PIPE LENGTH L	25° ANGLE		30° ANGLE	
	V	H	V	H
5	4.5	2.1	4.3	2.5
10	9.0	4.2	8.6	5.0
15	13.6	6.3	13.0	7.5
20	18.1	8.4	17.3	10.0
25	22.6	10.5	21.6	12.5
30	27.1	12.6	26.0	15.0
35	31.7	14.8	30.3	17.5
40	36.2	16.9	34.6	20.0
45	40.7	19.0	39.0	22.5
50	45.3	21.1	43.3	25.0
60	54.3	25.3	51.9	30.0
70	63.4	29.5	60.6	35.0
80	72.5	33.8	69.2	40.0
90	81.5	38.0	77.9	45.0
100	90.6	42.2	86.6	50.0
120	108.7	50.7	103.9	60.0
140	126.8	59.1	121.2	70.0
160	145.0	67.6	138.5	80.0
180	163.1	76.0	155.8	90.0
200	181.2	84.5	173.2	100.0

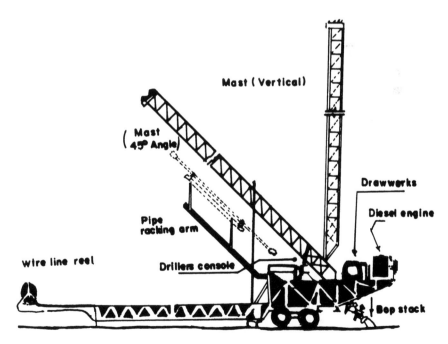

Fig. 3.26: Slant-hole drilling.

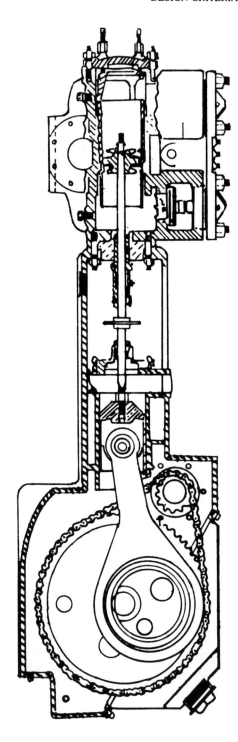

Fig. 3.27: Failing model FM-45 slush pump.

Table 3.6. Specifications of failing model FM 45 slush pump.

| Size, in. | | Max. Pressure lb per sq. in | Displacement | | Brake H.P. at Max. pressure | R.P.M. | | Max. Plunger Load, (lb) | Pipe Size | | Approx Wt of Pump (lb) | Approx. O.A. Dim., in. | | |
| | | | | | | | | | | | | | | |
Dia of Fluid Cylinder	Length of Stroke		Displacement U.S. Gal per Min/Rods deducted	Approx Barrels per Hr. (42 U.S. Gal.)		Crank-shaft	Jack-shaft		Suc-tion	Dis-charge		Height	Width of Mud end	Length
5		310	104-208	149-298	23-46									
4½		385	84-168	120-240	23-46									
4	6¼	492	66-132	94-188	23-46	50-100	236-473	6320	4	2	2630	27	32	80
3½		546	50-100	74-148	23-46									
3		895	36-72	52-104	22.5-45									

Model FM-45 —900 Pounds Working Pressure —1800 Pounds Test Pressure to a straight alloy steel shaft. These shafts are roller-bearing mounted and power is transmitted by a high-speed quadruple roller chain.

Connecting rods are one-piece steel with renewable eccentric bushings and bronze crosshead bushings. The one-piece skirt-type cast alloy crosshead is reinforced at all load areas.

The fluid end of the FM-45 is a one-piece casting with conventional pot-type valves. Steel studs secure valve covers, flanges and plates to the fluid end, and fluid and power end.

All bearings are sealed in individual pressure-gun lubricated housings. Eccentrics, chain, crossheads and pins are lubricated from the power-end reservoir through normal movement of parts.

Failing lubricator-type packing glands provide positive lubrication to piston rods. This pump is equipped with Failing Mudmaster pistons, hardened liners, rods and valves. Its specifications are illustrated in Table 3.6.

Modified Stable Foam Can Give Lower Drilling Costs

The advance of modified stable foam (MSF) is opening up a new perspective in flushing mediums. The aim, as always, is to attain the objectives, set but in so doing to achieve the lowest overall cost per foot. MSF offers assistance in loss of circulation zones, permafrost and high-temperature drilling areas.

Proven results in workover operations, water wells and shaft sinking confirm the arrival of MSF drilling. It is simple yet effective, ecologically safe and inexpensive. It should therefore be considered in drilling programmes for lower overall cost per foot.

Foam: Additions to air-flush media are certainly not new. However, the first driller to pour detergent down the drill-pipe to alleviate balling problems did not appreciate the changes he was instituting in drilling technique.

Air drilling has generallly been confined to dry, consolidated formations where the brute force of air velocity has cleaned the hole. The yardstick annular velocity is 3000 fpm although rates in excess of 5000 fpm are used.

When formation water is encountered, a foaming surfactant is injected into the high-velocity gaseous stream. The surfactant mixes with the formation water and increases lift efficiency of the high-velocity system. It also permits removal of water from the hole. However, large quantities of produced water obviously limit this approach.

Another drawback to the high-velocity system is that since the lifting ability is dependent on the annular velocity, air-volume requirements become prohibitive with an increase in hole size.

Cost, availability and space required (particularly offshore) are also mitigating factors.

Foam injection: A little over ten years ago, engineers in the US injected for the first time a bentonite-based mud containing a suitable foaming agent into the air stream. This resulted in an extremely viscous foam with excellent hole-cleaning and formation-stabilizing properties. The slurry was then circulated round the systems.

This mud system, cleaning by viscosity, was a recognisable advance. The air requirement necessary was that which just 'pushed' the columns around the system, rather like a slow moving materials conveyor.

However, general make-up water and formation contaminants limit application of this method, although excellent results were achieved by the Atomic Energy Commission. Later developments included a stable foam formed by the injection of a solution of water and a foaming surfactant into a gaseous phase on the surface and then circulating this foam down the hole. This method was considerably more contaminant-resistant than the gel-based stiff foam.

In general, drilling technology has improved dramatically within the past few years and foam technology has kept pace. The use of specific additives has extended the range of stable foam applications and clearly advanced the low-velocity flushing method to first place in optimised penetrating and hole-stabilising solids-free systems.

Foams and foaming surfactants: By definition a foam is a coarse dispersion of gas in liquid and two extreme structural situations can be recognised.

Dilute foams consist of nearly spherical bubbles separated by rather thick films of generally viscous liquid.

Concentrated foams are mostly gaseous and consist of polyhedral gas cells separated by thin liquid films.

The nature of thin liquid films is currently undergoing considerable research and improvement in foam stability has already been observed.

Emulsion chemistry tells us a great deal about foaming surfactant suitability. Good emulsifying agents are, in general, also good foaming agents since the factors which influence emulsion stability (against droplet coalescence) and foam stability (against bubble collapse) are very similar.

Therefore, the desired end-product circulating in the borehole is a concentrated foaming agent with emulsion and foam stability that results in a minute, tight, thin-walled dryer bubble concentration, much like an aerosol-type shaving soap lather.

This flushing medium has a carrying capability much in excess of a waterbased mud, and combined with a low hydrostatic head provides optimum penetration characteristics. Further addition of selected wall-stabilising high-molecular weight polymers and additives such as gilsonite will tolerate very high formation-produced water volumes and can stabilise swelling and sloughing shale sections.

Thus the MSF is justifiably considered as a basic mud system and merits consideration in all mud programmes.

There is only one drilling situation that precludes the usage of MSF, namely that in which hydrostatic pressure is deemed necessary. MSF densities range from 0.4 to 0.8 1b/gal.

How a modified stable foam works: Success of the MSF drilling concept is based on the following:

1. *Thickening the air* so that the hole can be cleaned at a slow annular velocity (about 45–150 fpm) compared to 3000–5000 fpm (or more) required when air alone is used. A substantial reduction in air requirement and low annular

pressure results, which in turn, reduce the chance of losing air to the formation. Furthermore, the low velocity reduces hole erosion and drilling problems consequent to hole enlargement.

2. *Casing the hole* with a thin impermeable mud cake. This film helps to confine the air to the borehole and limits the "pressuring up" of the wall and adjacent formation. This mechanism also coats sticky clays etc. and materially helps to prevent balling.

3. *Immediate clearing of the drilling face.* Cuttings are removed as rapidly as they are dislodged, usually in a marked increase in drilling rate and improved bit life.

This medium offers the penetration properties of a straight air-flushing system coupled with carrying capacity far in excess of a regular drilling mud.

Injection. Small amounts of a custom compounded mud slurry, with high efficiency foaming agent added, are injected into the air stream. The air stream is controlled to give an annulus velocity of 45–150 fpm 15–45 mpm.

The injection rate is dependent on hole size drilling rate annulus size, time on the hole, hole depth, type of formation and amount of water hole is producing.

Keep in mind that the cuttings and formation water are carried out of the hole, encapsulated in a thick foam of shaving lather consistency.

Expect to inject at least 7 gal/cu. ft of hole cut and adjusted to the above factors.

For injection, a low volume, variable-output, positive-displacement pump is required. The pump output is varied as required from 1 to 20 gpm at a pressure sufficient to allow injection into the rig stand pipe through a 'Y' arrangement.

It is usually best to start the injection pump before starting the compressor.

Modified stable foam slurry. For best results it is important that the MSF mixture be properly formulated. Care must be taken to obtain a compatible extra-high efficiency foamer. Not all foamers meet the requirements and give the desired results.

Mix selected additives in water to a uniform consistency, with viscosity in the range of 30–35 s/qt. The viscosity can be varied as required for best results. A low viscosity slurry is generally desirable; a slurry that is too viscous usually makes a foam that is too thick for efficient results.

Air-volume requirements for MSF drilling are modest compared to high-velocity types of air drilling. A ball-park estimate can be made using:

$$V = (D^2 \times .7854)868$$

where V is the air required in cfm and D is the bit diameter.

Field experience indicates that for moderate depth holes (2000 ft without unusual problems) requirements are in the range of:

26 in/660 mm hole:	400–600 cfm/11–17 cum/min
15 in/380 mm hole:	200–300 cfm/5–6 cum/min
10 in/254 mm hole:	90–130 cfm/2.5–3 cum/min
6 in/152 mm hole:	25–30 cfm/0.7–0.85 cum/min

Air-pressure requirements will vary widely with depth and other conditions. The desired range will usually allow for sufficient additional capacity (high volume and high pressure), based on the size and depth of hole, to unload the hole after a connection, or for handling large amounts of water.

The success of the MSF approach is absolutely dependent on not using too much air, and maintaining low velocity. If the limits are exceeded, the air will channel the foam, destroying its cuttings-carrying capacity and often cause wall packing and other hole problems.

The bubbles in the foam should be fairly small and as uniform as possible in size, which can be adjusted with injection rates (mud slurry and air).

The foam column is a slow-moving materials conveyor. The use of air usually results in high drilling-rate capability. Do not overload the foam column with cuttings as this will lead to problems. For efficient operations the drilling rate should be adjusted so as not to exceed the hole-cleaning capacity (rate) of the foam column.

When establishing the foam column, a steady rush of air at the return line indicates that a foam column is not being formed. Adjust by increased MSF injection volume or decreasing air volume. When proper downhole foaming action is taking place, a gentle pulling of air can be felt at the return line.

A steady smooth flowing column of returning foam is hard to maintain. Usually the foam will have a gentle surging action and there is reason to believe that this adds to its hole cleaning capability. If the foam is too stiff, this gentle surging action will not take place.

If the foam surges violently, or returns by heads at high velocity, this indicates too much air or drilling faster than the hole can be cleaned. These violent surgings can subject the hole to unnecessary pressure imbalance and that can cause sloughing of wall packing and unstable hole conditions.

The possibility of drill pipe corrosion arises when drilling with air and injecting water. When making a trip or laying down the pipe for storage, corrosion can be slowed up by coating the pipe inside and out with a corrosion inhibitor, or by incorporating a corrosion inhibitor in the MSF slurry.

Points to note in MSF drilling

Controlling MSF. Control of the MSF system is based on:

1) Surface injection pressure
2) Drill string torque
3) Condition of foam at the blooie line.
4) Regularity of foam at the blooie line.

Surface injection pressure should be the lowest possible for good foam drilling. Changes in the stand pipe pressure are the quickest way to determine when a problem exists. By observing pressure changes, adjustments in injection rate can be made that will rectify most problems.

Some typical situations are shown in Table 3.7.

Table 3.7. Pressure Change.

Foam condition at blooie line	Source	Corrective adjustments
Quick pressure drop	Air has broken through foam preventing formation of stable foam.	Increase MSF slurry injection rate and/or decrease air injection rate
Slow gradual pressure increase	Increase in amount of cuttings or formation fluid being lifted to the surface.	Increase air injection rate slightly.
Quick pressure increase	Bit plugged or formation packed off around the drill pipe.	Stop drilling and attempt to regain circulation by working pipe and slugging with MSF slurry.

Drill-pipe torque indicates the type of formation (sand shale, or clay) being penetrated, bit plugging and bridging of cuttings around the drill pipe or collars. Changes in drill-pipe torque can be easily detected when a power swivel is being employed. When a shale or clay formation is encountered, torque is excessive; the injection rate of both slurry mix and air should be increased in order to clean the hole. When bridging and bit plugging occur, the penetration rate should be reduced until the foam can clean out the cuttings.

Conditions of foam at the blooie line should be reduced until the foam can clean out the cuttings.

Conditions of foam at the blooie line should, under normal conditions, appear like shaving cream. If not, corrections should be made. Some typical conditions and corresponding adjustments are listed in Table 3.8.

Regularity of foam at the blooie line by regulating the MSF slurry and/or air-injection rates, foam returns at the blooie line may be continuous or "heading"

Table 3.8. Foam condition and adjustment.

Foam condition at blooie line	Source	Corrective adjustments
Air blowing free with line mist of foam.	Air has broken through liquid foam mix preventing formation of stable foam	Increase MSF slurry injection rate and/or decrease air injection rate.
Foam thin and watery (salt cuc)	Formation salt water entry.	Increase MSF and air injection rate and possibly increase percentage of foaming surfactant in injection slurry.

Sources:
1. How to Determine Your Rig's Depth Limit. by John L'Espoir, George E. Farling Co.
2. Drilling Technology-What Every Driller Should Know, parts I, II & III by John L'Espoir, George E. Farling Co.
3. Modified Stable Foam Can Give Lower Drilling Costs. by Bernard Higgins, Managing Director Drill—Aid
   (Methods & Materials) Ltd., London.

with foam coming out of the blooie line only part of the time. In most cases continuous foam returns are not necessary to ensure complete removal of cuttings. Noticeable heading occurs when constant circulation subs are not used to maintain circulation while making connections. As long as foam returns are regular and the time between unloading periods is no longer than 15 min, the foam system is generally operating satisfactorily.

If the time interval between unloading periods increases, the MSF injection rate should be increased and/or drilling rate decreased.

CHAPTER 4

DRILLING HOLES FOR MINING AND CONSTRUCTION INDUSTRIES

Drilling Selection Requires Value Judgements
Controlled Blasting
Bit Types and Applications

In mining and construction, earth excavation is an everyday part of the job. The most economical method for earth excavation is, of course, mechanical attack. However, when the rock material is too hard to succumb to such an attack, drilling and blasting must be employed. Standard procedure is to drill a pattern of suitably sized holes for the rock condition and formation, load the holes with a charge of explosive, clear the area of people and equipment, set off the charge, then move back in with mechanical equipment and 'muck out' the rock fragments. Rotary drills are especially suited for the blast-hole drilling these industries require.

Several basic strategies are employed in the drilling and blasting process. For example, in road construction projects production blast holes are generally drilled vertically. But presplitting holes are frequently used to establish a slope for a cut that will result in minimum erosion and land slide. In open-pit mining production blast holes are generally vertical, although in recent years some mines have used angled drill holes to great advantage. Occasionally, secondary drilling and shooting is required in either an open pit or a road construction job. This will occur during the mechanical 'mucking-out' process, when a large boulder, too big to handle with the equipment at hand, is uncovered. The practice is to break up this rock by mechanical means, utilising a crane-type piece of equipment with a heavy drop hall to shatter the boulder, or to move in some smaller drilling equipment and drill a few holes strategically in the boulder, place an explosive charge, and break the stone into smaller pieces which can be removed.

Hole depths for blast holes are generally drilled two or three feet/600 or 900 mm below the desired grade to obtain as much fragmentation of rock as possible above that grade line. This is generally true for all mining and construction projects. However, occasionally the contractor or miner will be fortunate enough to have the rock situated in layers or bedding planes in good alignment with the finished grade. In such cases minimal subdrilling is required. This is particularly true in the case of open-pit coal mining where, subgrade drilling would ruin the coal by mixing particles of rock and coal, making the product unusable. In most cases there is no hard rock cover immediately overlying the coal. Instead, it is covered with what miners refer to as 'slip-stone', usually a thin layer of shale or slate which can be removed without drilling and blasting. In these instances production hole depths are ended somewhere short of the coal body so that it remains undisturbed.

Hole diameters are generally determined by geological fractures in the native stone, as well as the density of the stone and the size of the 'mucking' equipment to be used on the job. For example, it would be considered unreasonable to drill a 3.5-in/89 mm diameter hole if the mucking equipment consisted of a 40 cu yd/30m^3 shovel. Usually the diameter of the drill hole must be determined by firing an initial shot pattern of just a few holes. The degree of fragmentation can then be observed. The mining engineer or the contractor can usually judge this very accurately.

The production drill must be the correct size for the job the contractor or mining engineer has to accomplish. For making a reasonably correct drill size recommendation, the manufacturer's representative has several sources of information available to him. In the construction industry huge dam projects are always core drilled. These core samples are available for inspection. Such samples reveal not only the type of material that will be encountered in the area to be excavated, but also give good information about the geological fractures in the ground. The same is true for any large new mining project. A quarry is only opened where a known deposit of suitable rock exists. New mines or quarries are usually located reasonably near an existing mine or quarry and the rock formation will thus be similar. When a road cut is to be made through a mountain, usually another road cut in the immediate vicinity will show pretty well what formations the contractor might expect. When an existing quarry or mining operation plans to buy new equipment, the rock face of the existing operation can be studied to help determine the correct drill size. In nearly all cases the size of the mechanical mucking equipment is known and usually is on order before any thought is given to production blast hole drilling. The size of the bucket on the loading shovel is a good indication of the degree of fragmentation that must be obtained by drilling and blasting. The same is true for the size of the crushing equipment to be used. All this information should be studied very carefully by the person who is to recommend drilling equipment. A thorough analysis of this information should provide the correct answer to a particular customer's drilling problems.

For the most part, the technique used to drill the first blast hole in a pattern will be used in every blast hole in the pattern. In a large mining operation requiring several patterns throughout the mine, the same basic technique will be used in drilling every blast hole. Today all these holes are drilled without a great deal of operator knowledge of the detailed drilling conditions. That is, the operator is concerned with moving the machine into position, levelling the machine, drilling the hole in a relatively solid rock formation with few worries about mud seams, lost circulation etc., retracting the drill string from the hole, raising and lowering the mast etc. These are, for the most part, purely mechanical operations on the part of the driller. In other words, the operator is only one of a team of men, each a specialist in his own field, and the entire team works together towards an end-result.

DRILLING SELECTION REQUIRES VALUE JUDGMENTS

Principles of Drilling

Selection of a particular machine for production drilling is the most critical drill evaluation the pit engineer is called upon to make. It is a true engineering design problem requiring value judgments. Generally, the procedure follows these steps:

1) Determine and specify the conditions under which the machine will be used, including such service factors as labour, site, weather etc.

2) State the objectives for the rock-breakage phases of the production cycle of operations—considering excavation and haulage restrictions, crushing capacity, production quota and pit geometry—in terms of tonnage, fragmentation, throw etc.

3) On the basis of blasting requirements, design the drill-hole pattern (hole size and depth, inclination, burden, spacing etc.).

4) Determine the drillability factors and select the drilling methods which appear feasible for the kind of rock anticipated.

5) Specify the operating variables for each system under study, considering drill, rod, bit and circulation fluid factors.

6) Estimate and compare performance parameters, including costs. Major cost items are bits, drill depreciation, labour, maintenance, power and fluids.

7) Select the drilling system that best satisfies all requirements with the lowest overall cost.

Probably the most difficult steps to accomplish in the entire design procedure are (4) and (6). This is because drillability determination and performance predictions are largely unreliable.

The various drillability factors may be grouped in six categories: (1) drill, (2) rod, (3) bit, (4) circulation fluid, (5) hole dimensions and (6) rock. Those factors in categories (1) through (4), components of the drilling system itself, are referred to as design or operation variables. They are dependent (controllable) within limits, although they are interrelated in some instances, being selected to match the environmental conditions reflected by category (6).

The hole geometry factors of category (5), drill hole size and depth, are dictated primarily by outside requirements and are independent (uncontrollable) variables in the drilling process.

The environmental factors of category (6) include the following: (a) rock properties (resistance to penetration, porosity, moisture content, density etc.), (b) geologic conditions (petrologic and structural, i.e., bedding folds, faults, joints etc.); and (c) state of stress (overburden pressure and formation fluid pressure; unimportant in shallow holes).

Percussion Drilling

There are two types of percussion drills in use today, the piston drill and the hammer drill. In the piston drill the drill steel is attached to the piston and both reciprocate and rotate.

The hammer drill is more commonly used. In this drill the piston or hammer reciprocates in a cylinder and strikes the drill steel, anvil block, or tappet on its forward stroke. The action of the percussion hammer drill incorporates two basic operating principles. The first is the principle that makes the piston reciprocate in the cylinder and the second that which makes the drill steel and bit rotate. In all drills piston movement is effected by a self-acting valve which admits compressed air at the proper instant, first to one end of the cylinder and then to the other. Rotation of the drill steel is accomplished by one of four methods: automatic rifle-bar rotation, integral independent rotation, external independent rotation and manual rotation.

A relatively new method of drilling which promises higher drilling speeds is the rotary percussion method, wherein drilling is performed either with a percussion bit and drifter drill or with a roller bit and downhole drill. Metallurgically much work has to be done on rotary percussion drilling but it may well be the next significant advance in the art of drilling.

A prime factor in percussion drilling is the type of mountings, of which there is a variety. Self-propelled mountings are by far the most predominant, but the type of mounting used depends on hole size, terrain and depth of holes. The most versatile is the self-propelled crawler type. This mounting, designed to tow the required air compressor, utilises drifter-type drills in the 4–6 in/100–150 mm bore size. The air compressors used are rubber-mounted, diesel driven and made in sizes up to 1200 cfm (35 m³m).

The next largest class of mountings are the crawler-mounted and self-contained. These are classified into two categories. Those in the first category are for 4.75–7.5 in/120–190 mm hole sizes. They are furnished with 500–900 cfm/14–25 m³m of compressor capacity and use 4–6 in/100–150 mm diameter drill steel and a 20–30 ft/6–9 m steel change drill tower. Two men are required to operate this class of drill and cabs are provided for the operators. Elaborate dust-collecting systems are furnished to protect the operators and also the air intake to the compressor and engine.

When conditions dictate, mine owners will use holes in the range of 7–9 in/175–225 mm. For such holes the larger crawler-mounted equipment of the second category will be used. As much as 1800 cfm/45 m³m of compressor capacity is required for these self-propelled mountings. Manufactures design these machines to be used with either percussion bits as downhole drills or roller bits, to add to their flexibility. The larger roller bits require high down pressures and the equipment is heavy. Drill towers capable of a drill-steel change are common. Rising labour costs dictate the need for more productive equipment. Hence, much work is being done on high-pressure (200 psig/14 kg/cm^2) drilling, improved button bits, improved flushing and improved sintered carbide.

High air pressure for drilling tools means more expensive air compressors and higher maintenance costs, both for the air compressor and percussion drill. Since the work of a percussion drill depends on air pressure available, this idea offers exciting possibilities. Metallurgy will play an important role in this development.

The button bit will undergo important changes as will the carbide used.

Improved flushing of drill holes may add to the art of drilling. Experiments have already been undertaken for flushing only with high-pressure air using drifter drills. The use of high-pressure air for downhole drill flushing would certainly add to performance by exhausting cuttings as soon as they are broken rather than recirculating them.

Rotary Drilling

Primary blasthole drilling by the rotary method was performed with drag bits until 1949, but its use was limited to soft formations. In 1949, it was found that air circulation to remove cuttings resulted in increased penetration rate and prolonged life for the rolling cutter-type bits (Fig. 4.1a). Prior to that time water circulation had been used with these bits. Dependence on water to clean the hole had delayed acceptance in mines of bits with rolling cutters due to water haulage and freezing problems as well as the loss of water through cracks in the formation. Once rolling cutters were introduced, subsequent improvements in tools made rotary drilling adaptable to harder formations.

Rotary drills attack rock with energy supplied to the bit by the rotating action and thrust of a drill stem. Supplemental energy is applied in oil-well rotary drilling by impingement of the circulation fluid (usually drilling mud) on the bottom through jets in the bit. The jetting action of air used in mining has little significance in breaking pieces of material from the mass but is useful in rapidly removing the material torn loose by the bit before these fragments interfere with subsequent drilling. Breaking of the rock is achieved by either a rubbing abrasive action, plowing scraping action, spalling action, chipping action or, more often, some combination thereof. Sufficient thrust must be provided so that stresses induced by the teeth prove effective in overcoming the compressive strength of the rock. Thrust force is obtained by the weight of the tools above the bita and by coupling to these tools a part of the weight of the drill rig through hydraulic cylinders, cable or chain pulldown.

Drag-bit drilling involves drag-bit bodies (Fig. 4.1b, c) which are castings or forgings in which the cutting blades, sometimes called bits, can be replaced when dull. These bits are occasionally hard-faced or may have sintered tungsten-carbide cutting edges brazed on them. Since drag-bit drilling is more popular in softer formations where shock resistance is not so important, replaceable bits are used more often than unit-body types. The cutting heads or bodies cost a few hundred dollars and will last through many tool replacements for thousands of feet of hole.

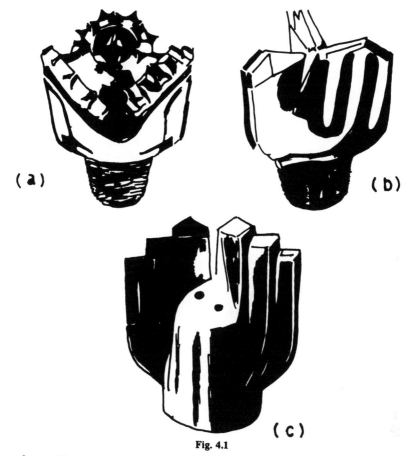

Fig. 4.1

a—3-cone roller rock bit; b—3-wing solid body insert drag bit; c—2-wing carbide insert drag bit.

Seven to 11 replaceable tools or bits are required to dress the most popular 6 and 8 in′ 150–200 mm bit sizes.

Rolling-cutter bits come in four types—for soft, medium, hard and very hard formations. They are made of alloy steel and specially processed for a hard car-burised surface and tough inner core.

Overall drilling costs are usually computed in cost per foot of hole. Except for cost of bits, cost items should first be converted to cost per hour, then to cost per foot/metre and then to cost per ton.

Rig depreciation is estimated by dividing the delivered price of the rig by 10,000 hours. Fifty percent depreciation will cover miscellaneous cost items, such as interest on investment, maintenance, power, supervision, overhead and adminis-tration. Labour costs will vary with the area and should be the total hourly wages of the operator and helper plus 25% to cover fringe costs.

Jet Piercing

Jet piercing is a patented thermal process dependent upon a characteristic of the rock termed spallability. Simply defined, spalling is decrepitation resulting from differential expansion of rock crystals due to thermally induced stresses.

In practice, flames from a multiflame burner consuming oxygen, and flux-bearing kerosene fuel are directed from a rotating blowpipe against a rock surface. The blowpipe is advanced as rapidly as the hole progresses.

The latest rotary-piercing machine is a 48-ton, self-propelled unit powered by electric motors and capable of manoeuvering over the usual open-pit mine terrain. Except for the blowpipe rotary-drive unit, all machinery components and electrical gear are housed within the insulated cab. The blowpipe consists of a long tubular steel member. At the upper end is an electric rotary drive and a joint through which the three process fluids are fed. The lower end of the blowpipe consists of a Kelly extension, header, fuel injector, burner and reamer shell. The three process fluids—oxygen, fuel and water—are fed from supply hoses to the rotary joint through separate conduits within the main drill pipe to the burner assembly at the lower end.

Two commonly used types of piercing equipment are the suspension-piercing machine and the manual-piercing blowpipe. All suspension-piercing machines presently in service are converted churn drills. In suspension piercing a burner with a single axial flame port is mounted at the lower end of a 1000 lb/454 kg blowpipe about 20 ft/6 m long, which is attached to the drill cable of the machine. Three hoses then carry water, fuel and oxygen respectively to the rear end of the blowpipe. The non-rotating blowpipe is oscillated up and down at approximately 45 cycles per minute with a stroke of 4 in/100 mm. As the blowpipe advances into the hole, process fluids hoses are attached to it and enter the hole with the drill cable.

The manual-piercing blowpipe weighs approximately 21 lb/10 kg and its standard length is designed to pierce holes $1\frac{1}{2}$-2 in/38-51 mm in diameter to a maximum of 5 ft/1.5 m. Needle valves for controlling the flow of process fluids are provided at the rear inlet of the blowpipe. An adjustable external spalling shield at the burner end protects the operator from the stream of gases, rock particles and steam emitted from the hole.

Jet piercing makes it possible to shape, enlarge or chamber the lower portion of the blast hole. Patented processes employing the jet burner have been widely applied to the working of dimension stone in the granite industry. The flame process for releasing large blocks of granite in the quarry is known as jet channelling and was introduced in the industry in 1955. The blocks are freed by channels or slots which are cut in the formation with jet burners in what is, in effect, a machining operation. The channels are approximately 2.5-3 in/63-75 mm wide, 10-100 ft/3-30 m long and up to 30 ft/9 m deep. Smaller blowpipes, both manually operated and mechanised are used to cut, shape and impart a flame-textured finish to granite building stone.

CONTROLLED BLASTING

The objective of controlled blasting is to reduce and better distribute the explosive charges in order to minimise stressing and fracturing of the rock beyond the neat excavation line. In other words, to blast to a line with as little 'overbreak' as possible. Three standard techniques are used to accomplish this objective.

Line Drilling

Line drilling consists of a single row of closely spaced, unloaded, small diameter holes drilled along the neat excavation line. These holes provide a plane of weakness which the primary blast can break. This plane also reflects some of the shock waves created by the primary blast, which reduces shattering of the finished wall. Line drill holes are generally 2-3 in/51-76 mm in diameter and spacings are generally two to four times the hole diameter along the neat excavation line. Depth of the line drill holes depends on how accurately the alignment of the holes can be maintained. Depths greater than 30 feet/9 m are seldom satisfactory. The primary blast holes directly adjacent to the line drill holes are generally loaded lighter and are more closly spaced than the other holes. The distance between line drill holes and the directly adjacent blast holes is usually 50% of the normal burden between primary blast holes. A common practice is to reduce spacing of adjacent blast holes by 50% with a 50% reduction in explosive load.

Best results with line drilling are obtained in homogeneous formations where bedding planes, joints and seams are minimal. These irregularities are natural planes of weakness that tend to promote shear through the line drill holes into the finished wall.

Cushion Blasting

Like line drilling, cushion blasting involves a single row of holes along the neat excavation line. Holes can be from 2-6-$\frac{1}{4}$ in/51-158 mm in diameter and are loaded with light, well-distributed charges of explosive completely stemmed and fired after the main excavation is removed. In cushion blasting the main cut area is removed, which leaves a minimum buffer or beam zone in front of the neat excavation line. The cushion holes can be drilled prior to any primary blasting or just before removing the final berm.

The burden and spacing will vary with the hole diameter being used. For example, with a 2-in/51 mm hole diameter, spacings are approximately three feet and burden approximately four feet. With a 4-in/101 mm hole diameter, spacing is approximately six feet and burden seven feet. With 6-$\frac{1}{4}$ in diameter holes, spacing is approximately seven feet and burden approximately nine feet. The holes are

string-loaded on primacord down lines with full or partial 1-1.5 × 8 in/25–38 mm
× 202 mm cartridges of dynamite, spaced one to two feet apart along the down
lines.

Presplitting or Preshearing

This technique involves a single row of holes drilled along the neat excavation
line. The holes are generally 2–4 in/50–100 mm in diameter and all are loaded.
Presplitting differs from line drilling or cushion blasting in that the holes are fired
before any adjoining excavation area is blasted or removed. With proper spacing
and charges, the fractured zone between the holes will be a narrow sheared area
which the subsequent primary blast can break. This results in a smooth wall with
little or no overbreak. Presplit holes are loaded similar to cushion blast holes, i.e.,
down-string loads of full or partial cartridges 1–1.5 × 8 in/25–38 mm × 202 mm
long, spaced at one to two ft intervals along the down string.

Spacings vary with the diameter of the hole. For example, for 2-inch diameter holes, spacings will be 1.5–2 ft/0.3–0.6 m; for 4-inch diameter holes, spacings
will be 2–4 ft/0.6–1.2 m apart. In extremely unconsolidated rock formations, results
are sometimes improved by using guide holes between the loaded holes to promote shear along the desired plane. These unloaded holes generally give better
results than increasing the explosive charge in the loaded holes. All loaded presplit
holes are completely stemmed along and between the explosive charges to prevent
gas venting into weak strata and causing poor results. Generally, presplit holes
are drilled to 50 ft/15 m depths or less in order to maintain good alignment. In
presplitting it is difficult to determine results until primary excavation has been
completed to the finished wall.

Theoretically, the length of a presplit shot is unlimited. In practice, however,
shooting far in advance of primary excavation can be troublesome if the rock
characteristics change. By carrying the presplit holes only one-half the primary
shot pattern in advance of the primary shot area, the knowledge gained from the
primary blast regarding rock formation can be applied to subsequent presplit shots.
In other words, the loads can be modified if necessary and less risk is involved
compared to shooting the full length of the neat excavation line before progressing
with the primary blast.

Combinations

In some instances it may be desirable to combine some of these techniques for a
particular result. For example, it is frequently advantageous to line drill or presplit
sharp corners where cushion blasting is normally employed. Occasionally, in difficult rock areas pre-splitting is done short of the neat excavation line, then cushion
blasting is used to clean the wall along this line. While far more expensive, this
technique assures holding the wall more accurately than a single technique alone.
Also, full knowledge of the formation will be available when removing the final
berm by cushion blasting.

Summary

Line drilling is generally unpredictable except in homogeneous formations. The closer spacings and large number of holes mean higher costs and more tedious drilling.

Cushion blasting offers several advantages over line drilling, including increased hole spacings, reduced drilling time and costs, and often better results.

BIT TYPES AND APPLICATIONS

By trial and error during the past 75 years, drilling bits have been developed to a point where a bit type and size will give good economy and performance for almost any drilling requirement. They are generally listed under four categories: blade bits, drill heads, tricone bits and down-the-hole bits.

Blade Bits

In most cases blade bits are used for drilling relatively small diameter holes (3–4.75 in/76–120 mm) in soft formations. However, they are also available in larger sizes. These bits can be used for drilling with either air or water/mud for flushing the hole. Blade bits are manufactured with two, three or four-blade configurations and usually have a slug of carbide on the cutting edge of each blade, which can be sharpened when the bit becomes dull.

Drill Heads

Drill head bits come in various sizes, from 6.75 in/170 mm through 12 in/300 mm diameter, and consist of a large, roughly round piece of steel for the bit body, with threads at one end for attaching to the drill stem and several holes drilled into the drilling end. Finger inserts, usually tungsten carbide-tipped are snap-locked into these holes, providing a toothed drill-head configuration. These bits are principally used in drilling coal overburden, or other very soft formation applications. The fingers are easily replaced when they wear down. In a typical coal mine application, the drill fingers are replaced at the beginning of each shift. These bits are suitable for air drilling and are very economical.

Tricone Bits

Tricone bits have been in use for a good many years and are now developed to a very sophisticated degree. They are used for drilling soft, medium-hard, hard and very hard rocks and can be used with either air or water drilling. The soft formation tricone bits have long thin teeth and each cone is slightly offset rather than pointing directly towards the centre of the bit. This combination provides for a tearing, cutting action and gives very fast penetration in soft material. Best performance requires high (100–120) rpm.

The medium formation bit has somewhat shorter, more widely spaced teeth and the cones are pointed directly towards the centre of the bit. This tooth arrangement

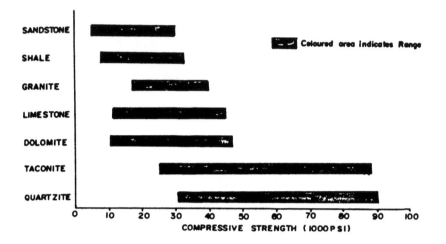

Fig. 4.2: Compressive strength of various rocks.

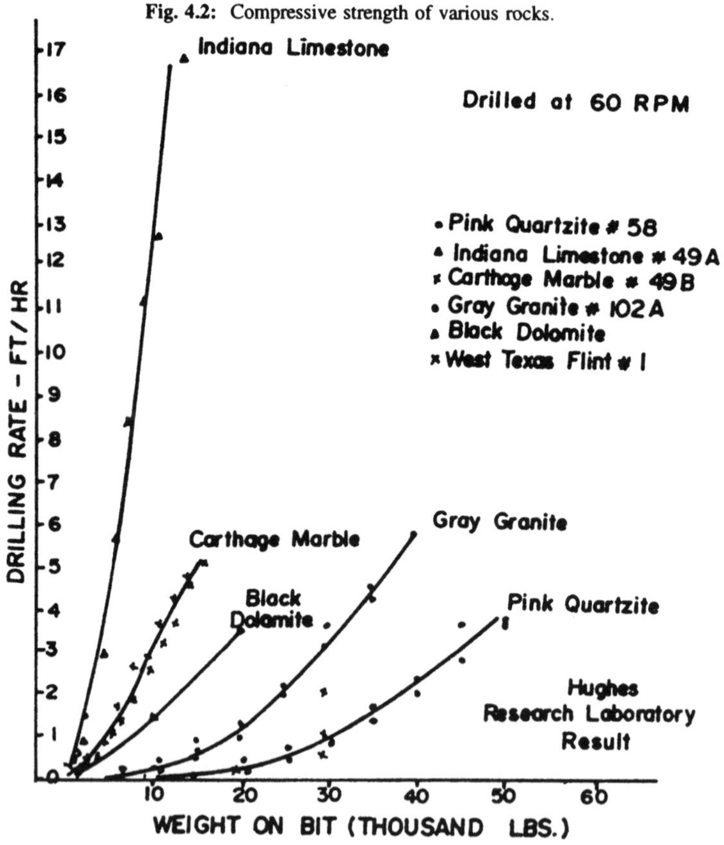

Fig. 4.3: Drilling rates for various formations.

provides for longer wear and the teeth are sufficiently sharp to give good penetration in medium-hard rock formations. Best performance requires medium (80–100) rpm.

The hard formation bit has short stubby teeth. The cones are pointed towards the centre of the bit and the gauge row of teeth on each cone is webbed to provide for longer wear. These bits should be rotated at 40–60 rpm and can be expected to give long wear and good penetration rates in hard rocks such as limestone, heavy sandstone, chert etc.

The teeth on all cone-type bits are surfaced with tungsten carbide.

Tricone Button Bits

Tricone button bits are used for drilling very heavy rock such as taconite, iron ore, granites, trap rock etc. and are constructed of three cones set with tungsten carbide inserts. These inserts have rounded points so as to present no sharp edges for rapid wear. They do not dig into the rock. Instead, heavy down pressure on this bit results in a crushing action at the face of the hole.

Other Cone Bits

Soft and medium-type cone bits can be obtained in a two-cone or a four-cone configuration.

Two-cone bit types are generally soft formation bits. They penetrate soft rock much faster than the tricone design but chatter violently when in contact with hard, broken rock formations.

Four-cone bit types should be used when drilling in broken ground that would pose rough drilling for a tricone bit. The four-cone bit action will be very smooth compared to that of the tricone but concomitantly penetration rates will be considerably slower.

Factors Affecting Bit Performance

The compressive strength and drillability of formations vary widely and prevent adoption of a universal drilling practice. The weight upon the bit, speed of rotation, and volume of air are all important factors in obtaining maximum penetration rate.

Figures. 4.2, 4.3 and 4.4 show the compressive strengths of various formations and the drilling rates of such formations with different loads applied on a rock bit. Rotary rock bits require weight to make them cut.

With weights greater than the foundering load, particularly in soft, low-strength formations, the teeth become fully buried and the cutter shell bears on the bottom and limits progress. Optimum weight, as well as other factors affecting drilling practice, can only be determined from actual field tests and experience.

In general, the rotary speed must be decreased as the applied weight is increased, otherwise the rock-bit bearings may be overloaded to the point of early failure. The types of formation being drilled and other factors that cause 'shock loads' on

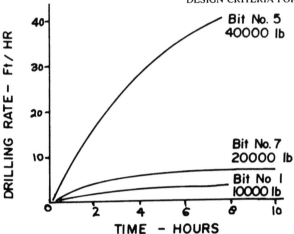

Fig. 4.4: Laboratory drill rig test—8.5 WR bits on ordinary sandstone with 10,000, 20,000 and 40,000 lb loads.

AIR VELOCITY DETERMINATION CHART

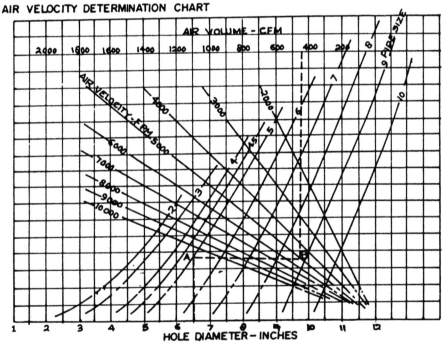

Fig. 4.5: To determine air velocity in the annulus when pipe size, hole diameter and air volume are known, follow vertical hole diameter line upwards to its intersection with pipe size. Move horizontally to intersect air volume line cfm air volume passing through the annulus, follow the hole diameter line to point A, at intersection with pipe size line. Move horizontally to point B, to intersect air volume line. Read annulus air velocity at point B (interpolating between 4000 and 5000 fpm) on 4700 fpm.

the rock bit bearings and teeth vary widely and sometimes dictate the operating practice with reference to rotation speed.

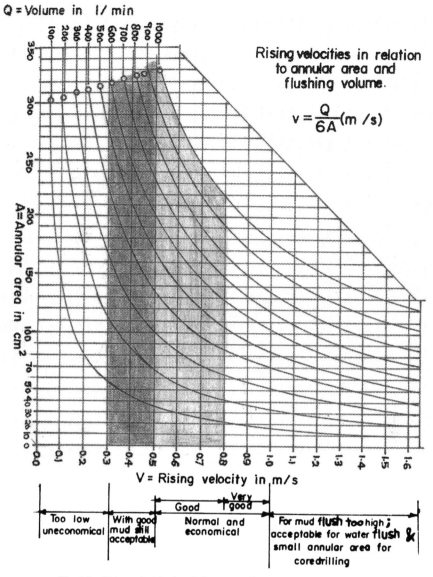

Fig. 4.6: Rising velocities in relation to annular area and flushing volume.

The efficiency of a rock bit depends on an adequate volume of air to remove cuttings (see Figs. 4.5 and 4.6), especially when drilling a softer formation or formations with a high water saturation. Very high air volume is frequently required to remove heavy cuttings such as coarse iron ore.

CLASSIFICATION OF ROCK MASSES

INTRODUCTION

Classification is defined as the arrangement of objects put into groups on the basis of their relationship. Classification have played an important role in engineering for centuries and are indispensable in engineering design.

The first major, classification was proposed 40 years ago by Terzaghi (1946) while suggesting steel supports for tunnelling. Since then many classifications have come into existence in civil and mining engineering, catering to different needs. Indeed, on many underground construction and mining projects, rock mass classifications have provided the only systematic design in an otherwise haphazard trial-and-error procedure. However, modern rock mass classifications have never been intended as the ultimate solution to design problems, but simply a means towards this end. While it is true that such classifications were developed to create some order out of the chaos in site investigation procedures and to provide much needed design aids, they were not intended to replace other means of engineering design. They should be used intelligently and in conjunction with observational and analytical methods to formulate an overall design rationale compatible with the design objectives and site geology.

ROCK CLASSIFICATION

Rock classifications can be broadly divided into two categories, i.e., (i) Intact classifications (ii) mass classifications.

Intact Rock Classifications

Based on laboratory tests of intact specimens of rocks, several classifications have recently been proposed. Figure 5.1 lists those based on uniaxial strength of rocks. The classification proposed by Deere and Miller (1966) includes compressive strength and modulus of elasticity. Their classification has been widely recognised as particularly convenient.

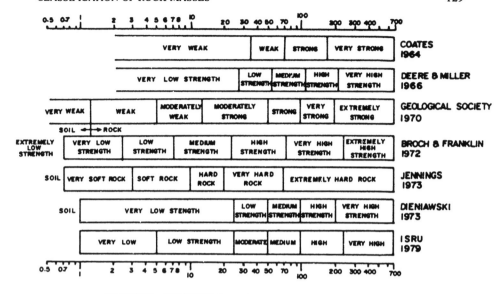

UNIAXIAL COMPRESSIVE STRENGTH, MPs

Fig. 5.1: Various strength classifications for intact rock.

These intact classifications, however useful, cannot provide quantitative data for engineering design purposes. Therefore, their value primarily lies in enabling better identification and communication during discussion of intact rock properties.

Rock Mass Classifications

Following the first classification of rock mass by Terzaghi, many classification systems were proposed (Table 5.1) which took into consideration the new advances in rock support technology. Among the many, the classifications proposed by Terzaghi (1946), Lauffer (1958), Deere and Miller (1967), Wickham et al. (1972) Bieniawski (1973) and Barton et al. (1974) are very important.

Rock Load Classification: Terzaghi formulated the first rational method of classification by evaluating rock loads appropriate to the design of steel sets. The main features of the classification are presented in Fig. 5.2. The revised version of Terzaghi's rock load coefficients is presented in Table 5.2. Though extensively used earlier, it is treated as too qualitative for modern tunnelling methods (Bieniawski, 1990).

Stand-up Time Classification: Lauffer's classification shows that an increase in unsupported tunnel span leads to a reduction in stand-up time. (Stand-up time is that period wherein a tunnel will stand unsupported after excavation and is affected by such factors as orientation of tunnel axis, shape of cross-section, excavation and support methods.)

This classification has been modified by a number of Austrian engineers, notably Pacher et al. (1974), leading to development of the New Austrian Tunnelling Method (NATM).

Table 5.1. Major Engineering Rock Mass Classifications Currently in Use.

Name of Classification	Originator and Date	Country of Origin	Application
1. Rock load	Terzaghi, 1946	USA	Tunnels with steel support
2. Stand-up time	Lauffer, 1958	Austria	Tunnelling
3. NATM	Pacher et al., 1964	Austria	Tunnelling
4. Rock quality designation	Deere et al., 1967	USA	Core logging tunnelling
5. RSR concept	Wickham et al., 1972	USA	Tunnelling
6. RMR system (Geomechanics Classification) *RMR system extensions*	Bieniawski, 1973 (last modified 1979-USA)	South Africa	Tunnels ***** foundations
	Weaver, 1975	South Africa	Rippability
	Laubscher, 1977	South Africa	Mining
	Olivier, 1979	South Africa	Weatherability
	Ghose and Raju, 1981	India	Coal mining
	Moreno Tallon, 1982	Spain	Tunnelling
	Kendoiski et al. 1983	USA	Hard rock mining
	Nakao et al., 1983	Japan	Tunnelling
	Serafim and Pereira, 1983	Portugal	Foundations
	Gonzalez de Vallejo, 1983	Spain	Tunnelling
	Unal, 1983	USA	Roof bolting in coal mines
	Romana, 1985	Spain	Slope stability
	Newman, 1985	USA	Coal mining
	Sandbak, 1985	USA	Boreability
	Smith, 1986	USA	Dredgeability
	Venkateswarlu, 1986	India	Coal mining
	Robertson, 1988	Canada	Slope stability
7. Q-system	Barton et al., 1974	Norway	Tunnels, chambers
Q-system extensions	Kirsten, 1982	South Africa	Excavatability
	Kristen, 1983	South Africa	Tunnelling
8. Strength-size	Franklin, 1975	Canada	Tunnelling
9. Basic geotechnical description	International Society for Rock Mechanics, 1981		General, communication
10. Unified classification	Williamson, 1984	USA	General, communication

In brief, the Lauffer-Pacher classification introduced stand-up time and span as relevant parameters in determining the type and amount of tunnel support, and also influenced development of more recent classification systems.

Rock Quality Designation (RQD) Index

Introduced by Deere et al. (1967), RQD is a modified core recovery percentage which incorporates only sound pieces of core that are 100 mm or greater in length. RQD has been extensively used as a quantitative measure for rock quality and incorporated as one of the important parameters in other classifications.

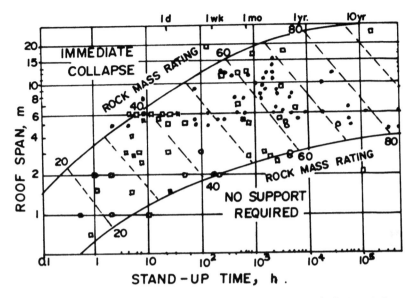

Geomechanics Classification of rock masses: Output for mining and tunnelling.

● = Case histories of roof falls in mining.

□ = Tunnelling roof falls.

Fig. 5.2: The tunnel rock load concept of Terzaghi, K. 1946.

Table 5.2. Terzaghi's rock load classification currently in use[a,b].

Rock Condition	RQD	Rock Load H_a (ff)	Remarks
1. Hard and intact	95 – 100	Zero	Same as Terzaghi (1946)
2. Hard, stratified or schistose	90 – 99	$0 - 0.5B$	Same as Terzaghi (1946)
3. Massive, moderately jointed	85 – 95	$0 - 0.25B$	Same as Terzaghi (1946)
4. Moderately blocky and seamy	75 – 85	$0.25B - 0.20(B + H_1)$	Types 4, 5, and 6 reduced by
5. Very blocky and seamy	30 – 75	$(0.20 - 0.60)(B + H_1)$	about 50% from Terzaghi
6. Completely crushed but chemically intact	3 – 30	$(0.60 - 1.10)(B + H_1)$	values because water table has little effect on
6a. Sand and gravel	0 – 3	$(1.10 - 1.40)(B + H_1)$	rock load (Terzaghi, 1946; Brokke, 1968)
7. Squeezing rock, moderate depth	NA	$(1.10 - 2.10)(B + H_1)$	Same as Terzaghi (1946)
8. Squeezing rock, great depth	NA	$(2.10 - 4.50)(B + H_1)$	Same as Terzaghi (1946)
9. Swelling rock	NA	Up to 250 ft irrespective of value of $(B + H_1)$	Same as Terzaghi (1946)

[a] As modified by Deere et al. (1970) and Rosa (1982)

[b] Rock load H_1 in feet of 10% on roof of support in tunnel width B (ft) and height H_1 (ft) at depth—of more than $1.5 (B - H_1)$.

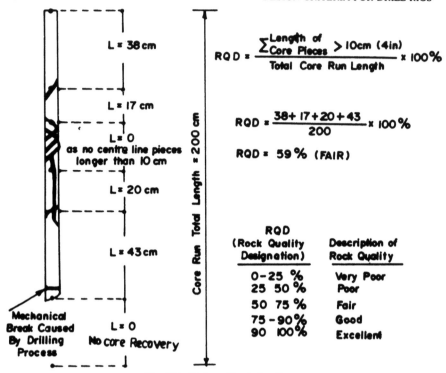

Fig. 5.3: Rock quality designation.

The procedure for measuring RQD is illustrated in Fig. 5.3. ISRM recommends that the drilling shall be of NX size with double-tube barrels. It includes only the pieces of sound core over 100 mm long, which are summed and divided by the length of the core run.

Fig. 5.4 illustrates the graphic borehole log presently being followed to describe the subject parameters. The double-tube core barrels should be of the NWG or NWM type depending on the formation to be drilled.

Cording and Deere (1972) related the RQD index to Tefzaghi's rock load factors and presented tables relating tunnel support and RQD.

Merritt (1972) correlated RQD with tunnel width and the compiled form is presented by Deere and Deere (1988).

Palmstrom (1982) presented the following relation between RQD and number of joints per unit volume, in which the number of joints per metre per each joint set is added.

$$RQD = 115 - 3.3J_n \qquad \qquad \ldots (5.1)$$

where J_n = total number of joints/m³.

Presently RQD is used as a standard parameter in drill core logging and forms a basic element of the two major rock mass classification systems, i.e., the RMR and the Q-system.

RECORD OF DRILL HOLE NO. DH 9

Drilling period from   Feb. 20th. to 15th March 1953   Rig – Longyear 38

Location — Hirakud

Angle with horizontal —

Name of Driller — Rajesh Bansal   Coller elevation in metres — 254

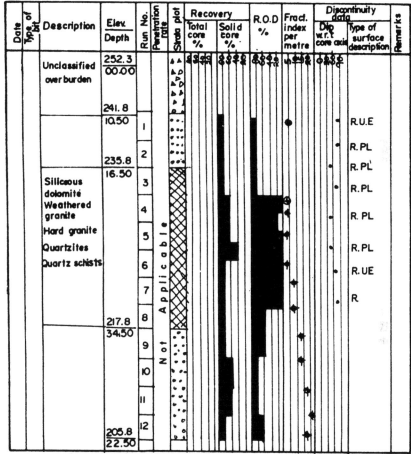

C – CURVED, CL- CLEAVAGE, F–FAULT, FL- FLEXURED, FR– FRACTURED
J – JOINT, P–POLISHED, PL–PLANAR, R– ROUGH, S – SLICKENSIDED, SH–SHEAR
SM– SMOOTH, ST– STEPPED, UE–UNEVEN, VN –VEIN, W –WAVY.

Fig. 5.4: Graphic bore hole log.

Rock Structure Rating (RSR)

The system proposed by Wickham et al. (1972) is a ground-support prediction model, which is complete in its own. The following parameters are included in the classification with varied ratings for each parameter.

Parameter A (general appraisal of rock structure) includes
— Rock-type origin (igneous, sedimentary, metamorphic)
— Rock hardness (hard, medium, soft, decomposed)
— Geological structure (thick, slightly faulted, folded, moderately, intensely).
Parameter B (joint pattern and direction of drive)
— Joint spacing
— Joint orientation (strike &/dip)
— Direction of tunnel drive
Parameter C (groundwater inflow)
— Overall rock mass quality due to parameters A and B
— Joint condition (good, fair, poor)
— Amount of water inflow

The RSR value can be obtained by summing the weighted numerical value determined for each parameter. Thus RSR = A + B + C with max. value of 100. RSR reflects the quality of the rock mass with respect to its need for support.

Further, RSR values are correlated with rib ratio (RR) of support installation. This ratio is defined as the ratio between theoretical support for a given rib size and spacing and actual support installed multiplied by 100. The RR for an unsupported section would be 100 (Bieniawski, 1990). This classification is very useful for selecting steel rib support for rock tunnels.

Rock Mass Rating (RMR) System

The rock mass rating (RMR) system, also known as the Geomechanical Classification, was proposed by Bieniawski (1973).

The following parameters, which can be measured in the field and also obtained from bore data, are used in the classification.
  i) Uniaxial compressive strength of rock
 ii) Rock quality designation, RQD
iii) Spacing of discontinuities
 iv) Condition of discontinuities
  v) Groundwater condition
 vi) Orientation of discontinuities.

Details of different categories of parameters with ratings are given in Table 5.3. To apply the classification, the rock mass along the tunnel route is divided into a number of structural regions, i.e., zones in which certain geological features are more or less uniform in the region.

In section A of Table 5.3, the first 5 parameters are grouped into five ranges of values with varied ratings. Influence of attitude of discontinuities is included in Section B of Table 5.3. Adjusted rock groups as shown in Sections C and D gives the practical meaning of each rock class by relating specific engineering problems. In the case of tunnels and chambers the output from the RMR classification is the stand-up time and the maximum stable rock span for a given rock mass rating as depicted in Fig. 5.4. The modified relation for tunnel-boring machine (TBM) is suggested by Lauffer (1988).

Table 5.3

A. CSIR geomechanics classification for jointed rock masses

Item	Class No. and its Description	1 Very good	2 Good	3 Fair	4 Poor	5 Very poor
1	Rock Quality RQD %	90–100	75–90	50–75	25–50	< 25
2	Weathering	Unweathered	Slightly weathered	Moderately weathered	Highly weathered	Completely weathered
3	Intact rock strength, MPa	> 200	100–200	50–100	25–50	< 25
4	Joint spacing	> 3 m	1 m–3 m	0.3 m–1 m	50 mm–300 mm	< 50 mm
5	Separation of joints	< 0.1 mm	< 0.1 mm	0.1 mm–1.0 mm	1 mm–5 mm	> 5 mm
6	Continuity of joints	Not continuous	Not continuous	Continuous, no gouge	Continuous, with gouge	Continuous, with gouge
7	Groundwater flow per 10 m	None	None	Slight < 25 lit/min	Moderate 25–125 lit/min	Heavy >125 lit/min
8	Strike and dip orientations	Very Favourable	Favourable	Fair	Unfavourable	Very Unfavourable

B. Rating adjustment for joint orientations

	Strike and dip orientation of joints	Very favourable	Favourable	Fair	Unfavourable	Very unfavourable
Ratings	Tunnels	0	−2	−5	−10	−12
	Foundations	0	−2	−7	−15	−25
	Slopes	0	−5	−25	−50	−60

C. Rock mass classes determined from total rating

Rating	100 ← 81	80 ← 61	60 ← 41	40 ← 21	< 20
Class No.	I	II	III	IV	V
Description	Very good rock	Good rock	Fair rock	Poor rock	Very poor rock

D. Meaning of rock mass classes

Class No.	I	II	III	IV	V
Average stand-up time	10 years for 15 m span	6 months for 8 m span	1 week for 5 m span	10 hr for 2.5 m span	30 min for 1 m span
Cohesion of rock mass	> 400 kPa	300–400 kPa	200–300 kPa	100–200 kPa	< 100 kPa
Friction angle of rock mass	> 45°	35°–45°	25°–35°	15°–25°	< 15°

Unal (1983) suggested the following relation for support load from RMR:

$$P = \frac{100 - \text{RMR}}{100} \gamma B \qquad \qquad \ldots (5.2)$$

where P = support load, kN;

B = tunnel width, m

γ = rock density, kg/m³.

The RMR classification has proved a useful method for estimating *in situ* deformability of rock masses, as shown below:

$$E_m = 2\text{RMR} - 100, \qquad \ldots (5.3)$$

where E_m is *in situ* modulus in GPa and RMR is > 50.

Serafim and Pereira (1983) provided the formula for predicting the range of RMR < 50:

$$E_m = 10^{(\text{RMR}-10)/40} \qquad \ldots (5.4)$$

Hoek and Brown (1980) and Rao et al. (1985) proposed a method for estimating rock mass strength using RMR rating.

This classification system also has been widely used in mining, slope stability and foundation design.

Q-System

Also known as the Norwegian Geotechnical Institute Classification developed by Barton et al. (1974), based on more than 212 tunnel case histories from Scandinavia. The Q-system takes into account the following parameters:

i) RQD
ii) Number of joint sets (J_n)
iii) Roughness of the most unfavourable joint or discontinuity (J_r)
iv) Degree of alteration or filling along the weakest plane (J_a)
v) Water inflow (J_w)
vi) Stress condition (SRF).

These six parameters are grouped into three quotients to give the overall rock mass quality, Q, as follows:

$$Q = \frac{\text{RQD}}{J_n} \times \frac{J_r}{J_a} \times \frac{J_w}{SRF} \qquad \ldots (5.5)$$

$$= (\text{block size}) \times (\text{shear strength}) \times (\text{active stress})$$

The Q can range from 0.001 to 1000 on a logarithmic rock mass quality scale. Table 5.4 gives the numerical values of each of the parameters. As per this, J_n, J_r and J_a play more important roles than joint orientation. However, orientation is implicit in J_r and J_a because they apply to the most unfavourable joints.

The Q value is related to support requirements by defining the equivalent dimension, which is a function of both the size and the purpose of the excavation, and is obtained by dividing the span, diameter or wall height of the excavation by a quantity called the excavation support ratio (ESR). Each excavation category has a separate ESR value (Barton, 1973). This relation determines the appropriate support measures. Barton et al. (1974) provide 38 support categories which give estimates of permanent support.

Correlation between RMR and Q Values: Based on the results of many case histories, a relation has been provided between RMR and Q (Bieniawski 1976):

$$\text{RMR} = 9 \ln Q + 44 \ldots \qquad \ldots (5.6)$$

Table 5.4. Descriptions and ratings for the parameters RQD, J_n and J_r.

1. Rock Quality Designation	(RQD)
A. Very Poor	0–25
B. Poor	25–50
C. Fair	50–75
D. Good	75–90
E. Excellent	90–100

Note 1.
  (i) Where RQD is reported or measured as $\leq$ 10 (including 0) a nominal value of 10 is used to evaluate Q in Eq. (1)
  (ii) RQD intervals of 5, i.e. 100, 95, 90 etc. are sufficiently accurate.

2. Joint Set Number	(J_n)
A. Massive, no or few joints	0.5–1.0
B. One joint set	2
C. One joint set plus random	3
D. Two joint sets	4
E. Two joint sets plus random	6
F. Three joint sets	9
G. Three joint sets plus random	12
H. Four or more joint sets, random, heavily jointed, 'sugar cube' etc.	15
J. Crushed rock, earth-like	20

Note 2.
  (i) For intersections use $(3.0 \times J_n)$.
  (ii) For portals use $(2.0 \times J_n)$.

3. Joint Roughness Number	(J_r)
(a) *Rock wall contact* and	
(b) *Rock wall contact before 10 cm shear*	
A. Discontinuous joints	4
B. Rough or irregular, undulating	3
C. Smooth, undulating	2
D. Slickensided, undulating	1.5
E. Rough or irregular, planar	1.5
F. Smooth, planar	1.0
G. Slickensided, planar	0.5
(c) *No rock wall contact when sheared*	
H. Zone containing clay minerals thick enough to prevent rock wall contact	1.0 (nominal)
J. Sandy, gravelly or crushed zone thick enough to prevent rock wall contact	1.0 (nominal)

Note 3.
  (i) Add 1.0 if the mean spacing of the relevant joint set is greater than 3 m.
  (ii) $J_r = 0.5$ can be used for planar slickensided joints having lineations, provided the lineations are favourably orientated.

4. Joint Alteration Number	(J_a)	τ_r **(approx.)**
(a) *Rock Wall contact*		
A. Tightly healed, hard, non-softening, impermeable filling, i.e., quartz or epidote	0.75	$(-)$

Contd.

Table 5.4. *Continued.*

4. Joint Alteration Number	(J_a)	τ_r (approx.)
B. Unaltered joint walls, surface staining only	1.0	$(25°-35°)$
C. Slightly altered joint walls. Non-softening mineral coatings, sandy particles, clay-free disintegrated rock etc.	2.0	$(25°-30°)$
D. Silty-, or sandy-clay coatings, small clay-fraction (non-softening)	3.0	$(20°-25°)$
E. Softening or low friction clay mineral coatings, i.e. kaolinite, mica. Also chlorite, talc, gypsum and graphite etc., and small quantities of swelling clays. (Discontinuous coatings, 1-2 mm or less in thickness)	4.0	$(8°-16°)$
(b) *Rock wall contact before 10 cm shear*		
F. Sandy particles, clay-free disintegrated rock etc.	4.0	$(25°-30°)$
G. Strongly overconsolidated, non-softening clay mineral fillings. (Continuous, < 5 mm in thickness)	6.0	$(16°-24°)$
H. Medium or low overconsolidation, softening, clay mineral fillings. (Continuous, < 5 mm in thickness)	8.0	$(12°-16°)$
J. Swelling clay fillings, i.e. montmorillonite. (Continuous, < 5 mm in thickness. Value of J_a depends on per cent of swelling clay-size particles, access to water etc.	8.0–12.0	$(6°-12°)$
(c) *No rock wall contact when sheared*		
K,L, Zones or bands of disintegrated or crushed	6.0, 8.0	$(6°-24°)$
M. Rock and clay (see G, H, J for description of clay condition)	or 8.0–12.0	
N. Zones or bands of silty-or sandy-clay, small clay fraction (non-softening)	5.0	
O,P, Thick, continuous zones or bands of clay	10.0, 13.0	$(6°-24°)$
R. (see G, H, J for description of clay condition)	or 13.0–20.0	

Note: (i) Values of $(\tau)_r$ are intended as an approximate guide to the mineralogical properties of the alteration products, if present.

5. Joint Water Reduction Factor	(J_w)	Approx. water pressure (kg/cm^2)
A. Dry excavations or minor inflow, i.e. < 5l/min. locally	1.0	< 1
B. Medium inflow or pressure occasional outwash of joint fillings	0.66	1.0–2.5
C. Large inflow or high pressure in competent rock with unfilled joints	0.5	2.5–10.0
D. Large inflow or high pressure, considerable outwash of joint fillings	0.33	2.5–10.0
E. Exceptionally high inflow or water pressure at blasting, decaying with time	0.2–0.1	> 10.0
F. Exceptionally high inflow or water pressure, continuing without noticeable decay	0.1–0.05	> 10.0

Note: (i) Factors C to F are crude estimates. Increase J_w if drainage measures are installed.
     (ii) Special problems caused by ice formation are not considered.

6. Stress Reduction Factor	(SRF)
(a) *Weakness zones intersecting excavation, which may cause loosening of rock mass when tunnel is excavated*	
A. Multiple occurrences of weakness zones containing clay or chemically disintegrated rock, very loose surrounding rock (any depth)	10.0
B. Single weakness zones containing clay, or chemically disintegrated rock (depth of excavation $\leq$ 50 m)	5.0

	σ_c/σ_1	σ_t/σ_1	
C. Single weakness zones containing clay, or chemically disintegrated rock (depth of excavation > 50 m)			2.5
D. Multiple shear zones in competent rock (clay free), loose surrounding rock (any depth)			7.5
E. Single shear zones in competent rock (clay free) (depth of excavation ≤ 50 m)			5.0
F. Single shear zones in competent rock (clay free) (depth of excavation > 50 m)			2.5
G. Loose open joints, heavily jointed or 'sugar cube' etc. (any depth)			5.0
(b) *Competent rock, rock stress problems*			
	σ_c/σ_1	σ_t/σ_1	
H. Low stress, near surface	> 200	> 13	2.5
J. Medium stress	200–10	13–0.66	1.0
K. High stress, very tight structure. (Usually favourable to stability, may be unfavourable to wall stability)	10–5	0.66–0.33	0.5–2.0
L. Mild rock burst. (massive rock)	5–2.5	0.33–0.16	5–10
M. Heavy rock burst (massive rock)	<2.5	<0.16	10–20
(c) *Squeezing rock; plastic flow of incompetent rock under the influence of high rock pressures*			
N. Mild squeezing rock pressure			5–10
O. Heavy squeezing rock pressure			10–20
(d) *Swelling rock; chemical swelling activity depending on presence of water*			
P. Mild swelling rock pressure			5–10
R. Heavy swelling rock pressure			10–15

Note:

  (i) Reduce these values of SRF by 25–50% if the relevant shear zones only influence but do not intersect the excavation.

 (ii) For strongly anisotropic stress field (if measured): when $5 \leq \sigma_1/\sigma_3 \leq 10$ reduce σ_c and σ_t to 0.8 σ_c and 0.8 σ_t; when $\sigma_1/\sigma_3 > 10$, reduce σ_c and σ_t to 0.6 σ_t where: σ_c = unconfined compression strength, σ_t = tensile strength (point load), σ_1 and σ_3 = major and minor principal stresses.

(iii) Few case records available where depth of crown below surface is less than span width. Suggest SRF increase from 2.5 to 5 for such cases (see H).

Rutledge (1978) developed the following relations for the three classification systems:

$$RMR = 13.5 \log Q + 43; \qquad \ldots (5.7)$$
$$RSR = 0.77\, RMR + 12.4; \qquad \ldots (5.8)$$
$$RSR = 13.3 \log Q + 46.5. \qquad \ldots (5.9)$$

In general, the Geomechanics Classification is more conservative than the Q-system.

CASE STUDIES

Though many case studies are available in the literature from different countries, Indian experience is scantly reported. Two cases of Indian mining industry have recently been in the limelight, one pertaining to coal mining and the other to metal mining. A brief account of the classification system adopted for these two cases is presented below.

Coal Mines

Venkateswarulu (1986) of CMRS modified the RMR classification for estimating roof conditions and support in Indian coal mines. This modification was called the CMRS Geomechanics Classification. It is a very simple and practical method of estimating roof conditions in a coal mine. The five parameters include (i) layer thickness, (ii) rock strength, (iii) weatherability, (iv) groundwater flow and (v) structural features with different ratings. The RMR values were correlated with support guidelines (Bieniawski, 1990).

The system has so far been tried in 47 Indian coal mines. Majority of the roof strata experiencing ground control problems came under the category of RMR class III (Fair) and Class IV (Poor).

Metal Mines

Santha Ram·and Jhanjhari (1986) presented stability studies of the underground Mocha Copper mines using rock classifications. The Mocha Copper deposit at Zawar of Hindustan Zinc Ltd. forms a part of the precambrian Aravalli group of rocks comprising phyllites, greywackes, quartzites and dolomites. For estimation of rock quality rating, four parameters were used, namely (i) discontinuity spacing, (ii) discontinuity strength, (iii) intact rock strength and (iv) RQD, and with the help of a power curve each unit was converted into weightage points.

Based on the total rating, rock mass classes were determined using Bieniawski's (1973) classification. Once rock quality was determined, the class boundary data was plotted on level plans and sections contoured at five-point intervals for delineating the areas of good to poor ground conditions. Poor ground conditions in mines were, in turn, corroborated with monitoring instruments installed in the adjacent areas. A broad agreement between instrumentation results and observed ratings was found. Rock mass strength was calculated for mine pillars based on rating and with the help of safety and stability assessment of pillars.

CONCLUSIONS

There is no unique classification system which caters for all types of rock masses and support systems. However, the RMR and the Q-systems seem to be useful for most situations. In fact, the better approach would be to ascertain the parameters which most influence a particular site and then evolve a classification by assigning ratings to these parameters. The existing classification systems may be used as general guidelines rather than blindly adopting a particular system which may not be relevant for site conditions.

REFERENCES

Barton, N., R. Lien and J. Lunde (1974). Engineering classification of rock masses for the design of tunnel support. *Rock Mech.* 6: 183–236.

Bieniawski, Z.T. (1973). Engineering classification of jointed rock masses. *Trans. S. Afr. Inst., Civil Eng.* 15: 335–344.

Bieniawski, Z.T. (1989). *Engineering Rock Mass Classification.* John Wiley & Sons, New York, 251 pp.

Cording, E.J. and D.U. Deere (1972). Rock tunnel supports and field measurements. *Proc. Rapid Excav. Tunneling Conf., AIME,* pp. 601–622.

Deere, D.U. and R.P. Miller (1967). Engineering classification and index properties of intact rocks. Air Force Lab. Tech. Report No. AFNL-TR, pp. 65–116.

Deere, D.U. and D.W. Deere (1988). The RQD Index in practice. *Proc. Symp. Rock Classif. Eng. Purp.* ASTM Publ. 984: 91–101.

Hoek, E. and E.T. Brown (1980). *Underground Excavations in Rocks.* Inst. Mining & Metallurgy, London, pp. 527.

Lauffer, H. (1958). Gebirgsklassifizierung fürden Stollenbau. *Geol. Bauwesen* 74: 46–51.

Merrit, A.H. (1972). Geologic prediction for underground excavations. *Proc. Rapid Excav. Tunneling Conf., AIME,* pp. 115–132.

Pacher, F.L., Rabcewicz and J. Golser (1974). Zum der seitigen stand der geibirgsklassifizerung in Stollen- und Tunnelbau. *Proc. XXII Geomech. Colloq.* Salzburg, pp. 51–58.

Palmstrom, A. (1982). The volumetric joint count—A useful and simple measure of the degree of rock jointing.

Rao, K.S., G.V. Rao and T. Ramamurthy (1985). Rock mass strength from classification CBIP Workshop, pp. 27–50.

Rutledge, J.C. and R.L. Preston (1978). Experience with engineering classification of rocks. *Proc. Int. Tunneling Symp.* pp. A 301–3.7.

Santha Ram, A. and V.K. Jhanjhari (1986). Rock mass classification system for stability of underground metal mines. *Ind. Geotech. J.* 16: 94–104.

Serafim, J.L. and J.P. Pereira (1983). Considerations of geomechanics classification of Bieniawski. *Proc. Int. Symp. Eng. Geol. Underground Constr. LNEC* 1:II, 33–42.

Terzaghi, K. (1946). Rock defects and loads on tunnel support. In: *Rock Tunneling with Steel Supports* (eds. Proctor & White). Commercial Shearing Co., pp. 15–99.

Unal, F. (1983). Design guidelines and roof control standards for coal mine roofs. Ph.D. thesis, Pennsylvania State Univ., 355 pp.

Venkateswarulu, V. (1986). Geomechanics classification of coal measure rocks vis-á-vis roof supports Ph.D. thesis, ISM, 251 pp.

Wickham, G.E., Tiedemann and E.H. Skinner (1972). Support determination based on geologic predictions *Proc. Rapid Excav. Tunneling Conf., AIME,* pp. 43–64.

CHAPTER 6

DRILLING AND ROCK MECHANICS

Of the many facets of rock mechanics, the more important ones in the context of drilling are concerned with rock properties such as hardness and abrasiveness, rock structures and water flow.

ROCK AND MINERALS

The term 'rock' refers geologically to an aggregation of materials which are commonly minerals but which may be liquids or organic materials such as peat. No distinction is made between hard and soft materials. Rock in the geological sense therefore relates both to 'rock' and 'overburden' in normal parlance. As it is a natural substance, infinite variations occur, a fact readily demonstrated by a single drill hole in a seemingly homogeneous material such as basalt.

The rock properties which most influence drillability are hardness, abrasiveness, toughness, grain size, fissility and water discontinuities. Each is briefly discussed below.

HARDNESS

Hardness of minerals is determined by Moh's Scale of 10 minerals, each of which is capable of scratching the next lower grade. Thus, diamond will scratch corundum and quartz will scratch feldspar. Moh's Scale is given below in a descending order. The absolute hardness values according to J. Murkes are given in parentheses. Certain common materials have been included for convenience.

MOH'S SCALE

10. Diamond	(15.1)	7. Quartz	(7.3)
9. Corundum	(8.9) Tungsten carbide	6. Feldspar	(6.5) Steel
8. Topaz	(7.9) Hardened steel	5. Apatite	(5.7) Glass

| 4. Fluorspar | (4.0) Brass pin | 2. Gypsum | (2.3) |
| 3. Calcite | (3.3) Finger nail | 1. Talc | (0.9) |

Hardness of rock depends not only on the hardness of the individual grains of which it is comprised, but also on the coherence between the particles, or strength of the bond. Quartz sand grains have a hardness reading of 7, and are harder than steel; quartzite, in which the quartz grains are bonded with additional silica could also be described as having a hardness of 7; calcareous sandstone, on the other hand, has a much weaker bond and can probably be cut with a knife, even though most of the material forming the rock is quartz.

In testing a rock for hardness, it should be scratched with a pocket knife if hard, or by a brass pin or the finger nail if soft. Should the rock remain unmarked by the pocket knife, a file or quartz fragment must be used. The resultant scratch should be carefully examined to determine whether the hardness is controlled by the mineral constituents or by the strength of the bond. When scratching hard minerals and rocks, care should be taken to differentiate between true scratches and the metallic fragments which are deposited from the knife blade on materials harder than the steel being used.

ABRASIVENESS

Abrasiveness may be defined as the ability of the rock to wear away drill-bit materials: steel, tungsten carbide and, in diamond drilling, the sintered matrix in which the stones are set. It is a relative quality and is intimately associated with hardness, but it is also affected by particle shape, tenacity of the mineral fragments and cleavage. In a manner similar to the wearing away of a stone by water, a degree of bit attrition can be observed after a period of drilling even in the softest rock, so that abrasion is not absolutely confined to the action of a harder, on a softer substance.

The size and shape of the particles of rock influence the type of abrasion, as do the torque and thrust applied to the bit. A relatively large, sharp-edged rock fragment held firmly in position is likely to cause a deeper scratch than would fine dust, provided that the shear forces applied to the bit are insufficient to break or dislodge the fragment. Dust has a polishing effect only, although it must be pointed out that ultimately this may prove to be more deleterious to the bit than a scratch; tungsten carbide which is polished in this way develops a high skin-hardness which may cause the material to spall. Similarly, diamonds rapidly lose their ability to cut if they are permitted to polish. Rounded sand grains, occurring in deposits formed from desert sands, apparently roll over the surface of a bit during drilling and act as miniature ball bearings, thereby causing less abrasion than sharp marine sand.

With percussive drilling, the only rocks which need to be considered as highly abrasive are those containing: (a) quartz, e.g. quartzite, sandstone, grit and acid-igneous rocks; (b) other forms of silica, e.g. flint, chert, jasper and wood-opal; (c) olivine, e.g. dunite and some forms of basalt, and (d) garnet, e.g. garnetiferous gneiss. With rotary drag-bit drilling, abrasion is a more critical factor than in percussive drilling and many other rocks must therefore be considered abrasive. It is a question of degree rather than of quality; a hard silty mudstone which might not normally be considered abrasive, could well prove too abrasive for successful rotary drilling with light equipment, the proportion of quartz dust in the rock being too high for the method used.

Very hard minerals such as topaz, corundum and beryl are only occasionally encountered.

Determination of hardness and abrasiveness of rocks

A number of methods are in use for determination of hardness and abrasiveness of rocks. The methods suggested by the International Society for Rock Mechanics, given in Appendix I, are recommended.

An improvement in interpretation of the Schmidt Rihand Hardness index as determined by the method suggested by the International Society for Rock Mechanics (ISRM) has been suggested by Goktan and Ayday (1993).

The recording technique described by ISRM considers only the upper rebound values. Goktan and Ayday have suggested a modification based on statistical techniques which allows integration of scattered data. This method is briefly described below.

Number of Samples

Sample statistics deal with relations existing between a population and samples drawn from the population. For conclusions of the sampling theory and statistical inference to be valid, samples must be so chosen as to be representative of a population. In statistics Chauvenet's criterion, a special case of the t-distribution, is used in selecting the observations which belong to a population and rejecting outlier observations. The criterion states: 'An observation in a sample of size N is rejected if it has a deviation from the mean greater than that corresponding to a $1/2\ N$ probability. The criterion is discussed in some detail elsewhere; rules for use of the criterion are given below (after Goktan and Ayday, 1993).

1. Compute the mean and standard deviation of all observations.
2. Determine the ratio of the 'suspiciously' large deviation divided by the standard deviation. Determine the limiting value P of this ratio from Table 6.1, for the corresponding number of determinations N.

Table 6.1. Sample size vs p-values.

n	p	n	p	n	p	n	p
2	1.15	7	1.80	15	2.13	50	2.58
3	1.38	8	1.86	20	2.24	100	2.81
4	1.54	9	1.91	25	2.33	250	3.09
5	1.65	10	1.96	30	2.40	500	3.29
6	1.73	12	2.04	35	2.45	1000	3.48

3. If the observed ratio is greater than the value found in the table, the observation may be rejected.

Example: Let us consider a set of actual Schmidt hammer test results obtained by a calibrated L-hammer. The test surface was inspected for microscopic defects and was made sufficiently smooth to take reliable readings with the hammer. Twenty single impacts separated by at least 30 mm were taken. Arranging the recorded data in rank order, we have:

$$37, 40, 40, 41, 41, 42, 42, 43, 44, 44, 44, 45, 45, 45, 46, 46, 47.$$

It can be seen that the distribution of these rebound values is normal about a mean value of 43,15. In order to arrive at an average value reflecting the hardness characteristics of this population, we have to discard those observations which do not belong to the population. For this reason, we are particularly interested in rejection of the lower rebound values. Therefore, in this case we want to know whether the rebound value 37 is representative of this population or not.

The mean of 20 rebound values is 43.15 and the standard deviation of all values is 2.41. The ratio of the suspected deviation to the standard deviation is:

$$(37 - 43.15)/2.41 = -2.55.$$

Since the observation 37 is to the left of the mean, this ratio correctly has a minus sign in front of it, although its absolute value is of interest for analysis. For 20 observations the tabulated value of p in Table 6.1 is 2.24. Since this is smaller than the ratio of the deviation to the standard deviation, the observation 37 should be rejected. Therefore, the best value for the test would then be the mean of the remaining 19 observations.

Toughness

This is related to hardness but in much more difficult characteristic to determine. In sedimentary rocks it would be correct to regard the strength of the bond as a measure of toughness, and limit hardness to the true hardness of the minerals. In igneous rocks it is the texture of the rock (i.e., the shapes and relationships of the crystals) and in metamorphic rocks often the structure of the rock (i.e., the relationships between two or more different textures, as in gneiss) which determine the toughness.

In granite, for example, many of the crystals making up the rock have dimensions which are approximately similar in each plane, and the faces of the crystals abut against one another. The minerals in dolerite, however, have an interlocking structure; although this rock is somewhat softer than granite, it is tougher. Amongst metamorphic rocks, siliceous schist can be difficult to drill because of the interlocking effect of the alternating layers, whereas coarse gneiss does not produce this result because the scale of the structures is disproportionate to the individual grains. Thus in this case the latter act as the controlling factor.

A measure of the toughness of rock can best be obtained by breaking pieces with a hammer. It should be remembered, however, that ease of fracture partly depends on the direction of the blow and, further, that interpretation may be subjective in part.

Generally, the uniaxial compressive strength of rock is considered to indicate its toughness. The suggested method for determining the uniaxial compressive strength of rock materials is given in Appendix II.

Grain size and uniformity

These factors have a bearing on drilling and should be considered when a hand specimen of rock is being examined. In general, coarse even-grained rocks drill faster than fine-grained rocks, or rocks with a very variable particle size.

Geologically, grain size is a useful factor in determining the method of formation. In igneous rocks a coarse grain indicates either that the material hardened under a thick cover of overlying strata and therefore cooled slowly, allowing ample time for crystal growth, or that the liquid from which the rock was formed (magma) was exceptionally rich in certain minerals and fluxes. Conversely, a fine grain indicates rapid cooling at shallow depth, or a small mass (as in dykes and stringers). In sedimentary rocks, particle size depends on the distance from source area to depositional area, on the contemporary topography and a number of other factors. Variable particle size can mean extremely rapid deposition, arctic conditions (in morainic deposits), or slumping of an original deposit (in greywacke).

Fissility

This refers to the ready separation of shales residing along planes parallel to the bedding (laminae), and also the cleavage of slates and schists. Although these arise in different ways, a common reason why the rocks can be split is that mica flakes or other crystals or particles are oriented about parallel axes. The laminae in shales may be at irregular intervals, or may be in a regular series of very fine partings aptly called 'paper shales'. Cleavage in slate is so familiar that it needs no description, although many people are surprised to find that slate in the average excavation, is much more irregular than the quality product of the famous roofing-slate quarries. Partings in schists and phyllites may be smooth, wavy or very contorted, depending on the history of the formation during and after metamorphosis.

Ease of drilling in rocks of this type is closely associated with direction of drill hole relative to the laminae.

Tectonics

A hand specimen of rock can thus yield a certain amount of information on drilling prospects, but this must be supplemented by information obtainable only by a study of the rock as it appears in the field. Many, but not all of the features at which the engineer must look are concerned with rock structures caused by earth movements. The following are of particular importance in rock drilling.

1. *Dip and strike:* Most sediments were deposited under water in horizontal layers which have hardened to form beds of rock, some of which were subsequently

disturbed and became tilted. On a tilted bed the direction in which the inclination is greatest is termed the true dip; this is expressed in terms of the angle from the horizontal. At right angles to the true dip, the beds are horizontal; this direction is termed the strike. An exposure of rock in a direction between dip and strike shows a lower angle of inclination, termed the apparent dip. These points are illustrated in Fig. 6.1. Dip is measured with a clinometer placed on the rock surface and rotated until the direction and degree of maximum inclination has been determined. Compass bearings are taken and the results noted on a map or drawing of the site by a small arrow whose tip is at the point of observation. The direction of the arrow must coincide with the dip and the angle should be appended as a numeral. A few such annotations will later yield much more useful and accurate information than would a vague impression that the 'rocks are dipping up the hill'. Where practicable, a number of readings should be taken as a check.

In block diagrams, dip and strike are simple geometrical expressions; conditions are rarely so simple in the field. The strike is not necessarily a straight line and the dip may be affected by major structures such as faults or folds, or by minor irregularities. Points to be remembered are: (a) the general dip (as opposed to minor variation) is sometimes identifiable by the eye after examining the 'grain' of the country; (b) sandstones and similar rocks are often associated with softer shales and it is the sandstones which are likely to be selected as the surface on which a reading of dip is taken as they are more resistant to weathering and therefore stand out. This is satisfactory provided that the exposure is of sufficient size to ensure that the reading taken is, in fact, the dip. Sandstones often exhibit current or false-bedding; (c) surface strata are sometimes disturbed by glacial action and other erosive agencies and (d) readings may not be taken from an isolated boulder.

The importance for the driller of dip and strike, as well as other features to be discussed, will be examined from various aspects in appropriate chapters.

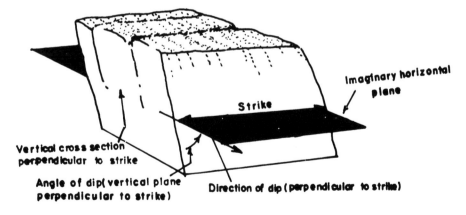

Fig. 6.1(a): Components of attitude of geological surfaces: dip and strike;

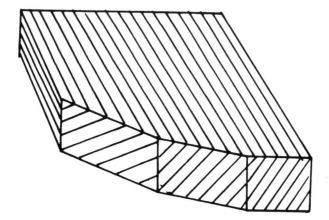

Fig. 6.1(b): Apparent dip. A strictly homoclinal sequence of beds is cut by vertical faces along different bearings. Only the face at right angles to the strike shows true dip.

2. *Folding:* The dip is an example of the general concept of folding. A little reflection will show that it is not possible for a rock stratum to continue in a given plane for more than a very limited distance. Dip relates, therefore, to a section (or limb) of a fold over a limited area. The terminology and classification of folds is of less importance to the engineer than to the geologist; but folding does become of interest to the former when it is on a relatively small scale, or when the excavation or tunnel cross-section is on a large scale. Two terms at least are important here: anticline and syncline. These are illustrated in Fig. 6.2 (C and D) which also shows a few other structures. An anticline is an area of weakness, the upper beds being stretched to give open fissures, shear faulting and collapse; this is compensated by a compression of the lower beds. In a syncline the beds tend to slide over each other, often accompanied by crumpling of upper beds and shear faulting.

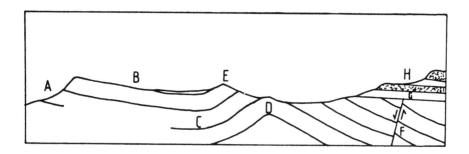

Fig. 6.2: Structure and scenery.
A—Scarp; B—Dip slope; C—Syncline; D—Anticline; E—Hog book ridge; F—Fault; G—Unconformity; H—Step topography.

3. *Faulting:* Faults are planes at which the rocks on one side are displaced relative to those on the other side. (Fig. 6.3). The angle of inclination of the fault plane is termed the hade and is measured from the vertical and not, as is the dip, from the horizontal. The vertical component of displacement is the throw and in faults having a throw, the two sides of the fault are termed upthrow and downthrow. Tear faults have lateral instead of vertical displacement. An overthrust fault is commonly termed a thrust (Fig. 6.4).

Faults are of vital concern to the mining engineer and to the civil engineering designer. In rock drilling faults create sudden changes in rock character which may be encountered in a drill hole or on a larger scale during the driving of a tunnel. They are also of interest because of the shattered rock which may accompany them. The faces of a fault may show slickensides, that is striations and high polish, caused by the grinding effect of movement.

Additionally or alternatively, the rock in the fault zone may be fragmented to angular material which forms fault breccia over the area of the 'shatter-belt'. Another and common possibility is the production of rock flour or fault gouge. This is one of the materials, varying in origin, familiarly known to the driller as clayback.

4. *Bedding:* The thickness of individual beds and whether they are regular or otherwise should be noted. Thin laminated strata can be drilled easily but not cleanly, and may present problems if drilling is parallel to the bedding. Thick-bedded rocks give more satisfactory results. Individual beds should be examined to see whether they remain uniform throughout their depth or become much finer as each bed is traced upwards. The succession of beds should also be inspected to determine whether they are constant, rhythmic or alternating.

Studies such as these not only ensure a better understanding of the rock, and consequently selection of the best method of drilling and blasting, but may also prove useful in deciding levels etc.

5. *Fissuring:* Joints and other fissures result from shrinkage, stresses, folding and relief of load. Joints often form a well-defined pattern in two or more planes which may be vertical or inclined. As joints have a marked influence on drilling, their frequency, direction and inclination should certainly be noted. Joints should also be examined to see whether they are tight or open, dry or wet.

Gouge, the clayey material often found in joints and cavities in rock, may derive from grinding action on a fault plane as a residue from the solution of limestone, as a deposit by hot mineral springs, or as a deposition of fine-grained material (clay or boulder clay) washed down from above. This material may have little effect on rotary drilling but is a constant source of difficulty in percussive drilling, causing jammed and broken steel. This is aggravated if the drilling is parallel to the jointing so that a drill hole may run for several feet in the gouge. A study of the joint pattern may suggest the advisability of altering the angle of drilling.

6. *Weathering:* Rocks are far from stable substances. To consider them synonymous with permanence is incorrect. They are subject to many forms of alteration, decay and denudation. Sedimentary rocks, for example, are the product of the

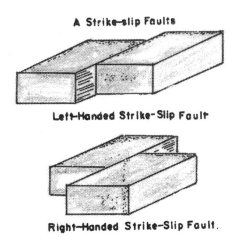

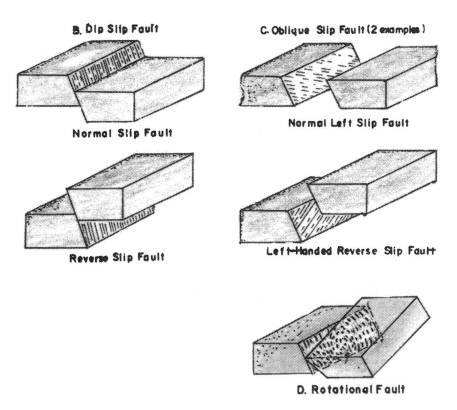

Fig. 6.3: Slip classification of faulting.

A—Block diagram showing left-handed and right-handed strike-slip faulting; B—Block diagrams showing normal slip, reverse slip; C—Examples of oblique slip faults; D—Schematic block diagram of a rotational fault.

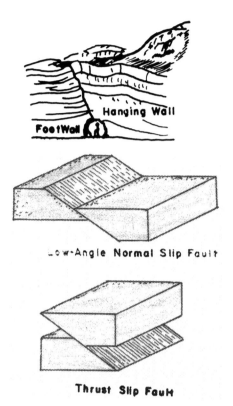

Fig. 6.4: Prospector in tunnel pauses to think about the difference between the footwall and the hanging wall of a fault. Low-angle normal slip fault and thrust slip fault.

prior erosion of earlier rocks; granite can alter to kaolinite, quartz sand and mica; underground water can carve valleys and caverns.

The rocks visible in most exposures show some degree of alteration, which must be discounted. An attempt should be made to obtain 'fresh' rock, otherwise misconceptions can occur. In general, weathered rock is softer and has a lower compressive strength than fresh rock, which affects drillability. Decomposition in igneous rocks containing feldspar is readily identifiable by the alteration of this mineral, as it becomes first cloudy and then opaque. Whitish, powdery patches in, say, granite are a sure sign of decomposition.

As an example of chemical changes, basalt is interesting. Under temperate weathering the feldspar crystals in basalt become kaolinised and at this first stage the rock is termed 'weathered basalt'. In the next stage of this process the dark (Fe · Mg) minerals also begin to break down. Under tropical conditions, a red clay-like substance called laterite is formed, and this may finally segregate to bauxite (aluminium ore derived from kaolinite), allophane (Fe · Al), and enriched laterite (iron ore).

DISCONTINUITIES

Discontinuity is a term used to describe all 'breaks' in rock masses. These could be joints, faults or bedding planes. A brief definition of each, as given by the International Society of Rock Mechanics, is given below.

Joint

This is a break of geological origin in the continuity of a body of rock along which there has been no visible displacement. A group of parallel joints is called a set and joint sets intersect to form a joint system. Joints can be open, filled or healed. Joints frequently form parallel to bedding planes, foliation and cleavage and may be termed bedding joints, foliation joints and cleavage joints accordingly.

Fault

This is a fracture or fracture zone along which recognisable displacement has occurred, from a few centimetres to a few kilometres in scale. The walls are often striated and polished (slickensided) as a result of shear displacement. Frequently the rock on both sides of a fault is shattered and altered or weathered, resulting in fillings, such as breccia and gouge. Fault widths may vary from milimetres to hundreds of metres.

Discontinuity

This is a general term for any mechanical discontinuity in a rock mass having zero or low tensile strength. It is a collective term for most types of joints, weak bedding planes, weak schistosity planes, weakness zones and faults. The ten parameters selected to describe discontinuities and rock masses are defined below:

1. Orientation: Attitude of discontinuity in space. Described by the dip direction (azimuth) and dip of the line of steepest declination in the plane of the discontinuity. Example: dip direction/dip (015/35).

2. Spacing: Perpendicular distance between adjacent discontinuities. Normally refers to the mean or model spacing of a set of joints.

3. Persistence: Discontinuity trace length as observed in an exposure. May give a crude measure of the areal extent or penetration length of a discontinuity. Termination in solid rock or against other discontinuities reduces persistence.

4. Roughness: Inherent surface roughness and waviness relative to the mean plane of a discontinuity. Both roughness and waviness contribute to the shear strength. Large-scale waviness may also alter the dip locally.

5. Wall Strength: Equivalent compression strength of the adjacent rock walls of a discontinuity. May be lower than rock block strength due to weathering or alteration of the walls. An important component of shear strength if rock walls are in contact.

6. Aperture: Perpendicular distance between adjacent rock walls of a discontinuity, in which the intervening space is air or water filled.

7. Filling: Material that separates the adjacent rock walls of a discontinuity and that is usually weaker than the parent rock. Typical filling materials are sand, silt, clay, breccia, gouge and mylonite. Also includes thin mineral coatings and healed discontinuities, e.g. quartz and calcite veins.

8. Seepage: Water flow and free moisture visible in individual discontinuities or in the rock mass as a whole.

9. No. of sets: The number of joint sets comprising the intersecting joint system. The rock mass may be further divided by individual discontinuities.

10. Block Size: Rock block dimensions resulting from the mutual orientation of intersecting joint sets and from the spacing of individual sets. Individual discontinuities may further influence block size and shape.

UNDERGROUND WATER

Some consideration must be given to this subject albeit one too large to be dealt with at any length here. However, it is of paramount importance both from the point of view of rock drilling per sec, and in the more general context of rock excavation.

Porosity is a measure of the voids in rock and their capacity for holding water; permeability, the capacity of rocks to let water pass through, is a measure of those voids which are of sufficient size and are in communication. Mud has a high level of porosity that diminishes during its conversion first to clay and later to mudstone or shale. Even then it has a certain porosity but because of the fineness of the clay particles and, therefore, the minute dimensions of the voids, it is impermeable to water. Permeability is thus a measure both of particulate size and degree of packing and can be defined as hydraulic conductivity.

Water flows most freely through unconsolidated gravels and sands, and through the softest limestones, such as chalk. Consolidated sandstones and grits, limestone and dolomite are to be considered highly permeable although, of course, impure versions of these rocks may contain a high clay fraction and so be that much less permeable. Clays, shales boulder clay, metamorphic and igneous rocks are per se impermeable but the two latter classes of rock permit water flow when fissured. In shattered igneous and metamorphic rocks with open joints water can flow most freely. This is very marked in basalt, less so in andesite.

Whether a given bed will contain water depends on various factors. At very shallow depths rocks and soils can become completely dry but this is very temporary; it is better to consider the upper layers as being aerated but composed of particles each surrounded with a water film. Below this level the ground is saturated with water, the upper surface of this saturation zone being the water table. This rises and falls with the amount of water which becomes available. The lower level of the saturation zone is its junction with impermeable material. In sloping beds the underground water will flow over the surface of the impermeable layer so that in a syncline composed of a sandstone bed sandwiched between shales, water will flow down each limb of the sandstone bed and collect in the synclinal axis. In repetitive alternations of permeable and impermeable rock a borehole may pass through alternate dry and wet layers, i.e., a series of perched water tables. On the other hand, the permeable layer may emerge at the surface at a contour level lower than the catchment area, giving rise to springs where it emerges and making the permeable bed and aqueduct rather than an aquifer.

After heavy rain or surface flooding, conditions are not in equilibrium and temporary high-level water tables may occur perched on aerated rock.

Prediction of underground water conditions requires considerable care, and also luck. The geology must be known and a three-dimensional picture built up by drawing sections or by mental exercise. The catchment, cover and relationship of aquifer to barriers (i.e., impermeable layers, intrusions etc.) must be considered.

At a more general level, wet conditions may be expected in limestones and sandy rocks if the beds are dipping towards the excavation from a higher level. Tunnels may be wet in any permeable or fissured rock which is inclined or in faulted areas.

REFERENCES

Goktan R.M. and C. Ayday C. (1993). A suggested improvement to the Schmidt Rebound Hardness: ISRM suggested method with particular reference to rock machineability. Technical Note. *Inter. J. Rock Mechanics & Mining Sciences & Geomechanics*, Abstracts 30 (3): 321-322.

ISRM (1978). Suggested methods for determining hardness and abrasiveness of rocks. *Inter. J. Rock Mechanics & Mining Sciences & Geomechanics*, Abstract 15: 89-98.

ISRM (1978). Suggested method for determing uniaxial compressive strength and deformability of rock materials. *Inter. J. Rock Mechanics & Mining Sciences & Geomechanics*, Abstracts (15): 113-114.

APPENDIX I: SUGGESTED METHODS FOR DETERMINING HARD- NESS AND ABRASIVENESS OF ROCKS

Part 1. Introduction and Review

The approach taken in this document is to review and reference those tests recently put to use. Those tests which have well-established usage are adopted as 'Suggested Methods' at the present time. Because of the active research underway, especially in the areas of drillability and machine boreability, it is anticipated that additional methods will be incorporated in the next revision of this document.

Definitions

The hardness and abrasiveness of rock are dependent on the type and quantity of the various mineral constituents of the rock and the bond strength that exists between the mineral grains. Tests for each property have been developed to simulate or to correlate with field experience. Many of the tests now used for rock have been adapted from highway materials, concrete and metals testing.

Considerable research has been conducted in the past and is now underway regarding these properties of rock. Many tests developed in a research study have not been evaluated by other organisations or have not been used in practical applications. Many tests which have been developed are used by only one commercial firm or governmental organisation, or are used only in a limited geographical area.

Abrasion and Abrasiveness

Abrasion tests measure the resistance of rocks to wear. These tests include wear when subject to an abrasive material, wear in contact with metal and wear produced by contact between rocks. Abrasiveness tests measure the wear on metal components (e.g. tunnelling machine cutters) as a result of contact with the rock. These tests can be grouped into three categories: (1) abrasive wear impact test; (2) abrasive wear with pressure test and (3) attrition test.

1. Abrasive wear with impact test: (a) Los Angeles abrasion test [1,2]. This test, developed for highway aggregates, subjects a graded sample to attrition due to wear between rock pieces and also to impact forces produced by an abrasive charge of steel spheres. (b) Sand blast test. The surface of the test sample is abraded by an air blast containing silica sand or aluminium oxide under specified conditions. The weight loss or depth of abrasion is a measure of the abrasive resistance of the rock. This method has its chief application in the evaluation of building materials [3]. (c) Burbank test. This test is designed to determine the relative abrasiveness of a rock sample on metal parts of mining and crushing equipment [4]. A single metal paddle of the test alloy is counter-rotated at 632 rev/min. This produces high-speed impact and rapid wear of the test paddle.

2. Abrasive wear with pressure test: (a) The Dorry test [5], ASTM test C-241-51 and the modified Dorry test (British Standard BS-812). These press the rock specimen against a rotating steel disc. A silica sand or aluminium oxide power is fed between the rock and steel surface and acts as an abrasive medium; (b) Bit wear tests. Several tests [6-8] have been devised to determine the abrasive resistance of rock by measuring the bit wear of a standard bit drilling for a specified length or time under specified conditions. These tests are also measures of drillability; (c) The abrasion resistance of a rock and the abrasive effect of the rock on other materials have been determined by use of a modified Taber Abraser Model 143 [9]. Each side of a 6 mm thick disc from an NX core is revolved 400 times under an abrading wheel which is forced against the disc by a 250 g weight. Debris is removed continously by vacuum. The weight loss of the rock is a measure of its abrasive resistance while the weight loss of the abrading wheel is taken as a measure of the abrasiveness of the rock. These values have been used in conjunction with hardness data to predict tunnel machine boreability [9].

3. Attrition tests: Attrition can be defined as the resistance of one surface to the motion of another surface rubbing over it. The wear is produced without impact, pressure or action of a third element of different and invariably higher hardness. The Deval test in which rock aggregates are tumbled at a slow speed without the abrasive charge of steel spheres used in the Los Angeles test, provides a determination of rock attrition. This test is not widely used at present.

Hardness

Hardness is a concept of material behaviour rather than a fundamental material property. As such, the quantitative measure of hardness depends on the type of test employed. Three types of tests have been used to measure the hardness of rocks and minerals: (1) indentation tests; (2) dynamic or rebound tests; (3) scratch tests.

1. Indentation tests: The Brinell and Rockwell tests are well-known tests used on metal but are not generally applicable to rock due to its brittle nature. The Knoop [10] and Vicker [11] tests determine the microhardness of individual rock minerals. A pyramidal-shaped diamond is applied to the surface with a specified force. The area of the permanent residual deformation divided by the applied force is a measure of the hardness. The Knoop test is capable of determining directional hardness of crystals.

2. Dynamic or rebound tests: These tests employ a moving indenter to strike the test specimen. Any plastic or yielding material behaviour produced by the impact will reduce the elastic energy available to rebound the indenter. The height of rebound is taken as a measure of the hardness of the material.

The Shore scleroscope is a laboratory test device that measures hardness by dropping a small diamond-tipped indenter on the specimen and measuring its rebound height. Because of the small size of the diamond indenter tip and the inhomogeneous nature of most rocks, it is necessary to conduct a large number of rebound tests to obtain an average for a particular material.

The Schmidt impact hammer, originally designed to determine the compressive strength of concrete, has been used for hardness determinations of rock. The device, which has both field and laboratory uses, consists of a spring-loaded piston that is projected against a metal anvil placed in contact with the rock surface. The height of the piston rebound is taken as an empirical measure of hardness.

3. Scratch tests: Scratch tests are widely used to determine mineral hardness. The hardness scale proposed by Mohs in 1822 is a scratch test still widely employed. In an attempt to provide a more quantitative measure of hardness, scratch sclerometers using a sharp diamond point to scratch the specimen have been developed. The Talmage and Bierbaum devices [12] are among the better known scatch sclerometers.

REFERENCES

ASTM Standard C131-69.

ASTM Standard C535-69.

ASTM Standard C418-68.

Burbank, B.B. (1955). Measuring the relative abrasiveness of rock minerals and ores. Pit Quarry, pp 114-118.

Obert L., S.L. Windes and W.I. Duval. (1946). Standardized tests for determining the physical properties of mine rock. U.S. Bur. Mines Rep. Invest. RI 389, 67 pp.

Selmer-Olsen, R. and O.T. Blindheim. (1970). On the drillability of rock by percussive drilling. *Proc. 2nd Congr. Int. Soc. Rock Mech.* Beograd.

White, C.G. (1969). A rock drillability index. *Col. Sch. Mines* 8.64(2): 1-92.

Goodrich, R.H. (1961). Drag bits and machines. *Col. Sch. Mines* 8.56(1): 1-21.

Tarkoy, P.J. (1973). A study of rock properties and tunnel boring machine advance rates in two mica schist formations. *15th Symp. Rock Mech. Custer State Park, South Dakota.*

Winchell, H. (1946). Observations on orientation and hardness variations. *Am. Miner.* 31 (3-4): 149-152.

Das B. Vicker's hardness concept in the light of Vicker's impression. *Int. J. Rock Mech. Min. Sci & Geomech.*, Abstr. (1974) 11: 85-89.

Williams, S.R. (1942) Hardness and Hardness Measurements. American Soc. for Metals, Cleveland, 101, 132.

IMPORTANT NOTE

The units stated in the document are the modern metric units in accordance with the Systeme International units (S.I.) which is an extension and refinement of the traditional metric system. The following should be noted:

Unit of length—1 metre (m) = 1000 mm:

Unit of mass—1 kilogram (kg) = 1000 g;

Unit of force—1 newton (N) = kg m/s;.

Unit of stress—1 pascal (Pa) = N/m.

Part 2: Suggested Method for Determining the Resistance to Abrasion of Aggregate by Use of the Los Angeles Machine

Scope

This method covers procedures for testing aggregate for resistance to abrasion using the Los Angeles testing machine. The abrasive charge and the test sample used are dependent on the aggregate size and grading.

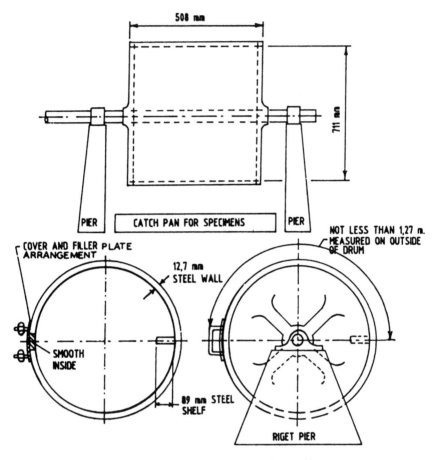

Fig. 6.5: Los Angeles abrasion testing machine.

Apparatus

(a) *Los Angeles Machine:* The Los Angeles abrasion testing machine, conforming in all essential characteristics to the design shown in Fig. 6.5 shall be used. The machine shall consist of a hollow steel cylinder, closed at both ends, having an inside diameter of 711 + 5 mm and an inside length of 508 + 5 mm. The cylinder shall be mounted on stub shafts attached to the ends of the cylinder but not entering it, and shall be mounted in such a manner that it may be rotated with the axis in a horizontal position within a tolerance in slope of 1 in 100. An opening in the cylinder shall be provided for the introduction of the test sample. A suitable, dust-tight cover shall be provided for the opening with means for bolting the cover in place. The cover shall be so designed as to maintain the cylindrical contour of the interior surface unless the shelf is so located that the charge will not fall on the cover, or come into contact with it during the test. A removable steel shelf extending the full length of the cylinder and projecting inwards 89 + 2 mm shall be mounted on the interior cylindrical surface of the cylinder, or on the inside surface of the cover, in such a way that a plane centred between the large faces coincides with an axial plane. The shelf shall be of such thickness and so mounted, by bolts or other suitable means, as to be firm and rigid. The position of the shelf shall be such that the distance from the shelf to the opening, measured along the outside circumference of the cylinder in the direction of rotation, shall be not less than 1.27 m. The shelf shall be made of wear-resistant steel and shall be rectangular in cross-section.

(b) *Balance:* A balance or weighing machine accurate within 0.1% of test load over the range required for this test.

(c). For coarse aggregate smaller than 38 mm the sample shall be recombined and the abrasive charge selected as described in Table. 6.3 (see Table 6.2 and 6.3).

Procedure

(a) Place the test sample and the abrasive charge in the Los Angeles abrasion testing machine and rotate the cylinder at a speed of 30–33 rev/min. The no. of

Table 6.2. Gradings of test samples*.

Sieve size, mm (square openings)		Weights of indicated sizes, g		
Passing	Retained on	1	Grading 2	3
75.0 mm	63.0 mm	2500 + 50	—	—
63.0 mm	53.0 mm	2500 + 50	—	—
53.0 mm	38.0 mm	5000 + 50	5000 + 50	—
38.0 mm	25.4 mm	—	5000 + 25	5000 + 25
25.4 mm	19.0 mm	—	—	5000 + 25
	Total:	10,000 + 100	10,000 + 75	10,000 + 50

*Coarse aggregate larger than 19 mm.

Table 6.3. Gradings of test samples*.

Sieve size, mm (square openings)		Weights of indicated sizes, g			
Passing	Retained on	A	Grading B	C	D
38.0 mm	25.4 mm	1250 + 25	—	—	—
25.4 mm	19.0 mm	1250 + 25	—	—	—
19.0 mm	13.2 mm	1250 + 10	2500 + 10	—	—
13.2 mm	9.5 mm	1250 + 10	2500 + 10	—	—
9.5 mm	5.6 mm	—	—	2500 + 10	—
5.6 mm	4.7 mm	—	—	2500 + 10	—
4.7 mm	2.3 mm	—	—	—	5000 + 10
	Total:	5000 + 10	5000 + 10	5000 + 10	5000 + 10

*Coarse aggregate smaller than 38 mm.

revolutions shall be 500 for aggregate smaller than 38 mm and 1000 for aggregate larger than 19 mm. The machine shall be so driven and so counterbalanced as to maintain a substantially uniform peripheral speed. If an angle-shaped steel member is used as the shelf, the direction of rotation shall be such that the charge is caught on the outside surface of the shelf.

(b) After the prescribed number of revolutions, discharge the material from the machine and make a preliminary separation of the sample on a sieve coarser than 1.7 mm (No. 12 US). Sieve the finer portion in a 1.7 mm sieve. Wash the material coarser than the 1.7 mm sieve, oven dry at 105-110°C to substantially constant weight and weigh to the nearest gram.

(c) Valuable information concerning the uniformity of the sample under test may be obtained by also determining the loss after 100 revolutions for cases in which 500 revolutions are specified or after 200 revolutions in cases for which 1000 revolutions are specified. The loss should be determined without washing the material coarser than the 1.7 mm sieve. The ratio of the loss after 100 or 200 revolutions to the loss after 500 or 1000 revolutions respectively, should not greatly exceed 0.20 for material of uniform hardness. When this determination is made, care should be taken to avoid losing any part of the sample; the entire sample, including the dust of abrasion, shall be returned to the test machine for the final 400 or 800 revolutions required to complete the test.

Calculations

(a) Express the difference between the original weight and the final weight of the test sample as a percentage of its original weight. Report this value as the percentage of wear.

(b) When the procedure described under (c) above is followed, the uniformity of wear ratio is the ratio of the loss after 100 or 200 revolutions to the loss after 500 or 1000 revolutions respectively.

Reporting Results

The report should include the following data:

(a) Source/location and geologic description of the sample tested.
(b) Grading of test sample.
(c) Grading of abrasive charge.
(d) The Los Angeles percentage of wear (see Calculations section (a) above).
(e) The Los Angeles uniformity of wear ratio (see Calculations section (b) above) if applicable.

Important Notes

1. This test method combines the essential features of ASTM standard test C131-69 and ASTM standard test C535-69. Aggregate in the size range of 19 mm to 38 mm can be tested by either one of the two procedures described in this Suggested Method. The specific procedure used for this size aggregate shall be reported with the results.

2. If the aggregate is essentially free from adherent coatings and dust, the requirement for washing before and after the test may be waived. Elimination of washing after testing will seldom reduce the percentage wear by more than about 0.2 percentage points.

3. Test sieves shall conform to ISO Standard 5650 1972 (E) 'Test sieves—woven metal wire cloth and perforated plate—nominal sizes of apertures', Series R 40/3.

4. Backlash or slip in the driving mechanism is very likely to furnish test results which are not duplicated by other Los Angeles abrasion machines producing constant peripheral speed.

Part 3: Suggested Method for Determination of the Schmidt Rebound Hardness

Scope

(a) This method is suggested for the Schmidt impact hammer test for hardness determination of rock.

(b) The method is of limited use on very soft or very hard rocks.

Apparatus

The apparatus shall consist of:

(a) The Schmidt hammer which determines the rebound hardness of a test material. The plunger of the hammer is placed against the specimen and is depressed into the hammer by pushing the hammer against the specimen. Energy is stored in a spring which automatically releases at a prescribed energy level and impacts a mass against the plunger. The height of rebound of the mass is measured on a scale and is taken as the measure of hardness. The device is portable and may be used both in the laboratory and the field.

Schmidt hammer models are available in different levels of impact energy. The type L hammer having an impact energy of 0.74 Nm shall be used with this suggested method.

(b) A steel base of minimum weight of 20 kg to which specimens should be securely clamped. Cored specimens should be tested in a steel "cradle" with a semi-cylindrical machined slot of the same radius as the core, or in a steel V-block (Fig. 6.6).

Procedure

(a) Prior to each testing sequence, the Schmidt hammer should be calibrated using a calibration test anvil supplied by the manufacturer for that purpose. The average of 10 readings on the test anvil should be taken.

(b) Specimens obtained for laboratory tests shall be representative of the rock to be studied. When possible, use larger pieces of rock for the Schmidt hardness test. The type L hammer should be used on NX or larger core specimens or on block specimens having an edge length of at least 6 cm.

(c) The test surface of all specimens, either in the laboratory or in the field, shall be smooth and flat over the area covered by the plunger. This area and the

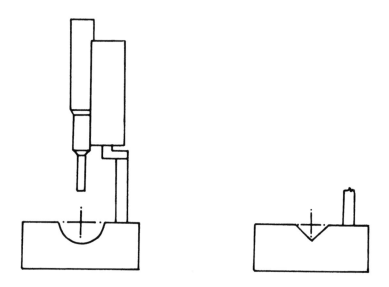

Fig. 6.6: Core specimen holders.

rock material beneath to a depth of 6 cm shall be free from cracks, or any localised discontinuity of the rock mass.

(d) Small individual pieces of rock, whether tested in the laboratory or in the field, shall be securely clamped to a rigid base to adequately secure the specimen against vibration and movement during the test. The base shall be placed on a flat surface that provides firm support.

(e) The hardness value obtained will be affected by the orientation of the hammer. It is recommended that the hammer be used in one of three positions: vertically upwards, horizontally, or vertically downwards with the axis of the hammer +5 from the desired position. When use of one of the three orientations is not feasible (e.g. *in situ* testing in a circular tunnel), the test should be conducted at the angle necessary and the results corrected to a horizontal or vertical position using the correction curves supplied by the manufacturer. The hammer orientation for the test and any corrections applied to non-vertical or non-horizontal orientations should be recorded and reported in the results.

(f) At least 20 individual tests shall be conducted on any one rock sample. Test locations shall be separated by at least the diameter of the plunger. Any test that causes cracking or any other visible failure shall cause that test and the specimen to be rejected. Errors in specimen preparation and testing technique tend to produce low hardness values.

Calculations

(a) The correction factor is calculated as:

$$\text{Correction factor} = \frac{\text{Specified standard value of the anvil}}{\text{Average of 10 readings on the calibration anvil}}$$

(b) The measured test values for the sample should be ordered in descending value. The lower 50% of the values should be discarded and the average obtained of the upper 50% values. This average shall be multiplied by the correction factor to obtain the Schmidt Rebound Hardness.

Reporting of Results

The following information shall be reported:

(a) Lithologic description of the rock. Source of sample, including geographic location, depth and orientation.

(b) Type of specimen (core, blasted or broken sample *in situ*). Size and shape of core or block specimen.

(c) Date of sampling, date of testing and condition of storage (i.e., exposure to temperature extremes, air drying, moisture etc.).

(d) Orientation of the hammer axis in the test.

(e) Method of clamping sample (V-block or clamps).

(f) The Schmidt Hardness value obtained as per the Calculations section above.

Part 4: Suggested Method for Determination of the Shore Scleroscope Hardness

Scope

This laboratory method is suggested for hardness determination of rock minerals using the Shore scleroscope and for verification of other scleroscope hardness instruments. Rock hardness may be obtained as an average of readings taken at random on individual mineral grains.

Apparatus

The instrument used for determining scleroscope hardness numbers is supplied in two models, designated Model C and Model D. Model C-2 is recommended for use with rock.

(a) The Scleroscope Model C-2 consists of a vertically disposed barrel containing a precision bore glass tube. A scale graduated from 0 to 140 is set behind the barrel and is visible through the glass tube. A pneumatic actuating head affixed to the top of the barrel is manually operated by a rubber bulb and tube. A hammer drops from a specified height and rebounds within the glass tube. The hammer for Model C-2 shall have the following dimensions:

Diameter	5.94 mm
Mass	2.3 + 0.5 g
Overall length	20.7–21.3 mm
Distance hammer falls	251.2 + 0.13–0.38 mm.

(b) The diamond must be shaped to produce a correct reading on reference bars of known hardness. In profile, the diamond is convex, having a radius terminated by a flat striking surface, as shown in Fig. 6.7. The flat striking surface

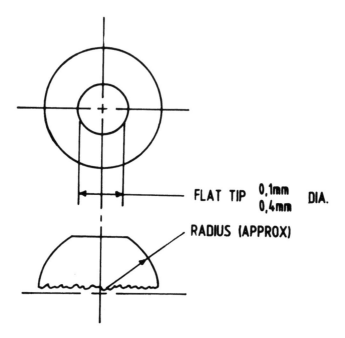

Fig. 6.7: Profile of scleroscope diamond showing range of diameters of flat tip.

is approximately circular and from 0.1 to 0.4 mm in diameter, depending on the hardness and other physical characteristics of the diamond.

Procedure

(a) Before each day's use, make at least five hardness readings on the standard test block furnished by the manufacturer at the hardness level at which the machine is being used. If the values fall within the range of the standardised hardness test block the instrument may be regarded as satisfactory; if not, the machine should be verified using procedures recommended by the manufacturer.

(b) Tests shall be made on flat surfaces ground smooth using a No. 1800 grade aluminium oxide abrasive powder. An excessively coarse surface will yield low and erratic readings.

(c) Specimens should have a minimum test surface of 10 cm and a minimum thickness of 1 cm. Small specimens should be clamped securely with the flat test surface perpendicular to the scleroscope axis.

(d) To perform a test hold or set the instrument in a vertical position with the bottom of the barrel in firm contact with the test specimen and normal to the surface of the specimen. Bring the hammer to the elevated position by squeezing the rubber bulb, allow it to fall and strike the test surface and measure the height of rebound. The height to which the hammer rebounds on the first bounce indicates the hardness of the material.

(e) To prevent errors resulting from misalignment the instrument must be set or held in a vertical position, using the plumb bob or spirit level on it to determine verticality. The most accurate readings of the scleroscope are obtained with the instrument mounted in a clamping stand. Lateral vibrations must be avoided since they tend to free fall of the hammer, impede which results in a low instrument reading.

(f) An error may result if the hammer indentations are spaced too closely together. Space indentations at least 5 mm apart and make only one test in one spot. At least 20 hardness determinations should be taken.

Calculations

The Shore Scleroscope Hardness test shall be the average of not less than 20 measurements made on the same specimen, using the above method.

Reporting results

The report should include the following information on each specimen tested:

(a) Lithologic description of the rock source of the sample including geographic location, depth and orientation.

(b) Approximate mineral composition and grain size of the rock specimen.

(c) Date of sampling, date of testing, storage conditions, and specimen preparation procedures.

(d) Orientation of the test surface with respect to bedding or foliation planes when these are significant characteristics of the rock.

(e) The number of tests conducted and the average Shore hardness.

The test procedures described above are based on the ISRM suggested test procedures.

APPENDIX II: SUGGESTED METHODS FOR DETERMINING THE UNIAXIAL COMPRESSIVE STRENGTH OF ROCK MATERIALS

Part 1: Suggested Method for Determination of the Uniaxial Compressive Strength of Rock Materials.

Scope

This test method is intended to measure the uniaxial compressive strength of a rock sample in the form of specimens of regular geometry. The test is mainly intended for strength classification and characterisation of intact rock.

Apparatus

(a) A suitable machine shall be used for applying and measuring axial load to the specimen. It shall be of sufficient capacity and capable of applying load at a rate conforming to the requirements set in Section 3. It shall be verified at suitable time intervals and shall comply with accepted national requirements such as prescribed in either ASTM Methods E4: Verification of Testing Machines of British Standard 1610, Grade A or Deutsche Normen DIN 51 220, DIN 51 223, Klasse 1 and DIN 51 300.

(b) A spherical seat, if any, of the testing machine, if not complying with specification 2(d) below, shall be removed or placed in a locked position, the two loading faces of the machine being parallel to each other.

(c) Steel platens in the form of discs and having a Rockwell hardness of not less than HRC58 shall be placed at the specimen ends. The diameter of the platens shall be between D and $D + 2$ mm, where D is the diameter of the specimen. The thickness of the platens shall be at least 15 mm or $D/3$. Surfaces of the discs should be ground and their flatness should be better than 0.005 mm.

(d) One of the two platens shall incorporate a spherical seat. The spherical seat should be placed on the upper end of the specimen. It should be lightly lubricated with mineral oil so that it locks after the dead weight of the cross-head has been picked up. The specimen, the platens and spherical seat shall be accurately centred with respect to one another and to the loading machine. The curvature centre of the seat surface should coincide with the centre of the top end of the specimen.

Procedure

(a) Test specimens shall be right circular cylinders having a height to diameter ratio of 2.5–3.0 and a diameter preferably of not less than NX core size, approximately 54 mm. The diameter of the specimen should be related to the size of the largest grain in the rock by the ratio of at least 10:1.

(b) The ends of the specimen shall be flat to 0.02 mm and shall not depart from perpendicularity to the axis of the specimen by more than 0.001 radian (about 3.5 min) or 0.05 mm in 50 mm.

(c) The sides of the specimen shall be smooth and free of abrupt irregularities and straight to within 0.3 mm over the full length of the specimen.

(d) The use of capping materials or end surface treatments other than machining is not permitted.

(e) The diameter of the test specimen shall be measured to the nearest 0.1 mm by averaging two diameters measured at right angles to each other at about the upper height, the mid-height and the lower height of the specimen. The average diameter shall be used for calculating the cross-sectional area. The height of the specimen shall be determined to the nearest 1.0 mm.

(f) Samples shall be stored, for no longer than 30 days, in such a way as to preserve the natural water content, as far as possible, and tested in that condition. This moisture condition shall be reported in accordance with 'Suggested method for determination of the water content of a rock sample', Method 1. ISRM Committee on Laboratory Tests, Document No. 2, First Revision, December 1977.

(g) Load on the specimen shall be applied continuously at a constant stress rate, such that failure will occur within 5–10 min of loading; alternatively the stress rate shall be within the limits of 0.5–1.0 MPa/s.

(h) The maximum load on the specimen shall be recorded in newtons (or kilonewtons and meganewtons where appropriate) to within 1%.

(i) The number of specimens tested should be determined from practical considerations but at least five are preferred.

Calculations

The uniaxial compressive strength of the specimen shall be calculated by dividing the maximum load carried by the specimen during the test, by the original cross-sectional area.

Reporting results

(a) Lithologic description of the rock.

(b) Orientation of the axis of loading with respect to specimen anisotropy, e.g. bedding planes, foliation etc.

(c) Source of sample, including geographic location, depth and orientation, dates and method of sampling, and storage history and environment.

(d) No. of specimens tested.

(e) Specimen diameter and height.

(f) Water content and degree of saturation at time of test.

(g) Test duration and stress rate.

(h) Date of testing and type of testing machine.

(i) Mode of failure, e.g. shear, axial cleavage etc.

(j) Any other observations or available physical data such as specific gravity, porosity and permeability, citing the method of determination for each.

(k) Uniaxial compressive strength for each specimen in the sample, expressed to three significant figures, together with the average result for the sample. The pascal (Pa) or its multiples shall be used as the unit of stress and strength.

(l) Should it be necessary in some instances to test specimens that do not comply with specifications as stated above, these facts shall be noted in the test report.

REFERENCES

Obert L., S.L. Windes and W.I. Duvall. (1946). Standardized tests for determining the physical properties of mine rocks, U.S. Bur. Mines Rep. Invest. No. 389, 67 pp.

International Bureau for Rock Mechanics. Richtlinien zur Durchfuhrung von Druckversuchen an Gesteinen im Bergbau. Bericht, 5, Landerireffen des I.B.G. Akademie-Verlag. Berlin 1964, pp. 21-25.

U.S. Corps of Engineers. Strength parameters of selected intermediate quality rocks—testing procedures. Missouri River Division Laboratory Reports No. 64/493. July 1966, pp. 1A-6A: IB-7B.

ASTM Standard D-2938-71a.

Hawkes, I and M. Mellor. (1970). Uniaxial testing in rock mechanics laboratories. *Engng. Geol.* 4: 177-285.

CHAPTER 7

DRILLING SPEEDS

Drill and Air Supply
Variables Relating to Material Properties
Variables Relating to Operation
Variables Relating to Site Organisation and Management

The published literature on the theoretical aspects of rock drilling is of great value in understanding the process of rock drilling, but of little help in predicting the outputs.

Since a large number of factors affect drilling speeds, some of which are still unresolved theoretically, prediction is currently based simply on experience. Hence it is necessary to understand these factors as a means towards more logical prediction.

Some definitions generally used in this connection are as follows:

Penetration Speed is the actual rate of drilling at any given time.

Drilling Rate is the average speed of drilling in metres per hour over one or more holes inclusive of all drilling operations but excluding delays etc.

Gross Drilling Rate is the average speed of drilling in metres per hour, inclusive of all delays, over a series of shifts.

The term 'overall drilling speed' has been discarded as ambiguous. Manufacturers often use the term to refer to drilling rate whilst estimators use it as a synonym for gross drilling rate, which can lead to misunderstandings.

Variables in drilling can be grouped into those relating (a) to the drill and air supply, where applicable, (b) to the hole (size, depth and inclination), (c) to the rock (toughness and abrasiveness), (d) to the rock structure (dip, fissuring and faulting), (e) to the operation of drilling (setting up, ability of operators) and (f) to the organisation (lost time). By considering as many of these variables as practicable, a reasonable forecast can be made.

Applied thrust is an important factor in penetration speed but is difficult to determine for percussive drilling, particularly in the field and specially in handheld drilling where the thrust varies with every shift of the operator's weight. Also, applied thrust means little unless correlated with other factors relating to the rock. However, it may be stated that the ability to determine and select the optimum thrust level at any given time is the main characteristic of the good operator and thus the factor can be expressed in terms of the operator's skill. This is variable and although difficult to analyse mathematically, can be estimated empirically. On the other hand, this is not true for rotary drilling, where thrust has to be taken as one of the major factors.

DRILL AND AIR SUPPLY

Air Consumption

It is known that work output per minute is related to the air pressure, area of the piston and the stroke. If the work output is plotted against penetration speed it should approximate to a straightline graph, but in practice the results are inconsistent, particularly when different drills are compared.

A simple approximation can be achieved by plotting air consumption at a given pressure against penetration speed. In the relationship mentioned above air consumption is expressed in terms of air required to maintain pressure. The use of air consumption as a guide to penetration speed presupposes that all rock drills make equally efficient use of the compressed air which flows through them. This is not strictly true but it is better to have a datum than not.

Air Pressure

Penetration speed varies with absolute pressure of compressed air to the power of 1.5; the basic validity of this statement can be demonstrated. An increase in pressure from 5.0 kg/cm^2 to 6.0 kg/cm^2 improves penetration speed by 31% and an increase from 5.0 kg/cm to 7.0 kg/cm by 65%.

Air pressure can be used as a factor provided that the air consumption is taken at a common pressure level.

WEAR AND TEAR OF MACHINE

A new rock drill will give optimum performance after having drilled a few hundred metres and, provided that it is correctly and adequately lubricated, will continue to give a performance which can be considered as 100%. Penetration speed will deteriorate, as a function of compressed air consumed, with increasing wear of the piston and valve and, as the rifle bar and/or rifle bar nut and chuck become worn. The former is of importance economically but affects penetration speed mainly if the air supply is limited. Wear on the rifle bar and nut reduces speed of drilling by a drastic reduction of rotation. Loss of speed due to chuck wear results from the eccentric impact of the piston on the rod shank.

Lubrication is of great importance, both because it reduces wear and hence minimises the above effects, and because it decreases friction. The loss of transmitted energy in an unlubricated machine is shown by overheating of the rock drill cylinder and front head.

Strict routine maintenance, replacement of partially worn components and thorough lubrication can reduce the loss of drilling to a minimum. On many sites this surveillance is considered unnecessary where an overall reduction in penetration of say 10% may be allowed. Much higher losses can be experienced with badly maintained drills. Although these lead untimely to breakdown and therefore replacement and renewed efficiency, the average value would need to be adjusted.

Rigidity of Mounting

Drills mounted on a heavy carriage or rigid structure give a higher penetration speed than a similar drill if hand-held or mounted on light frames. However, this factor is less critical in percussive drilling than with rotary drills. The loss is largely due to friction of the rod or rod-couplings on the wall of the hole as the drill moves away from the hole axis. Obviously the effect increases with length of the drill string.

The gross drilling rate is not necessarily affected because an increase in rigidity is obtained at the expense of increased handling time.

PARAMETERS RELATING TO THE HOLE

Size

Generally the penetration speed varies inversely with the cross-sectional area of the hole. Thus for a given percussive value, a 100 mm dia hole will be drilled at 25% of the rate for a 50 mm hole. However, the energy transmitted to the rock per unit length of cutting edge of the bit would only be approximately equal in the two instances, if the smaller bit was of the cross-type and the larger bit a single chisel.

Depth

The rate of penetration decreases with increase in hole depth. This loss is due to several factors: loss of energy through couplings, loss of energy in the drill rods, and increased compression of the rock with increase in weight of overlying strata.

This can be seen by comparing the results of percussive and down-the-hole drilling. In the latter technique there is no loss of energy in drill rods or couplings because the percussive hammer is virtually at the rock face all the time; nevertheless a loss in penetration speed occurs with increase in depth. In percussive drilling with the drill at the surface and using coupled rods, the loss, expected, is even greater.

A large proportion of rock drilling is done in stress-relieved strata, e.g. blast holes subparallel to a quarry face, and also many rock structures have undergone partial relief naturally. In cases where this is not so, penetration rates are likely to be lower.

Rod Diameter

The resultant effect of differing diameters of drill rod is an interaction of a number of factors. In relation to bit diameter, decrease in rod size gives (a) more flexing and therefore greater energy loss; (b) decreased flushing-hole size but increased annulus between rod and hole wall and therefore lower return velocity and less efficient flushing; but (c) decrease in absorption of energy within the rod.

The optimum cross-sectional area of the bit appears to be 2.25–3 times the cross-sectional area of the rod. Acceptable results, however, are obtainable with bits having up to seven times the area, provided the drilling conditions are not unduly difficult.

Average return velocities are low when large bits are used in percussive drilling. If these figures are compared with down-the-hole drilling, assuming the bit does not commonly exceed twice the rod area and an average return velocity of 1520 cm/s is considered minimal, it is apparent that adverse ground conditions will be reflected more readily by percussive drilling than by down-the-hole drilling for an equivalent sized hole. The load-carrying capacity of the air stream is related to the velocity; it is probable that the larger particles cut by the bit would remain in partial suspension at the bottom of the hole, absorbing a proportion of the impacts and preventing clean rotation. This would be aggravated in fissured ground. An example of the limited load-carrying capacity of flushing air is the build-up on coupling ledges.

The component of the loss of penetration speed which is associated with length of drill strings is represented in the hole depth factor already mentioned. For reduction in penetration rate due to flexing, low return velocity of up to 20% is suggested, whenever the area of the bit is more than three but less than seven times the rod cross-sectional area.

Hole Inclination

The drilling rate is likely to be slower in inclined holes with light wagon drills or hand-held drills; with heavier track-mounted drills the factor is relatively unimportant. Reduced speeds, where they do occur with light wheeled rigs, are due to loss of alignment.

Large rotary rigs are more affected by hole inclination. The necessarily heavy thrusts exerted are not normal to the surface plane and therefore tend to push the rig along the surface It may, not be practicable, therefore, to use the full power of the rig.

One factor is that angled holes are more likely to be at an adverse angle to the bedding of the rock than are vertical holes.

VARIABLES RELATING TO MATERIAL PROPERTIES

The various factors which concern the properties of the material being drilled can be determined in the laboratory. Of these, 'abrasiveness' can be isolated, whilst the remainder are best described as 'drillability'.

Drillability

A test for determining the toughness of rocks involves the drilling of small diameter holes by rotary means. This appears to place undue emphasis on the shear strength of the rock. Alternatively, applying theories of wave action tests can be conducted with a Schmidt concrete test hammer. This consists of a piston which is released against an anvil held in contact with the rock surface at a constant pressure. The rebound of the piston, after striking the anvil, is a measure of the compressive component of the reflected wave. The harder the rock, the more the piston rebounds: the softer the rock, the greater the tensile component of the wave reflection and the less the rebound.

An alternative method is to simulate rock-drilling conditions in the field on a miniature scale. Some of the material in the form of hand specimens can be embedded in concrete blocks.

More accurate indices can be obtained by using fresh specimens only, carried out *in situ*. Because of the small scale, no account is taken of the dynamic condition of the rock; this can only be incorporated by carrying out similar small-scale sampling at regular intervals at the bottom of a large drill hole or trial excavation. Similar objections also apply to more subtle laboratory tests involving cores or prepared blocks of stone.

It is usually possible to obtain data from manufacturers regarding penetration speed or net penetration rate of their drills in a standard granite. By carrying out a drillability test using the equipment suggested on both granite and the rock type for which information is required, then relating to this the data obtained from the manufacturers, a useful indication of drillability can be obtained.

Abrasiveness

The ability of the rock to abrade the bit affects the penetration speed because of the reduction in depth of penetration with each blow as the bit becomes increasingly blunted. The loss is progressive; wear rate increases as penetration falls.

The degree of abrasiveness can be determined by laboratory test but a convenient method is to use, as a scale, the Silica (SiO_2) content. This method should be reasonably sound because the proportion of other highly abrasive minerals (e.g. corundum AlO) present in most rocks is small, but it is not strictly accurate because silica combines to form silicate minerals of varying hardness. Thus on Moh's Scale, the common silicon-bearing minerals are:

6.5–7.5	Tourmaline
	Quartz
	Chalcedony (including Flint and Chert)
	Garnet, varieties
	Olivine
5.5–6.5	Augite
	Hornblende
	Orthoclase feldspar
	Plagioclase feldspar
1.5–3	Chlorite
	Mica, varieties

These compare with tungsten carbide (about 7.8) and steel (usually about 6).

Ideally the silica content of any chlorite or mica present should be deducted but the error may be regarded as insignificant.

VARIABLES RELATING TO GEOLOGICAL STRUCTURES

Dip

The angle of drilling in relation to grain of the rock can affect the drilling rate either directly through penetration speed, or by causing bit-sticking and similar troubles. By 'grain' in this context is meant bedding planes in sedimentary rocks, foliation and slaty cleavage in metamorphic rocks, and flow-banding and preferential orientation in igneous rocks.

Figure 7.1 shows in exaggeration (a) the typical profile produced by a single-chisel bit in thick rock (b) the wide, flat crater when drilling normal to thin bedding, and (c) the deep, narrow crater when drilling parallel to thin bedding. Obviously (b) favours relatively heavier percussion and (c) requires relatively strong rotation with a four-point bit rather than a single-chisel bit.

This is only one aspect of the matter. Some minerals (e.g. 'kyanite') have a hardness which differs on each crystallographic axis. Many minerals (e.g. the micas) will cleave easily in one specific plane only. In other minerals cleavage is

Fig. 7.1: Base of percussive drill hole. Diagrammatic representation of craters produced in a) hard homogeneous rock, b) thin bedded rock drilled normal to bedding and c) rock drilled parallel to bedding.

in several planes (e.g. parallel to the three axes of the cube) and so perfect that if struck a tangential blow, a crystal will cleave to show a structure resembling a flight of steps.

Laminated rocks tend to have mineral components arranged parallel to the laminae. Thus in laminated rocks containing platy minerals, cleavage is often perfect to the laminae but negligible in other planes. Where a rock is substantially composed of a mineral having several planes of perfect cleavage (e.g. fluorospar, which cleaves parallel to the octahedron), direction is of less importance than the degree of crystallisation.

Some segregation of the mineral components or the particle size of the rock may also result in certain layers being easier to drill than others. 'Parallel' holes therefore give a wider scatter of results than 'normal' holes.

The practical result of these causes is that when drilling normal to the grain, results differ little from those in massive rock; but when drilling along the laminae penetration speed is improved. This may be masked by increased difficulty of drilling so that, on balance, holes parallel to the grain give a slower drilling rate. This is most marked when drilling at a small angle to the laminae, when a hole will tend to deflect as the bit follows the easiest path, giving a combination of initial fast penetration speed but an overall slow net penetration rate.

Even where drilling is difficult in a given plane, the average results over a contact may be little affected if it is possible to alter the angle of the holes to suit rock conditions or where the dip is variable.

Fissures

Joints, fissures and planes of weakness are commonly present, their frequency and extent depending on the geological history of the area. Related to rock drilling, as distinct from blasting, three points must be considered: frequency, degree of weathering and angle in relation to the drill hole.

A clean tight fissure (for example a blasting crack) will have only a minor effect on speed of drilling, although in theory it forms a limiting plane affecting the transmission of energy from the bit through the rock. Other things being equal, however, a close network of cracks (as in a shatter zone) will give slower drilling than will relatively unfissured rock. This is partly due to reduced flushing and partly to the wedge action exerted by partially displaced angular rock fragments.

'Open' fissures are of much greater importance. The term is used to include cracks which have been open at some stage in their history, permitting weathering, decomposition by the passage of underground liquids, or infilling from above where the interstitial material (boulder clay, earth rock flour) is markedly softer than the country rock. The material, broadly speaking, can be termed clayback.

The greater the differential in hardness between rock and clayback, the greater the drilling problem. Such rocks as Carboniferous limestone, which are hard but capable of solution, are amongst the worst offenders. The applied thrust required for drilling in hard rock causes the bit to plunge forward in clayback, which this commonly results in a jammed bit and/or blocked flushing holes.

Open joints are often wet and may cause partial or complete loss of flushing at a time when flushing is vital. The closer the angle of drill holes to the plane of fissuring, the greater the delays.

The techniques required to avoid trouble need not be discussed here but an experienced operator has to be assumed. There is no doubt that in open-fissured rock a considerable loss of footage will result from abandoned holes, from the time taken for extra precautions by the driller and so on. The gross drilling rate in such circumstances will be considerably reduced.

If this is compared with large-hole rotary drilling, a marked difference will be apparent. Provided that flushing can be maintained, as for example by the use of large quantities of water-injected air, the presence of claybacks may actually improve rotary drilling performances.

An accurate calculation of percentage losses due to fissuring is obviously impracticable, but estimates based on observation may be of interest. In bad conditions the net penetration rate may drop by 70–75%. Bad conditions, however, although they may occur for the full depth of a drill hole, represent only a proportion of the total drilling in fissured rock when considered over a series of holes. For example, when drilling a pattern of 9-m blast holes in fissured Carboniferous limestone, it was found that each hole had passed two or three claybacks varying in length from 0.15 to 3 m. The latter figure occurred when a drill hole coincided for a distance with a vertical joint. Calculation showed that only 15% of the gross footage was affected.

Percentage loss on gross drilling rate due to fissuring varies from 5 to 30% in hard limestone, down to 0 to 10% in less critical rock types.

VARIABLES RELATING TO OPERATION

Setting up etc.

The process of drilling includes a number of essential operations which are independent of the net penetration rate.
These include:

Operation	Factors per hole include	Frequency

1 Setting up
   Moving, lining up
   Moving compressor
	Distance between Holes
	Ease of drill Movement
	Conditions of terrain, Necessary accuracy (e.g. angled holes)
		Once per hole

2 Collaring	Rock conditions	Once per hole
3 Blowing, cleaning out, redrilling	Rock conditions No. of rod changes necessary. Uphole velocity	Once per rod change and as necessary

4 Adding Rod or Changing Steels	No. of additions required	Once per rod addition
	No. of changes required	Once per rod change
5 Removing Rods	No. of rods added Type of rod coupling Use of grease	Once per rod addition
6 Adjusting Leg	Spatial restrictions	As necessary

It has been found that setting up with a self-propelled Tracdrill takes 2–4.5 min for 3.0–6.0 m holes, 3–5 min for 12–18 m holes and 5.6 min for 30 m holes. This increase in time with depth of hole is partly due to extra care taken to line up deeper holes, which in turn necessitates more frequent movement of the compressor. The variation shown for each depth largely relates to the condition of the quarry top. Wagon drills, which have to be manhandled, are much more sensitive to bad terrain, requiring 3–7 min for holes up to 6 m and 4–8 min for 12 m holes.

Collaring, the cautious penetration of the first few inches of rock, varies little with the type of drill, usually requiring 15–25 s/hole.

Blowing and redrilling is a highly variable factor intimately associated with flushing. Depending on the latter, the average time spent for a large proportion of holes varies between 0.5 and 1.0 min. rod change until the last change, which averages 0.66–1.8 min.

The proportion of time lost in rod handling is largely a matter of the relationship of hole depth to length of steel. Thus with a Tracdrill having a 3-m feed, holes up to 3 m deep can be drilled with no rod handling, holes 3.3–6 m deep with one rod added and removed, and so on. Obviously the time lost on a series of 3.3 m holes will be considerably greater (proportional to the total footage) than on 6 m holes. The time taken to add a 10 ft (3 m) coupled rod is 0.75–1.25 min and to remove it (after the thread has been tightened during drilling) 1.5–3.0 min.

With hand-held sinkers and feed-leg drills, the time required for both blowing and changing steels is of the order of 1.0 min/0.6 m or 0.8 m change.

Obviously, there are many departures from these figures but nonetheless they offer a basis for estimation.

Human Factor

The assumptions made so far have been based on averages obtained by knowledgeable but not necessarily expert handling. It is naturally true that a lethargic person will take longer on each operation, especially under difficult conditions. It is equally true that a keen man· will achieve better drilling rates than the lazy or unenthusiastic.

Inept handling (correct amount of thrust not selected or not maintained) will reduce penetration speed. With hand-held drills the fault is commonly underthrusting since it is easier to let the machine bounce than to hold it to the rock; but with mounted drills 'crowding' is commoner. In the former much of the energy produced is not transmitted to the rock, resulting in poor penetration rates, overheated and broken front heads, chipped pistons and drill shanks and rapid loss of bit gauge. Crowding impedes rotation and leads to a reduced stroke and less power; it also causes bit jamming, misaligned holes and so on.

In quantitative terms, the good driller will achieve net penetration rates within 10% of the maximum but may deliberately select a lower rate in order to avoid possible trouble. In many cases experience of the specific rock conditions will permit even better figures but obviously, there must always be a margin between that which is possible and that which is advisable and can be maintained. Underthrusting can reduce net penetration rate by 50% or more in extreme cases, but 10–20% is the more usual figure. Overthrusting can reduce the rate by about 25% before stall conditions become so marked as to be obvious even to the untrained.

It is also common to find that the untrained driller makes only a perfunctory attempt to clean out a hole during and after drilling. The small amount of time saved on the operation is negated by reduced length of usable hole, by jammed steels and by a poorer quality of hole, which necessitates more time during loading and stemming.

Operations such as setting up can take two minutes for one man and four minutes for another.

Thus the human factor acts as a modifier in both penetration speed and drilling rate. Whether it is justifiable to treat it as a function of the drilling rate is arguable, but facility in one aspect of drilling will usually be coupled with general ability. So the concept of a good (or mediocre or poor) driller is acceptable and general ability can be regarded as being reflected in all operations.

Climatic conditions also affect the rate of drilling. In adverse physical conditions (e.g. tropical climate) outputs will necessarily be low. The untrained indigenous driller in the tropics will often achieve only 50–60% of the footage obtained by a good driller in temperate conditions. As the local drillers become more experienced this differential reduces.

VARIABLES RELATING TO SITE ORGANISATION AND MANAGEMENT

To arrive at a gross drilling rate, due allowance must be made for both recurrent and intermittent delays. Meals and evacuation for blasting are examples of the former; machine breakdown and adverse weather represent the latter. No driller has control over the weather but in a general sense these variables are organisational and reflect upon managerial skills.

Models for Drilling Prediction

A few attempts have been made in the recent past to evolve models which could predict the advance rate of drilling through boreholes. Worth mentioning is the effort of G. Wijk (1989, 1991), who developed the Stamp Test for rock drillability classification. This test yields the rock strength parameters required for rotary drilling rate prediction with the aid of dimensional analysis, whereby a single, non-dimensional constant is fitted to an experimentally determined drilling rate for a particular combination of rock, drill bit, thrust force and rotation rate. The drilling rate equation so obtained is then shown to predict the drilling rate for all other combinations of rocks, drill bits, thrust forces and rotation rates covered by the experiments. If the thrust force in rotary drilling is too low, the roller bit buttons will not break the rock and the drilling rate is vanishingly small. This particular situation is more adequately described by a direct physical model (instead of the dimensional analysis mentioned above) in very close analogy with the Stamp Test itself. It is concluded that the model for rotary drilling thus described yields the accuracy required for practical applications. A simple model for rotary drilling economics is defined. With a reasonable assumption for the rotary bit life as a function of the thrust force and the rotation rate, it is shown that the drill rig should be utilized either at the maximum available rotation rate and a certain optimal thrust force, or at the maximum thrust force and a certain optimal rotation rate.

CHAPTER 8

DOWNHOLE HAMMER DRILLING

DTH Rock Bits
Operating Criteria
Holddown Pressure
Water and Foam Injection
Correct Assembly Procedures for DTH Hammers
DTH Drill Resharpening Procedures
Investigation of Failure Modes
DTH Hammer Componnts

Since the commercial introduction of the down-the-hole (DTH) hammer in the 1950s, the technique has developed into what is arguably the most widely used drilling system in use today. The hammer is used in mineral exploration drilling, mining production, both surface and underground, water-well drilling, construction for anchoring, piling, dewatering and oil and gas exploration. In modified forms the DTH system is used in multiple mounts for large-diameter hole forming with cluster drills and in raised drilling applications in both civil engineering and mining.

The hammer has influenced drill rig development, has permitted the development of new piling techniques and designs, was largely instrumental in providing the hole conditions necessary to exploit the then new slurry techniques in blasting, accelerated development of small portable air booster packages to drill ever deeper holes in a wider range of conditions and in many areas introduced the concept of air drilling into the oil and gas industry.

In the alternative form of percussion drilling with a tophole hammer or drifter, the technique suffers severe limitations in that the percussive energy generated at the surface must be transmitted through the drill string to the drill bit at the rock face. The total impact energy is dissipated by losses at each rod joint and elasticity of the drill string. These losses are cumulative with depth to a point where the impact effect generated at the surface is severely reduced at the bit face.

The tophole hammer suffers further difficulties as a technique in that large volumes of air are required as a dedicated flushing medium, the holes have a tendency for deviation, particularly in long-hole applications, strata changes tend to accentuate the deviation, fractured formations are problematical since the transmission of the percussive forces interact with the loose material in the borehole wall and the drill string components suffer a high rate of attrition, increasing unit cost in terms of drilling consumables and hence cost per foot drilled.

In areas where noise is of environmental consideration, percussion hammers require the exhaust to be muffled, which can lead to a reduction in operational efficiency.

Introduction of the independent rotation mechanism in the 1960s and subsequently the tophole hydraulic hammer has certainly helped to improve the effectiveness of the technique. But though this has widened the options for shallow drilling in smaller diameters (the expensive technology has increased the power output, blow count and impact energy with reduced exhaust noise), the problems of blow energy dissipation, deviation, separate flushing and relatively higher consumable costs still remain.

The drilling operation can be regarded as a chain of disintegration processes in that to produce advance in a borehole, the rock must be subjected to stress levels which exceed the natural strength of the formation, causing it to break away and be evacuated to the surface as drilled cuttings or debris.

In the case of downhole hammer drilling, the hammer generates the forces and the drill bit transmits them to the rock face. Since the hammer is located immediately behind the bit at the bottom of the borehole, the drill string carries no forces other than a comparatively light feed pressure and the cushioned reactive forces reflected through the hammer. In this form of drilling therefore it is the hammer and the bit which are under the greatest stress.

THE DOWNHOLE HAMMER

The downhole hammer is essentially a free-floating piston running in a hollow cylinder. Compressed air is fed into the hammer through the drill pipe and by use of one of a variety of valves or porting arrangements, the piston is caused to reciprocate up and down in the cylinder.

The drill bit is fitted into the base of the hammer, the piston strikes the shank of the bit at the end of each down stroke and the impact energy is transmitted by the bit to the rock face.

The live air used to drive the piston is exhausted at the face of the bit and acts as the flushing medium to both cool and clean the cutting edges of the bit and to transport the drilled debris back to the surface.

The downhole hammer has an air strainer fitted into the top connection to trap any major introduced foreign material in the air stream. A check valve is fitted in the back end to seal the hammer against back pressure and ingress of water into the system, should water be present in the borehole.

The hammer is fitted with a choke, which can be varied in size, to bypass live air through the hammer direct into the exhaust stream. This is used in circumstances wherein the head of water present is such that it could otherwise create back-pressure problems when the hammer is turned back on after making a drill pipe connection. Adjustment of the choke size can be used to regulate the available air supply and ensure optimal conditions between the air activating the hammer and that which is bypassed to the face of the bit for direct flushing. In circumstances wherein either the volume or pressure is on the low side of the hammers designed requirements, the fitting of a blank choke will ensure maximum performance from the hammer in the prevailing circumstances. When the formations are such that water or foam slurry must be injected into the live air supply, the choke enables the fluids to bypass the internal components of the hammer and to be delivered to the point at the face of the bit where they are required.

Various designs are employed by different manufacturers to achieve reciprocation of the piston in the cylinder. These take the form of valves in either a tubular or flap design which, depending on the open or closed position, divert the air supply to either the top of the piston for the power stroke or the underside for the return stroke.

In another hammer design there are no valves (the valveless hammer) and control is achieved by a hollow control tube with ports. The control rod runs through the centre of the hammer and a ported piston is employed which slides over the control rod. The relative position of the porting in the control rod with the porting in the piston controls the direction of the air flow to pressurise the top chamber for a power stroke or the bottom chamber for the return stroke.

DTH hammers are designed to work under a variety of air-supply volumes and pressures ranging from 100 psi through 600 psi air boosters are available.

As a general observation, a valved hammer has a slightly increased piston area vis-à-vis a similar-sized valveless hammer since the valveless piston must sacrifice area to accommodate the central control tube and the internal porting. This in general leads to a higher impact energy per blow from the valved hammer and in circumstances wherein the volume of the air supply is towards the lower end of the register, the valved hammer provides a possibly higher penetration rate.

It will be shown, however, that the factor exerting greater influence on the performance of a DTH hammer is in fact air pressure. The absence of piston mass in a valveless design is adequately compensated by an increased strike count or blows per minute whenever the air supply is adequate in both volume and pressure. In practice, the valved hammer has proved more efficient when only low pressures in the order of 100–125 psi are available; it has a tendency to start of 'fire up' better in such conditions compared to its valveless counterpart.

The valveless hammer (Fig. 8.1) has fewer moving parts and thus a much simpler design compared to the valved hammer alternative, but the final selection of a DTH hammer for a specific application must be a balance between the restrictions of air supply available coupled with the normal considerations of unit and operating cost, operating efficiency, availability of back-up supplies etc.

The drill bit is fitted into a splined chuck and the bit is retained in the chuck by either a key or split ring assembly, depending on the individual manufacturer's design and type of bit shank employed.

A foot valve is fitted into the top of the drill bit to channel the exhausted air from the hammer to the face of the bit for flushing.

The DTH hammer is designed to provide what is termed a 'blow mode', which comes into operation when the bit is in the extended position ahead of the

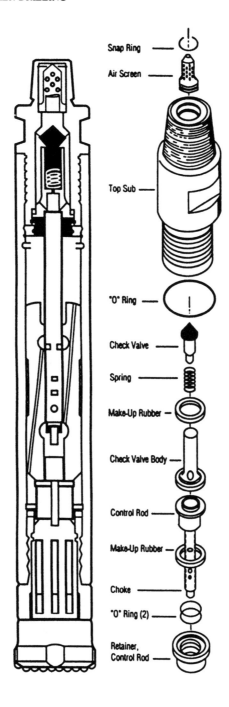

Fig. 8.1: Section through DTH hammer.

hammer. This condition is used when lowering the hammer into a hole prior to commencement of drilling and during the drilling process. If it is necessary to clean cuttings from the bottom of the hole, the hammer is raised a few inches off the bottom and the bit slid forward in the splines. The piston in this situation ceases to reciprocate and travels to the bottom of the cylinder, causing the entire supply of air to be channelled to the face of the bit to clean the bottom of the hole.

The most significant influence on the performance of a DTH hammer is the air pressure supplied and the volume available. Valved hammers are designed to operate at pressures between 100 and 150 psi whereas valveless are designed to run from 125–350 psi or 600 psi if boosters are in use.

The specific performance characteristics of a hammer can be altered at the design stage to meet either a higher or lower operating pressure, the number of blows delivered per minute or the kinetic energy provided by the piston within certain tolerance bands. However, in practice a hammer design is set to provide optimum performance across a wide variety of conditions. This factor should be borne in mind when running comparative performance tests since it would be undesirable to compare the performance of a low-volume hammer against a standard hammer in a low-volume environment. Change the restrictions on the available volume and the outcome would be reversed.

In general terms, provided adequate volume is available, air pressure exerts the greatest influence on a hammer's performance.

The choke assembly is used to match the consumption of the hammer to the available air supply. As the hole depth is increased, the back pressure in the hole increases. This back pressure reduces the effective operating pressure available for the hammer and thus reduces efficiency. The situation is exacerbated when water is present in the borehole, resulting in increased back pressure which must be overcome if the drill hole is to progress. Chokes of different sizes are fitted to the hammer either to bypass air through the system to help lift the water out of the hole in the exhausting air stream or to help raise the operating pressure in the hammer itself.

The DTH hammer will produce a blow count or strike rate (i.e., the number of times the piston strikes the bit) in the order of 1400–3000 blows per minute depending on the design and the air supply available. It is therefore inevitable that wear will take place in the component assembly.

In order to maintain tolerances and alignment as such wear takes place, hammers are fitted with a variety of make-up devices, which differ between individual manufacturers, but perform the same basic function. The design of these make-up systems can be in the form of belville springs fitted into the back end assembly or steel make-up rings with alternating neoprene rings fitted between as an assembly.

When a new hammer is introduced into service or a used one reintroduced, the stand off between the back head and the casing should be in the order of .030–0.050 inches prior to machine tightening. This make-up should be done under power to set up both the correct compression and alignment in the assembly.

Theoretical Design Parameters

Considerable research has been brought to bear on the operation of downhole pneumatic tools to establish the mathematical relationships which govern the theoretical performance of a hammer, given changes in either operating conditions (air volume and pressure) or the physical characteristics of the tool in the form of piston weight, length of stroke, area of piston etc.

Early work carried out in the 1950s established the basic equations and a great deal of empirical fieldwork done to verify that the mathematically deduced formulae were valid in practice.

There is little to be gained from an operational standpoint in providing deep analysis of the mathematical relationships which can be established between the physical properties of downhole hammer designs by way of cylinder diameters, length of stroke, weight of piston etc., since these are variable solely within the ambit of the design engineer. However, it is worth noting that laboratory observations have established the limitations of design parameters in seeking higher performance and efficiency as follows.

Within the design of any pneumatic piston tool there are four variable factors: weight of the piston (W), area of the working face of the psiton, length of stroke of the piston and mean effective operating pressure working on the face of the piston (Fig. 8.2).

Certain assumptions must be made in developing the work output to be derived from the piston, namely:

a) Air pressure during the power stroke is uniform and constant over the working face of the piston;
b) Air pressure on the working face of the piston is the mean effective pressure acting for the full stroke of the piston; and
c) The piston accelerates at a uniform rate during the power stroke.

$$\text{Force on the working face of piston} = PA \text{ (lb)}$$

$$\text{Work on power stroke:} \quad e = \frac{PAS}{12} \text{ (ft/lb)}$$

Acceleration of piston on power stroke:

$$a = \frac{PAg}{W} \text{(ft/s}^2)$$

$$\text{Time of power stroke:} \quad t = \left(\frac{WS}{6PAg}\right)^{1/2} \text{ (s)}$$

Terminal velocity of piston on power stroke:

$$v = \left(\frac{PASg}{6W}\right)^{1/2} \text{ (ft/s)}$$

$$v \alpha \left(\frac{PAS}{W}\right)^{1/2}$$

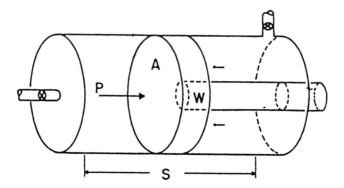

P = PRESSURE (air, mean effective) ON WORKING FACE OF PISTON

W = WEIGHT OF PISTON

A = AREA OF WORKING FACE OF PISTON

S = STROKE OF PISTON

Fig. 8.2: Typical pneumatic tool

Time of piston round trip:

$$T = (1 + k) \left(\frac{WS}{6PAg} \right)^{1/2} (s)$$

BLOWS per minute:

$$n = \frac{60}{(1 + k)} \left(\frac{6Pag}{WS} \right)^{1/2}$$

$$n\alpha \left(\frac{PA}{WS} \right)^{1/2}$$

TOTAL work output per minute:

$$E = \left(\frac{WORK}{BLOW} \right) \cdot \left(\frac{BLOWS}{MINUTE} \right)$$

$$E\alpha(PAS) \cdot \left(\frac{PA}{WS} \right)^{1/2}$$

$$E\alpha \frac{P^{3/2}A^{3/2}S^{1/2}}{W^{1/2}} \left(\frac{ft/lb}{min} \right) \cdot$$

Examining the interrelationships established in the formulae it can be seen that the energy of the piston per power stroke is variable and is modified by changes to either the area of the piston face, the stroke or the air pressure.

The weight of the piston does not affect the piston energy since the relationship between the weight and the terminal piston velocity is $c\ WV;$ as the weight is changed the terminal velocity V is varied inversely $(1/W)^{1/2}$. The energy per blow therefore remains constant.

For any given hammer size there is a working relationship between the cylinder inside diameter and hence piston diameter which can be accommodated. Hence any major adjustments to the equation must derive from either the length of the piston stroke or the air pressure.

Increasing the stroke length decreases the number of blows per minute.

Air pressure thus is the major factor influencing the energy output of a DTH hammer. The $\frac{3}{2}$ exponent of the air pressure means that as the air pressure increases, the piston energy output increases in direct proportion, i.e., double the pressure = double the power output. If the pressure is doubled, the number of blows per minute increases in the ratio of the square root of 2 or 1.41. So, if the pressure is doubled, the number of blows per minute rises 2.82 times.

Although the foregoing mathematical approach establishes the relationships which exist between the design components and their effect on the energy output derived from the DTH hammer, the relationships are in fact far more complex in practice and need to take into account the relative exhaust and inlet areas of the valves or, in the case of a valveless hammer, the relative configuration and location of the ports. Since the high-pressure air is fed into a chamber and expanded, the exhaust area must exceed the feed area by a minimum 60%, rising to 75%. Alternatively the exhaust ports must be open for a longer time period during piston travel to compensate.

When a hammer is run on the surface the exhaust is vented to atmosphere and there is minimal restriction in the air-flow system and the hammer can produce it's maximum rated power output.

As a hole progresses the pressure losses in the system are subjected to a variety of factors which increase their significance to the hammer's performance.

The transmission and frictional losses through the surface equipment (airlines etc.) are constant and can be readily measured by the gauges. The increase in frictional losses through the drill pipe are cumulative with depth and serve to decrease the operating pressure at the hammer.

Table 8.1. Back Pressure.

PRESSURE DROPS ACROSS ORIFICES (DISCHARGE DIAMETER)			
CFM free air	$\frac{3}{8}$ inches	$\frac{5}{8}$ inches	$\frac{3}{4}$ inches
100	35	5	1
150	60	12	5
200	65	21	10
250	110	30	14
300	135	36	22
350	150	48	30

The pressure drop across the orifices in the drill bit (Table 8.1) can be as high as 60 psi and this coupled with the pressure required to produce an air velocity in the order of 3–4000 ft per in the external annulus to transport the drill debris to the surface combined with the increased demand to overcome any head of water present, imposes a high back pressure on the system.

If a hammer is to continue to run at its rated energy output, the cumulative back pressure must be overcome by increasing the primary air supply volume/pressure accordingly to provide the optimum working pressure at the piston face in the hammer.

Energy Transfer

An important factor to be considered in the engineering design of a DTH hammer is the energy transfer between the piston and the strike face of the bit. In the drilling process it is essential that the DTH hammer develop and transmit sufficient energy through the cutting tool (the bit) to generate levels of stress which exceed the minimum strength of the formation.

Exceeding the minimum strength level of the rock produces fragmentation resulting in cuttings which must be immediately removed by the air exhausting through the face of the bit to expose fresh rock to the cutting edges of the bit with the subsequent impacts.

It is important to consider the time period over which the piston and strike face are in contact and the time intervals required for the forces to be transmitted and dissipated prior to the next blow being encountered. If the time interval between blows is less than the time for absorption, the overlapping stress waves could be reflected back into the components of the hammer, creating immense metallurgical problems.

It has been established that energy transfer takes place in fact with considerable rapidity in an order of magnitude measured in milliseconds and that the pattern of rise and fall in stress levels created by the impact dissipates in intervals likewise measured in milliseconds.

The time/stress relationship developed during the energy transfer is variable, however, and is influenced by velocity of the blow (Fig. 8.3). In a high-velocity blow the stress/time relationship indicates a high but rapidly decaying level of stress whilst a lower velocity blow gives rise to a lower level of stress but longer contact time. If the stress level generated by the lower velocity blow exceeds the minimum rock strength, the rock will fragment and an increased amount of rock will be removed due to the increased period of contact.

In practice it is common to so regulate the design as to hold the terminal velocity of the piston at around 45 ft/s. With the stress/time decay measured in milliseconds the hammer blow count could in theory be of the magnitude of 6000 blows per minute without stress wave overlap and in practice 1500–3000 blows per minute would be the more normal hammer count.

There is always a great difficulty in observing with a high degree of precision the specifics of an individual design compared to another since in actual fieldwork we are dealing not in the precise world of the laboratory, but in a highly variable

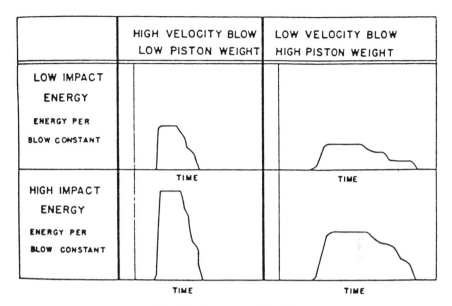

Fig. 8.3: Time/stress relationship.

natural environment where changes in geological formation, the condition of related equipment, climatic environment and individual skill of the operator have a wide ranging effect both individually and collectively. Given this situation it is essential that field data be collected over as wide ranging a set of conditions as possible if objective judgments are to be made. However, the theoretical interrelationships for basic design parameters remain valid

Specific Design Features

The mathematical relationships are very important when a manufacturer designs a DTH hammer with a specific application in mind. For instance, in open-cut mining it is possible to tailor hammer performance such that it will perform more efficiently across a relatively narrow band of drill depths and for a specific set of matching drill bit sizes to optimise performance and production within a given set of parameters.

In exploration or water-well operation the depths encountered may well exceed those of a blast-hole situation. High-pressure air-packs or boosters may well be required and hammer performance characteristics developed accordingly.

Similarly, if production in a reasonably consistent formation is envisaged and the stress levels of the predominant formation can be predetermined, it is possible to ensure a piston weight, piston velocity and blow count that will accurately match that required to provide maximised output.

Such specifically designed products are economic for a manufacturer only when the number of specialised applications is sufficient to provide a financial return. In general terms the tendency is to produce a hammer which is classed as a blast-hole hammer, deep-hole, exploration, high-pressure hammer, general purpose tool etc.

Certain manufacturers produce variations of their specific hammer types with the dimensions of the cylinder outside diameter increased to accommodate increased wear. Alternatively, the cylinder's metallurgy can be changed, either by the choice of material or by heat treatment to provide a superior abrasion resistance. The chuck and back-head assembly can have the outside diameter increased to just below the drill bit size to act as a stabiliser to the DTH hammer and bit assembly, hence leading to straighter holes in long-hole drilling.

It should be borne in mind that the use of such techniques can tend to increase the back pressure in the hole by limiting clearances, which in the case of low air supplies can affect hammer performance.

In some manufacturers specifications the outer cylinder or wear sleeve is designed to be reversible in assembly. The reason for this is an acknowledgement that the major area of wear in the hammer is located at the chuck and the bottom part of the cylinder's external diameter. This is due to proximity to the highest air velocity in combination with the maximum quantity of freshly cut and therefore irregularly shaped cuttings to create maximum abrasion. As the cuttings rise up the hole they are more often milled in the process and gradually assume a more rounded profile. When the cylinder has reduced in diameter at this bottom point, it should be reassembled in the reverse position when the hammer is next stripped out. Care should be taken to ensure that any such wear which is not that excessive that it will weaken the internal thread causing splits due to incidental hoop stresses in the reduced diameter.

The practice of building up the external diameter subsequent to heavy wear is always to be avoided since with even the most rigorous pre-heat and post-heat procedures, which are seldom carried out in the field in any event, any applied heat will change the carefully controlled microstructure of the metallurgy of the cylinder and will inevitably produce an alteration in the roundness tolerances in the inner cylinder, leading to a binding piston and premature failure of both the piston and the cylinder.

DTH ROCK BITS

The operating conditions experienced by the percussion rock bit and the task it is required to perform present one of the most demanding work environments for any manufactured component. Trapped between the percussive hammer and the rock face, the drill bit is required to transmit all the drilling forces to the face of the drill hole, penetrate the rock and to endure the abrasion of the drill debris and high-velocity air working its way past the body of the bit into the outer annulus for transport back to the surface.

To perform in this environment requires that the bit be correctly designed and constructed and that the materials be of high and consistent quality. The bit body is produced from a high alloy steel forging, which was developed to withstand the heavy impact loads imposed on the landing gear of large jet aircraft. The forging ensures good grain flow lines within the microstructure of the material; it is also heat treated to produce the toughness and durability required from the finished product.

All machining must be of the highest standards. Surface finishing is of paramount importance since any surface flaws or machine marks will introduce stress risers in the product during operation.

The cutting edge of the drill bit consists of tungsten carbide inserts located at the face of the bit. Depending on the bit type, the inserts may be in the form of single slabs of carbide cross-mounted in an 'X' form but cylinders of tungsten carbide with their ends preshaped to a hemispherical profile are gaining in popularity.

In the case of the cylindrical form the DTH bit is termed a button bit and the physical composition and characteristics of the buttons can be regulated to provide specific design features (Fig. 8.4).

In the the button bit cemented carbide inserts are fitted into the bit body by pressure. The hole in the bit body to accommodate the insert must be perfectly parallel and the internal finish of mirror quality. Any departure from this high standard of finish would produce an interactive fit in which the tungsten carbide inserts would be pinched at the high point, leading to a point of deformation and bending stresses at the point of contact. This in turn would lead to premature failure in operation since with the extreme hardness of the tungsten carbide there is an associated brittleness and low tolerance to bending stress.

Tungsten carbide is an alloy of tungsten powders together with a cobalt medium which is fused together under extreme temperature and pressure to form a material of uniform through hardness and abrasion resistance. The characteristics of the material can be regulated to provide specific and consistent qualities by specifying the mesh sizes of the carbide, density of the mix and additional alloying mediums used with the cobalt in the production. Manufacturing should employ the hot isostatic pressing or "hipped" carbide process to ensure a consistent quality in the material best suited to those conditions found in the drilling operation. Cemented carbide always suffers some adverse effect on its mechanical properties due to residual porosity during the manufacturing process. Hot isostatic pressing helps in overcoming this difficulty. The process involves reheating the material after initial infiltration and recompacting to some 2000 bars in an argon-enriched atmosphere. The result is a product with a far greater wear resistance and a transverse rupture strength.

Borrowing from a technology developed for the tungsten carbide inserts used in the roller bits for the oil industry, there are new insert materials being introduced into the DTH drill bit industry in an endeavour to increase the inhole life of the inserts.

In the latest developments a new element is introduced into the equation to improve the wear characteristics in one case by adding a cap of synthetic diamond to the outer cutting surface of the insert, and in another new technology a synthetic abrasive in the form of cubic boron nitride is introduced as a component part of

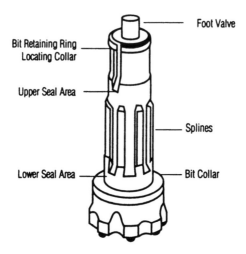

Foot Valve

Bit Retaining Ring
Locating Collar

Upper Seal Area

Splines

Lower Seal Area

Bit Collar

Fig. 8.4: DTH hammer bit.

the alloy construction of the tungsten carbide to produce greater wear resistance qualities to the material.

Drill Bit Design

Within any drill bit design there are certain features which can be varied either in whole or in part to "tune" the performance of a drill bit to meet specific conditions. The flexibility achieved in the design features of the insert button bit has led to improved economies in drilling costs, totally displacing the "X" bit in many areas, and expanding the areas of application for DTH hammer application.

The ability of the bit designer to tune the performance of the bit to suit a given set of conditions has led to individual mines being able to specify a 'mine standard' design which from field testing has been developed to match specific requirements. These requirements can be set by both geological conditions or known difficulties, such as low air supply, excessive water flows or a host of other circumstances which cannot be readily changed at the mine site.

The design variables of the the insert bit encompass such features as the following:

Tungsten Carbide Inserts

Type of insert (including new technology)
Grade of Material
Size (diameter of insert)
Shape (spherical or domed section profile)
Position (location in face of the bit)
Protrusion
Body Length (determines length of grip in the bit body)

Bit Body Configuration

The profile of the bit face can be varied to suit specific rock types and operational conditions and manufacturers offer a range of shapes:

Flat Face
Concave
Drop Centre
Dome
Gauge Protected (a feature used in long-hole drilling where undersizing is the predominant wear pattern rather than insert wear)
Relieved (used to assist larger fragmented cuttings to clear the bit face)

Waterways

Number
Shape
Location

Drill Bit Shank And Bit Face

Carburised.
Tempered.

In order for any bit design to work correctly and provide cost-effective performance every bit must have the right balance and combination of the above-listed variables.

The carbide inserts protrude above the bit body. This exposure promotes efficient rock breakage and hole cleaning. Ideally in most ground conditions, good bit design will give a flow of rock cuttings over the face of the bit. This flow of cuttings will erode the base metal away at a rate similar to the wear of the carbide buttons. However, if the protrusion is too great, the inserts might shear off due to the side stresses involved.

Conversely, if a button has too little protrusion from the face of the bit, the airflow across the face of the bit is restricted, which has a detrimental effect on hole cleaning and leads to accelerated wear on the face of the bit. This in turn can cause premature failure by overloading the bit and the carbide inserts.

Bit Face Design

Manufacturers offer a range of face designs with their bits to meet a wide variety of drilling conditions. Where an individual mine site has sufficient consumption, it is possible that this could justify the development of a specification tailored to provide optimum performance at that location. In general terms, however, the more normally available off-the-shelf product would fall into four basic designs.

The flat-face design (Fig. 8.5) is designed to work in hard solid rock formations with medium to high compressive strengths.

The concave design (Fig. 8.6) is more of a general purpose bit and the more generously proportioned airways permit fast evacuation of any larger cuttings generated away from the face of the bit.

Flat Face or Flat Bottom Button Bits

Recommended for:

- banded formations
- consistent penetration
- deep hole drilling
- hard formations

Fig. 8.5: Flat face.

Concave Button Bits

Recommended for:

- general ground
  conditions
- excellent penetration
- bottom hole cleaning
- hardest formations

Fig. 8.6: Concave.

The drop-centre design (Fig. 8.7) is effective in softer type ground where clay bands exist. This face configuration will provide substantially higher penetration rates than other designs in the recommended formations but it should not be regarded as a general purpose bit for all grounds. In this type of design the protrusion of the carbide is often greater than in other designs, making it vulnerable to high side loadings if run in formations of high compressive strength.

The double-gauge row design is used in areas where gauge wear is a major factor and in situations in which the continual tripping of bits to maintain both hole diameter and adequate penetration is required.

Hole openers can be provided in various sizes. These are intended to be used in situations wherein a pilot hole has been previously drilled and the diameter has to be opened out. Their success varies according to the ground involved and the efficiency of the hammer used. If could be accommodated, this would often be more cost effective as a full size hole could be drilled in the first instance. However, circumstances do not always permit employing this technique and the hole opener is a valid method of creating a larger diameter hole with a smaller hammer.

It should be noted that if the suggested technique is used, mismatch of hole size/drill pipe ratio will be apparent in terms of uphole velocities. This problem does not arise with a hole opener as the cuttings are permitted to fall into the pilot hole.

In practical terms the greatest danger is the use of too large a diameter of bit face on too small a diameter shank design. This can set up bending stresses which will quickly lead to shank failure and the problem of having to fish the oversized head of the bit and pilot from the base of the hole.

Drop Center Button Bits

Recommended for:

- medium air pressure
- excellent flushing
- maximum penetration in soft to medium formations
- broken formations

Double Gauge Bit

Recommended for:

- high air pressure
- abrasive formations
- medium to hard formations
- deep holes

Step Gauge Bits

Recommended for:

- high air pressure
- medium hard formations
- abrasive banded formations
- broken faulted formations

Fig. 8.7: Drop centre

Depending on size ratio, the head may be a parabolic profile or be fitted with a blank dolly which acts as a socket to follow the predrilled pilot hole. Generally speaking, the parabolic configuration is equally as accurate in following the line of the pilot and provision of tungsten carbide inserts around the centre ensures break up any bridges or hang-ups which may occur in the pilot hole.

OPERATING CRITERIA

It is imperative that sufficient air be provided at all times to ensure efficient hole cleaning. If a bit is continually running on a cushion of cuttings, the progress of the hole will be impeded and harmful stresses set up in both the drill bit and the hammer system. Failure to blow a hole clean before pulling a bit can lead to bit blockages in the subsequent bit run.

The DTH hammer is so designed that as the bit is lifted off the bottom it slides out of the chuck housing and with the piston at the bottom of the cylinder, the hammer bypasses all the air into the face of the bit to clean the bottom of the hole.

For this reason it is essential that once the hammer and bit are being tripped either into or out of the hole, the air supply should be reduced to avoid damage to the wall of the borehole by the air stream.

In all formulas for hammer design air pressure is the most important criterion for efficient operation and maximised production from the hammer. However, the weight on the bit and the rotational speed provided by the drill string carry an equal significance in the overall performance. If a bit is run in a lightly loaded condition, production is lost and the impact energy output of the hammer is absorbed into the other drill component parts and the drill string. This will cause metal fatigue, which in turn results in damage to the hammer components, causes inserts to pop out from the face of the drill bit and inflicts damage on both the face and shank of the drill bit.

Rotational speed is another area which requires accurate control in order to avoid damage to the DTH hammer and bit. Too fast a rotational speed will accelerate the bit wear and increase the risk of gauge row carbides shearing. Too slow a rotational speed will lead to excessive bit wear and peening of the bit.

Every bit should be inspected for wear and damage after each run. Any problem, such as flatted or cracked buttons, should be attended to before further problems arise.

If broken carbides are left in the hole following a bit run, these must be fished out before running another bit into the hole or premature failure of the new bit will occur.

Air Pressure Demand

It has already been demonstrated that air pressure exerts a significant influence on the performance of the DTH hammer, be it valved or valveless in design. Provision of a high-pressure compressor in the operating range from 250 to 350 psi is virtually an industry standard wherever efficient performance is essential. Considerable effort has been expended in developing DTH hammers which harness the production potential of high-pressure air whilst concomitantly reducing the total volume of air demanded.

It is important to recognise in practice that gauge pressure at the compressor is not the same as pressure at the hammer. A high-pressure compressor still requires attention to the basics in good maintenance and design in surface equipment.

Moving air requires energy to move through the system; measurements of air pressure at any two points in the system will never be identical.

Air hoses should retain as large a diameter as practicable and bends should be as smooth as can be accommodated. Small-diameter hoses and fittings rob the system of pressure energy and should be minimised regardless of cost.

Air pressure in the system can be read at various points either by provision of separate gauges set into the system or hypodermic needle gauge in the airline. It is necessary to check the system with the hammer running in order to ascertain where pressure drops are occurring within the system.

For efficient hole cleaning and transportation of the cuttings back to the surface, it is essential that a minimum annular velocity of 3000 ft per minute be obtained.

This minimum figure can rise to to 7000 ft per minute as the hole size increases or in certain formations with a specific gravity value in excess of 3.

The uphole or bailing velocity can be determined from the formula:

$$\text{Uphole Velocity} = \frac{\text{CFM} \times 183.4}{\text{BIT DIA}^2 - \text{PIPE DIA}^2}$$

Alternatively, most manufacturers of DTH hammers will provide charts in their brochures which permit direct reading of the air volume requirements based on various combinations of bit sizes and drill pipe dimensions.

In general, all hammers will pass sufficient air during the drilling cycle to maintain an adequate uphole velocity. However if the air demand to obtain the required bailing velocity exceeds the operating air demand of the hammer, an appropriately sized choke must be fitted to bypass the additional demand through the hammer to maintain the uphole velocity.

Field observation of the cuttings returning to the surface will provide a reasonable indication as to correct bailing velocity.

Having established the theoretical demand and fitted an appropriate choke, the cuttings returning to the surface should be observed. These should exit the collar of the hole in a steady flow and with sufficient velocity to cause a stinging sensation to the hand. An irregular return which ebbs and flows is an indication of insufficient volume.

The size and shape of the cuttings will also provide an indication of both bottom hole activity and the overall efficiency of the cutting system. The cuttings returned should be of reasonable size and their shape should indicate their having been freshly spalled from the formation. Rounded cuttings and powdered fines indicate regrinding by the bit. This means that penetration rate is being sacrificed and bit life reduced due either to an inappropriate bit type, a dull bit requiring attention or insufficient air volume to lift the cuttings away from the face of the bit as they are spalled.

Air is an easily compressed medium and in designing the combination of the drill string and hammer assembly this should be taken into account. The fitting of stabilisers either to the hammer or further up the drill string causes a physical restriction which in turn creates a back pressure.

Back pressure in the external annulus should be held at as low a value as possible if maximum uphole velocity is to be maintained. A 15 psi increase in back pressure reduces annular velocity to almost 50% of that which would be obtained if no back pressure were present.

Similarly, hole enlargement through too high an annular velocity reduces the uphole velocity substantially. It may be recalled that in the equation to determine the bailing velocity, the square of the hole diameter is the divider.

Even in the most arid areas it is common to strike water in hole of some magnitude. The volume of water present will affect the back pressure in the hole and requires consideration.

Calculation of the back pressure of water assumes the following:

A 230 ft column of water exerts a back pressure of 100 psi at the bottom of the hole, i.e., every 2.3 ft = 1 psi of back pressure, or 1 ft = 0.43 psi back pressure. If therefore a hole is making water to the extent that when a drill pipe connection is made 75.25 psi is developed in the hole, and the amount of pressure required to blow the hole clear would be:

$$\text{Water column 150 ft} \times 0.43 = 64.5 \text{ psi.}$$

This is termed the unloading pressure and equates to the pressure which must be overcome by the flushing air from the hammer before the latter will 'fire' or commence operation. If the water inflow is less severe than the evacuation rate, the operating pressure demand will fall away as the water is removed from the borehole. If the water inflow is such that the head of water is continually fed, the unloading pressure will increase with depth accordingly and must be allowed for in determining the capacity of the compressor.

When the cuttings are in suspension in the water, this increases the water density and the back pressure accordingly.

The back pressure and peak unloading pressure may well prove to be determining factors in how deep a hole may be drilled using standard methods. Further progress may require the introduction of either a booster into the system or a foam injection into the air stream to help lift the water from the hole.

Holddown Pressure

When a bit and hammer assembly is run into the hole, the bit extends from the chuck and assumes what is known as the 'blow mode' in which any air supplied to the hammer will bypass the latter's internal operating ports and pass straight through the face of the bit. When the hammer reaches the bottom of the hole, the bit will nest back into the chuck closing the bypass path with the foot value and the hammer will go into cycle or the hammer mode.

The weight required by the drill bit to operate correctly is relatively low compared to bit weight demand of other drill systems drilling a similar sized hole. The general rule for a DTH hammer bit is in the order of 500 lb per inch diameter. Any increase in holddown pressure above this level serves only to increase wear on the bit and leads to a higher rotational demand without increasing the penetration rate.

As the drilling depth increases and hence the weight of the pipe in the hole, it will become necessary to compensate by withholding weight on the bit to avoid overloading.

In relatively shallow holes the operator may detect a low bit weight condition by a 'bounce' in the drill string. With increasing depth and in larger pipe sizes this effect is less easily detected and the operator must rely on the rotation pressure gauge to provide an indication of overloading.

When drilling with the drill pipe in tension to avoid overload, it is essential that the bit not be allowed to become extended into the blow mode position by extending out of the chuck. Running a bit in this lightly loaded condition will create problems at the bit due to heat generation, which can be sufficiently severe as to cause changes in the metallurgical structure of the bit body and crack the shank or alternatively the buttons may be loosened and pop from the seating in the bit face due to distortion or expansion of the grip areas.

As a guide to the efficient application of bit weight the following procedure is recommended:

1) Increase the down pressure until the air pressure ceases to increase.
2) Note the hydraulic feed pressure at this point.
3) Reduce the down pressure until the air pressure starts to decrease and note the hydraulic feed pressure at this point.
4) Maintain the operating parameters between these two settings.

Observing the above procedure will keep the hammer operating with sufficient weight to keep the bit closed during drilling and should provide an optimum reading on the rotational pressure gauge.

Table 8.2 provides bit weight indications for individual bit sizes as a general guide.

Rotational Speed

The correct rotational speed is a most important factor in achieving both optimum penetration speeds and a good bit life. Bits are readily damaged when either too high or too low a speed is used. The selection of too high an rpm produces bit scrubbing and premature breaking of inserts by imposing high side loads on the extremely tough but brittle cemented carbide inserts. High rpm increases gauge body wear, thus exposing the gauge buttons at an inequitable rate.

Too low an rpm results in premature insert failure due to their bogging or burial. Burying of the insert leads to slow penetration due to failure to attack fresh rock at each blow of the hammer. Too slow an rpm results in an erratic and irregular rotation imposing side loadings on the inserts with consequent premature failure,

Table 8.2

Bit Diameter (Inches)	Weight on Bit (Pounds)
$3\frac{1}{2}$	1750
4	2000
$4\frac{1}{2}$	2250
5	2500
$5\frac{1}{2}$	2750
6	3000
$6\frac{1}{2}$	3250
7	3500
$7\frac{1}{2}$	3750
8	4000
$8\frac{1}{2}$	4250
9	4500
$9\frac{1}{2}$	4750
10	5000

damage to drive splines of the bit and accelerated wear on the chuck components of the hammer.

A combination of high or low feed with high rotational speed will rapidly damage or dull the bit. The rpm employed should be adjusted to suit drilling conditions and the hardness and drillability of the formation. Generally, the harder the formation, the lower the rpm required for efficient performance.

Allowance should be made for the bit diameter in use. Bits with a larger diameter may experience high rates of wear and sustain damage to the inserts if run at too high a peripheral speed.

The proper rotational speed is always a compromise between the rate of penetration and the life of the bit. The operator must establish an optimum speed which provides a smooth rotation and an economic bit life.

As a guide to ascertaining the correct rpm range it is recommended that the bit should complete two revolutions per inch of penetration. In other words, if the penetration is 12 inches per minute, the indicated rpm would be in the order of 24 rpm.

Observation of the drill and the drill string performance is the best guide for an operator who understands the drilling process. The drilling pressure employed will interact with the rotational speed. Too high a pressure causes a jerky and uneven rotation and increasing the rpm may only aggravate the condition. The final selection of the combined rpm and feed pressure must always be that which provides the smoothest running consistent with an acceptable rate of penetration and bit wear.

Lubrication

As has been illustrated, the DTH hammer is a closely toleranced piston running at high frequency in a hone finished cylinder. It carries out the task of converting the supply of compressed air into work energy at the bit under the most stressed conditions imaginable. To ensure that the hammer continues to function under such circumstances it is essential that it remain lubricated at all times with the correct formulation of rock drill oil at the prescribed rate. Failure to observe this requirement will result in premature and expensive breakdown in the equipment.

Rock drill oil is specially formulated for this purpose and is the only type of oil recommended. Rock drill oil is characterised by a high film strength which will not break down under heavy loads. The correct viscosity ensures that the oil will flow freely under a wide range of temperatures. Good emulsibility precludes the oil film being washed off the running components by the inevitable moisture present in the air stream.

The oil must have good adhesive properties to protect metal surfaces which would otherwise tend towards corrosion in the air/moisturised conditions. Finally, the oil must have a high flash point to avoid dieseling when the drill is operating.

Most hammers designed to drill in the 6-inch hole size range require a continuous minimum oil supply of 2 pints per hour or 2 gallons per 8-hour shift. Oil consumption will vary as the hammer size increases. The general rule of thumb is injection of $\frac{1}{3}$ pint of oil per hour for each 100 cfm of air consumed by the hammer.

In practice, the oil should be evident as a light film on the bit shank whenever a hammer is pulled and at each joint as a drill pipe is broken out.

On most rigs designed to run with DTH hammers the oil is supplied through either a dedicated injection plunger pump or a venturi-type oil bottle. Both types of injection system incorporate adjustable delivery which can be utilised to increase the supply should circumstances demand it in, say, low ambient areas where the natural flow rate may be retarded or when injecting water or foam into the air stream.

A point which is often overlooked in the periodic maintenance schedules is the filter and metering orifice associated with the oil delivery system. After 1000 hours the system should be drained and the filters and needle valve cleaned. In the case of the venturi-type system, the oil bottle should be drained and the bottle steam-cleaned to remove the accumulation of oxidised oil sludge on the bottom.

It should be noted that with the venturi-type system, the oil container is under constant pressure when the compressor is in operation and on no account must the top be removed without first evacuating the air pressure.

WATER AND FOAM INJECTION

The practice and application of injecting either water or foam into the air stream either for environmental reasons as a dust suppressant in the case of water or as a means of increasing the efficiency with which water is removed from the borehole, may be safely carried out whilst drilling with a DTH hammer system provided certain basic rules are observed.

In injecting either medium, it is necessary that it be introduced into the system downstream of the oil injection.

It is preferable that an open choke be fitted on the hammer to provide a more direct route to the bit than running through the hammers internal ports. It is essential that the water or water phase of the foam solution be as pH neutral as possible to avoid initiation of corrosion in the hammer and the drill pipe. The oil injection level should be raised to ensure retention of the lubrication film, and the hammer and the drill pipe must be blown out at the end of the drill shift if an accelerated rate of corrosion is to be avoided.

Use of a straight water injection is normally applied whenever difficulties are experienced in drilling due to a slightly moistened formation that tends to ball the cuttings and cause intermittent circulation stoppages. Injection of sufficient water into the system to create a wet rather than moistened environment will assist circulation of cuttings up the hole.

In areas where dust is of major concern or where health policy mandates that dust be suppressed, a simple water injection will suffice to bring the situation under control.

In areas where high water inflows are experienced, which reduce the efficiency of the air-circulation system, the use of foam is advocated.

Drill foam is a high molecular strength polymer formulation which, when mixed with water in low concentrations and injected into the air stream, is transported to the expansion point at the face of the bit. As the air expansion/pressure drops the mixture is transformed into tightly compacted bubbles of foam with a good surface tension. The effective back pressure of the water is reduced by the introduction of foam bubbles and the water is more readily carried to the surface on the interactive surface tension of the foam, which replaces velocity with viscosity in acting as a transportation medium for the water and drill cuttings.

Drill foam is both biodegradable and non-polluting. It may be safely used in areas which are later to be developed for water supplies or in areas where primary aquifers may be present.

As a rule, the bailing velocity required is reduced to approximately half that required for a straight air-circulation hole.

The foam should be injected through a separate injection pump to ensure correct mixture and control of air/foam mix ratio.

The rate of injection is varied according to observation of the returns and will vary according to circumstances and combinations of hammer size, bit size, formation, rate of penetration and volume of water in the hole.

It should always be remembered that the reaction of the airborne fluid phase and transition into a foam phase take time to develop and stabilise. It is important not to overreact to an immediate lack of indicated response, therefore. Too much foam injection can be detrimental and may flood the hammer.

When the use of foam is anticipated, it is advisable to set a blooey line to conduct the foam away from the top of the hole collar since it takes time to dissipate once back on the surface.

Do not attempt to step up the rates of penetration; remember the surface indicators are time lagged and not as immediate as when using direct air circulation. Do not be tempted to bury the hammer and bit as a consequence of this delay.

Maintain the air supply at as low a level as is required by the hammer; an oversupply merely serves to channel the foam and destroys its carrying capacity.

Although the foam will stabilise a hole over a short period of time, the bubbles will decay over time. So sufficient material must be premixed to ensure continuity of injection.

If the surface barriers do break down in the bubbles due either to too weak a mix or a prolonged period of dormancy in the borehole, the cuttings in transportation will fall back into the hole and create a potential blockage. It is customary this provides after each pipe connection to increase the injection for a short time; a slug of new foam in the hole which helps clear any such blockages.

The mixture of air and water is a highly corrosive one and good housekeeping practice dictates that all the equipment be thoroughly purged and the hammer stripped and lubricated at the end of each foam-drilling operation.

In any form of DTH drilling it is important to pull a hammer back from the bottom prior to standing over night; this usually means at least two pipe stands.

In foam drilling it is important that the hammer be tripped out totally since the corrosive attack of the combined air and water could damage it. When a stabilised foam is in use it acts to keep a potentially caving hole open. As the bubble phase decays, it is possible that the hole side may become unstable and fall back on any equipment left in the hole. Removing equipment may appear an inconvenience but the short time spent can preclude devoting an entire shift to recovery of circulation and stuck equipment.

CORRECT ASSEMBLY PROCEDURES FOR DTH HAMMERS

Before a hammer is put into operation, it is important to carry out an inspection of the internal assembly. It is normal practice for a manufacturer to supply a hammer with blank chokes in place to ensure the best performance from the available air supply. However, a blank choke may not be the most appropriate under the circumstances in which the hammer is to be operated.

Connections at the back head and hence the make-up assembly will normally be supplied handtight only to facilitate the prerun inspection. The hammer should be thoroughly lubricated and reassembled and the back head torqued prior to running in the hole. Similarly, the new bit fitted into the chuck should be checked to ensure it slides freely, that the bit retainers are present and correctly fitted, and the chuck should be torqued again prior to entering the borehole.

Any time a hammer is returned to a base or a workshop for repair the top and bottom connections should be broken out on the rig prior to transporting. Although many major workshops may have breakout facilities, this may not be so in all areas and even if such facilities are present, their capacity may be below that available on the rig. In the assembly and disassembly process it is essential that the correct tools be used and correct procedures adopted.

In the following visual assembly an SDS Digger Tools valveless hammer is used as a model but other manufacturers will require a similar approach:

A With the cylinder lying on its side, use a pair of 90-degree circlip pliers to insert the piston retainer ring, raised lip first into the cylinder. Ensure that the raised lip sits in the groove against the shoulder.

B From the other end of the cylinder insert the piston, nose first (small diameter) but do not slide to the other end of the cylinder.

C Place 'O' rings on the outside diameter and inside diameter of the control tube mount.

D Slide the control tube mount make up rubber over the control tube and assemble the control tube make up rubber and mount together.

E Holding the mount assembly together, slide the control tube stem into the piston and push the mount into the cylinder assembly until it sits against the shoulder in the cylinder.

F The control tube mount assembly shown sitting on the shoulder in the cylinder.

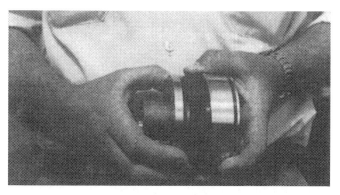

G Place the check valve guide "O" ring onto the check valve case O.D then the rubber make up ring into the recess in the check valve case, with the steel make up ring fitted over the top.

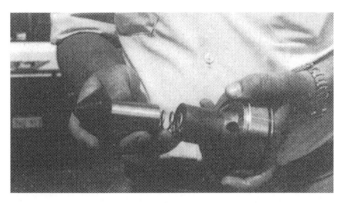

H Install the check valve spring and check valve plunger inside the check valve guide.

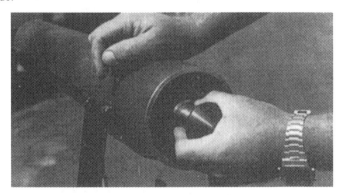

I Place the check valve guide assembly into the cylinder until it sits on top of the control tube head.

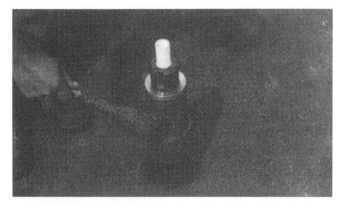

J Ensure that all threads are coated with proper thread compound to prevent galling.

K Screw in the top sub and tighten by hand until snug. The stand off between the shoulder of the top sub and the end of the cylinder should fall in the range of (0.040″–0.090″. 1 m/m-2.29 m/m). Failure to achieve this correct stand off will cause damage to internal parts.

L Assemble drive sub, split rings and "O" rings onto the shank of the bit.

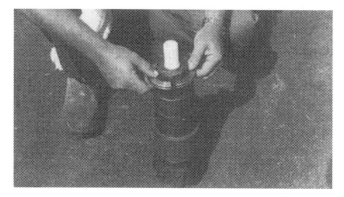

M Install bit/drive sub assembly into the lower end of the cylinder using proper thread compound.

It is important to note that heat should never be applied to any hammer surface as a means of freeing up a connection. The materials selected for use in DTH hammers and their subsequent heat treatment are carefully controlled in the manufacturing process and any subsequent application of heat will result in premature failure.

Good housekeeping requires that all inhole equipment be carefully inspected both prior to and after conducting any drilling operation, especially the DTH hammer. A hammer is a high-precision instrument required to perform its task in the most arduous and hostile conditions for hours on end. It will repay any care and attention with increased life. In dry holes the hammer should be stripped out every 250 hours and in wet holes every 100 hours. When stripping out for inspection, look for wear in check valves, springs, exhaust tubes, make-up rings etc. The policy should always be 'tear out before wear out'.

All internal components should be carefully washed down, dried and inspected for wear or heat marks before lubrication and reassembly in the hammer.

Good practice requires similar attention to all other components in the drill string.

The drill pipe should be stored with the threads greased and the end caps in place. Before the pipe is put back into operation after storage, it should be blown through to remove any accumulations of scale or rust. When temporarily stored on the rig whilst drilling is taking place, the ends should be raised from the floor level and if no carousel or rod rack is used, care must be taken to avoid damaging the threads.

DTH DRILL RESHARPENING PROCEDURES

To maintain drill bits in a condition that will continue to provide maximum performance, it is essential that attention be given to the state of both the bit body and the cutting inserts, be they of the 'X' design or insert button bit variety.

A major cause of component failure in both the DTH hammer and the drill bit can be directly attributed to overrunning the drill bit, i.e., operating the bit after the cutting edges have gone beyond the limits of their cutting efficiency. Continuing to drill with a dull bit reduces the penetration rate and imposes high stress loads on the system.

If a drill bit is kept in service much beyond its efficient cutting condition, the damage will rapidly achieve a level at which the entire bit is an unrecoverable proposition.

Wear on the drill bit is manifested in two ways: the gauge of the bit and the chipways are reduced in size and the inserts become flattened.

Reconstituting the bit to a useful condition requires close attention and an understanding of what wear has taken place and hence the requisite treatment.

In the case of 'X' bits it is the 'roof angle' which wears down and the cutting edge presents a gradually widening flatness. This should never exceed $\frac{3}{32}$ inches at the extreme outer edge.

Gauge wear of the insert is generally indicated by gradually rounding corners. Associated with both types of wear, the inserts may show a crowning effect, wherein the sectional angle of inserts falls away from the centre to the outside diameter (Fig. 8.8). If the crown angle exceeds 5°, the bit will fail to cut.

Gauges are available which facilitate measurement of the wear areas and enable an operator to restore cutting angles back to condition.

The 'roof angle' for the insert should be 110° and the inserts should all be flat and level when returned to service.

When the antitaper on the bit body (Fig. 8.9) has reached a height of $\frac{1}{4}$ inch, the body will require grinding back. After restoration, all the inserts should be of equal length from the centre point to the outside diameter and the chipways should be enlarged back to their clearance dimensions as presented on a new bit.

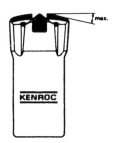

Fig. 8.8: Crowning angle.

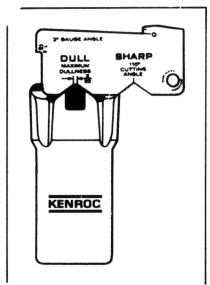

Roof angle sharpening is required if cutting edge wear is 3/32" or greater.

Fig. 8.9: Roof angle

The grinding should be carried out in a controlled and careful manner whilst observing the usual safety precautions required when using hand-grinding equipment. The grade of the grinding wheel should be neither too hard nor too fine-grained as this can cause clogging of the cutting structure of the wheel, leading to a rapid heat build-up and consequential damage to the metallurgy of the insert.

The use of too coarse a wheel can leave scratch marks which in turn will become stress risers in the carbide and promote rapid failure.

It is generally accepted that a button bit will operate at a level of efficiency some 3 to 4 times that produced by an 'X' bit. However, the insert buttons will wear and require regular inspection and attention to maintain the drill bit at maximum efficiency.

With the insert bit the buttons themselves wear to a flat point. The buttons located in the gauge areas will by and large develop flats faster than the more centrally located buttons.

For softer formation bits a wear of $\frac{1}{4}$ inch flats can be tolerated, in medium formation bits this is reduced to $\frac{3}{16}$ inches whilst in hard formations $\frac{1}{8}$ inch is the maximum (Fig. 8.10).

In reconditioning an individual button, the requirement is to bring the surface section back to a rounded profile. The smaller the flat allowed to develop, the less stock required to be removed and the greater the overall extended life.

For insert buttons to be re-ground on site, hand-held grinders are now available with both specially profiled grinding wheels and an in-built coolant supply arrangement to ensure that stock removal does not cause detrimental heat build-up.

Special grinding tools are available for various button sizes as are wheels which enable manoeuvering the button grinder around the button to obtain the required spherical profile.

In extremely hard drilling conditions, the heat generated at the bit face can cause a breakdown in the basic structure of the carbide, creating a snake-skin appearance on the surface of the button or insert. If this condition is noted, it is essential that the surface be ground back to good condition even if this leaves an individual button below the level of the others. A snake-skin appearance is an early sign of a breakdown in the cemented carbide and if left unchecked the surface will flake off and cause subsequent damage to the other inserts.

Similarly, if an individual button is cracked through or has a piece missing, it is better to grind it flush with the metal surface than to allow it to crack further or break away completely.

The basic method of reconditioning is to mark a spot in the centre of the flat and then grind down from the centre point using a rolling motion to restore the rounded configuration.

The waterways and chipways should be ground back to full depth to ensure smooth release of cuttings when the bit is returned to service.

The diameter of the steel head is always slightly smaller than the gauge diameter, which is measured across the outer gauge buttons. The head is also tapered back towards the shank of the bit. This gauge relief, or difference in gauge between

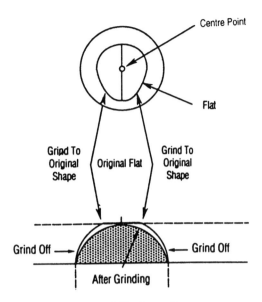

Fig. 8.10: Button flats.

the bit head and the outer row of gauge buttons, must be maintained to prevent bit binding in the hole. To do this, grind a band about $\frac{3}{16}$ inch wide around the bit, just below the gauge buttons, until the gauge buttons stand proud by about $\frac{1}{32}$ inch. Then grind a smooth taper back towards the shank. This taper must not run back into the drive splines or preload areas. (Fig. 8.11).

Most bits can be reconditioned a number of times, but the limit is reached when the the original gauge diameter is reduced by $\frac{1}{2}$ inch (Fig. 8.12). Any further regrinding will reduce both the thickness of the gauge button and the metal holding it in place.

It is essential to have a series of bits available on the drill rig; these should be ringed and marked with any reduction in diameter. The bits should be put into service in descending order in the case of long holes in areas where an individual bit will not last for the duration of the hole. Obviously, running a full-diameter new bit into a hole previously drilled by one of reduced outside diameter should always be avoided as the pinching of the new bit in reaming to the bottom will severely reduce its life and may cause the buttons to pop out of the face when hammering in a non-loaded condition.

INVESTIGATION OF FAILURE MODES

All major manufacturers operate with an extremely high level of in-house quality control to ensure a consistency in componentry, product and product performance. Although failure in a component can creep into the system and indeed batch failure could occur if one area of the control system were faulty, such failures are rare.

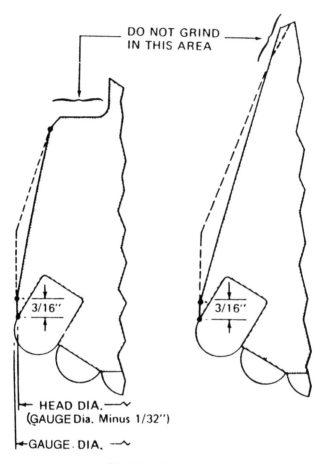

Fig. 8.11: Bit body taper.

It is more probable that component failure will be due to some external factor. Understanding some of the failure mechanisms which can be engendered in the drilling process, provides an explanation for many of the more commonly seen component problems.

It should be recognised from the outset that no driller will admit to equipment abuse. This is evident in an examination of some of the parts offered for warranty investigation. The term abuse is relative, however, and the objective of every manufacturer is that this equipment wear out rather than fail. Every drilling component has a finite life and this life can be extended or foreshortened by the drilling practices employed. By understanding the drilling process and recognising the classic trouble signs, field personnel can correct the commonly expressed view that 'it just broke'.

The drilling process represents one of the most destructive working environments it is possible to encounter and the drilling equipment is in the direct line

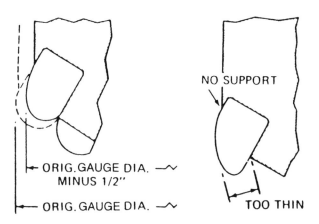

Fig. 8.12: Limit of bit reconditioning.

of fire since it is the mechanism by which stresses are created and transmitted in sufficient magnitude to induce excess stress levels at the rock face. It is this excess stress energy level which causes the rock to fracture and hence the borehole to advance. It follows that continual exposure to high stress energy levels will ultimately lead to a breakdown in the equipment itself. It is the timeframe and the externally imposed stress levels that one must try and isolate in order to understand why a component fails.

Examining the mechanism of failure in rotating components under compressive loads, as in a typical drill string, it can be shown that metal fatigue is the most common cause of failure.

Fatigue is set up by a concentration of bending stresses. This is best illustrated by considering a pipe rotated and operated under a compressive load in an unsupported condition. As the pipe is rotated and the compressive forces increased, it will buckle at that point at which the fibres of the metal are successively stretched or compressed depending on their location relative to deflection of the pipe. In other words, successive bending and stretching produces fatigue, which in turn leads to failure. The more severe the bending due to deflection or the greater number of times alternation takes place between bending and stretching due to rotary speed, the sooner cracks will appear in the metal surfaces.

Fatigue is accelerated by a concentration of bending stresses or stress risers at any change of section since a thick section will not bend as readily as a thin one. Any nicks or notches caused either by tool marks, machining or corrosion pits create a discontinuance in the fibres of the metal and thus potential areas of failure through stress risers.

In percussion drilling the forces are not merely rotational but compounded by impact forces generated by the percussion source, i.e., the DTH hammer.

The materials selected and the heat treatment processes are designed to produce components capable of transmitting and absorbing high levels of stress energy.

If, however, a stress riser is introduced into a component by either poor-quality machine finish, badly designed change of section or a corrosion pit mark, the forces involved in the percussive blow are not transmitted across but reflected back at the discontinuance, which results in a high stress area which will rapidly propagate into a crack. In a similar manner, if high energy levels are generated in the system through overrunning of a dull bit, the impact energy will be reflected back into the system rather than transferred to the rock face and again failure will occur. This is generally the reason for bits breaking in the shank but the potential for damage is by no means confined to the bit shank in such circumstances.

Hot spots introduced either by friction due to a breakdown in lubrication or an external influence such as welding or 'building up' the outside diameter of a component, will change the microstructure and render it incapable of withstanding the impact energy involved in the drilling process.

Torsional failures more commonly occur in the areas of threaded connections and tool joints and can normally be ascribed to overtorquing the joint, leading to a swollen box connection or a cracked pin. Running with a loose or undertorqued connection will invariably give rise to flexing and a typical bending failure will result.

Failure in Tungsten Carbide Inserts

Cemented tungsten carbide is a superhard composition material and like all hard materials, has an associated brittleness factor. This must be treated with respect at all times. Inserts should never be subjected to shock loads by striking with a hammer or other implement when fitting or removing a bit from a DTH hammer. It is imperative that they do not strike against metal objects such as the conductor casing or deck bushings when being run into a hole and, of course, they should never be dropped onto a hard surface.

Correctly used, tungsten carbide inserts will provide a long service life. In operation, flats will develop on the inserts and if a bit is to continue to be used to cut rock, it is imperative that these flats be removed by grinding and the button returned to its domed shape before the flats become excessive. Should the flats be allowed to develop too far and the buttons become dull, the inserts will not effectively transmit the energy of the hammer but simply reflect the shock waves of energy back from the increased surface area in contact with the rock face into the bit shank and the hammer. These reflected shock waves lead to failure in the button inserts, the shank and the components of the hammer. This can be likened to striking a rock face with a chisel. If the chisel is sharp, the rock will splay off. If the chisel is blunted, it will bounce back from the undamaged rock face.

It is not uncommon in practice for a bit of much larger diameter than that designed to be used by the hammer to be pressed into operation. Unless a pilot hole is drilled first to relieve some of the area to be cut by the bit, the latter will ultimately shank prematurely. The buttons may still appear to be in good condition but the total contact and strike area being too great, will give rise to reflected shock waves in the system and premature failure.

Overrunning of dull bits represents the most common form of failure in percussion rock bits. Being totally within the control of the operator, it constitutes bad drilling practice.

A second condition which can be readily observed in viewing dulled bits is thermal shock due to drilling hard formations with insufficient air circulation through the bit. This is manifested in the 'snake-skin' appearance of the bit (so termed because of the reptile-like scaling on the surface of the carbide). Once established, the subsurface of the carbide will have been structurally altered to a point where rapid failure is inevitable.

Bit Body Failures

In inspecting dull bits for damage, attention should be given to the shoulders of the bit at the back of the head. In drilling long holes, if a bit is becoming dull towards completion of the hole, the temptation to overrun it rather than trip out the drill string for a short distance is strong. Such a practice leads to slow laboured penetration and the chuck will often be in prolonged contact with the bit, resulting in the back of the head of the hammer being peened or mushroomed. A shanked bit is the usual consequence.

In the case of a broken shank on a bit a similar peening on the strike face is a manifestation of overruning.

If on inspection of a bit which has seen a short service life the gauge buttons are shattered or show high wear whilst the face buttons are either in 'new condition' or popped from the face of the bit, it is probable that the bit has been run into a severely undergauged hole in an attempt to ream to the bottom.

DTH HAMMER COMPONENTS

The ultimate mode of piston failure is its breaking into two pieces. Commencement of failure will manifest itself at a much earlier stage, however, and can be detected by careful visual inspection.

The general cause of piston failure is heat checking due to a breakdown in lubrication film. The heat check gives rise to a change in the metallurgical microstructure causing enbrittlement; this in turn reduces shock resistance and gives rise to the ultimate failure of cracking.

There are obviously several other factors which can lead to piston galling, such as the introduction of foreign matter if the hammer is assembled in the field without due care, or material entering the hammer in the air stream due to internal hose shredding, or scales or old bentonite from the internal diameter of a badly stored drill pipe passing through the air screen in the hammer, or in the worst case an air screen which has been removed or cracked.

As in any other close toleranced mechanism, it is essential to observe all basic precautions of cleanliness when assembling a hammer in the field. The introduction of foreign matter, irrespective of its nature or origin, will quickly lead to problems in either the piston, the cylinder or control tube area, or the threads of the individual components. The evidence of heat galling or trapped material can be spotted either with the naked eye or a magnifying glass. Galling creates score marks on what is an otherwise polished surface. Heat checking will often show local discoloration of the metal and inclusions will often leave a polished spot or area where the material grinds away in the threaded points, leaving a highly polished isolated area.

Most cylinder failures can be attributed to mishandling. As with the piston, cleanliness is essential in fitting parts together and the evidence of foreign matter having been introduced is the same as that described for the piston above.

Excessive force in either disassembly or assembly of components can lead to difficulties. Cylinders can become squeezed or deformed; this is readily detected by measuring concentricity or looking for signs of crimping.

As stated earlier, at no time should heat be introduced into the component by using a welding torch, say, leads to changes in microstructure and premature failure.

External hammer blows with a hard-faced hammer will lead to deformation and ultimately a cracked cylinder.

It is essential that the correct grade of thread grease be applied to the threads when assembling since the components will be expected to perform many hundreds of hours between inspections also the thread should be correctly torqued before running into the hole. Threads should be inspected for galling caused by metal-to-metal contact or washing at the faces of the threaded components due to escaping pressure.

Cylinders are subjected to high rates of external wear in the drilling process due to the passage of high-velocity air and drilled debris in transportation. It is important to continually monitor this wear by observation and measurement. The wear will not be uniform along the cylinder nor indeed will it be consistent around the diameter since the configuration of the chip grooves in the bit or stabilised chuck will set up eddy currents in the discharge air stream. It is essential to measure the cylinder at several points to establish the site of maximum wear. The major wear will always be at the bottom end of the cylinder closest to the drill bit in hammers in which the cylinder is a reversible feature. Rotation of ends during the assembly process will aid overall cylinder life.

Uneven wear on the control tube is invariably caused by lack of make-up in the overall assembly. The control tube is in effect no longer centralised by the mount and thus is no longer held concentric in the assembly. The effect is a pronounced wear area established on one side of the tube towards the bottom choke area.

Obviously, galling can only be attributed to failure of the lubrication film or ingress of foreign matter.

Loss of make-up is a progressive factor and can be detected by feel and observation of the 'stand-off' when assembling the hammer. Shims are available to restore the make-up value to prolong the life of the control tube. Lack of attention to lubrication and over - or undertorquing are the usual problems associated with threaded components, such as the top and bottom threaded components of the back head and chuck. The comments made earlier regarding the correct tools for assembly are pertinent to prolonging the life of these components. Corrosion is the enemy of all highly stressed components and none more so than in DTH hammers. A hammer must be stripped out and cleaned immediately upon withdrawal from service and the internal components well lubricated prior to any prolonged storage. Attention to these minor details will ensure longer-lived and more cost-effective drilling tools.

CHAPTER 9

ALTERNATIVE DOWNHOLE HAMMER APPLICATIONS

The versatility of the downhole hammer (DTH) has been well illustrated by the diversity of its applications. Conventional hammers are now available in larger sizes, making it possible to drill holes up to 36 inches in diameter using direct flushing methods. Reverse circulation hammers in smaller sizes (4.25 inch bit dia) concomitantly accommodate many exploration programmes.

Hammers are used in oil and gas drilling operations in conjunction with high-pressure air packages to produce higher productivity than could be achieved by rotary rock bits under certain conditions. However, it is in the more novel applications that hammer versatility can best be illustrated. These range from drilling large diameter holes for piling and hole formation for the construction industry to large diameter raise drilling for the mining industry.

In the construction industry there are many applications which require the insertion of micropiles through variable strata into a rock base. The technique now widely practised uses a downhole hammer in which the bit cutting system has been developed to underream the pile. The DTH hammer and drill string is inserted into the steel pipe which will ultimately form the pile and the underreaming bit drills ahead of the steel pipe which acts as a casing to keep the hole open around the drill string. When the rock head has been reached and drilled to the required depth, the DTH hammer with the specially developed underreaming bit working on an eccentric principle (Fig. 9.1) is retracted and the drill string is removed. The outer steel pipe remains *in situ* to be re-enforced internally and concreted in place to form the pile. Pile driving by percussive means is therefore eliminated, providing a substantial reduction in noise levels during the drilling process, a most important point in urban areas. At the same time the new DTH technique provides a more precise method for both locating and founding piles than can he achieved in driving from the surface.

For larger hole formation for the construction industry, large-diameter hammers have been developed which are capable of drilling a 36-inch diameter in a single pass. But a more innovative system of cluster drills in which multiples of smaller

DTH in hammers are assembled to drill as a single unit to drill holes up to 60 inches diameter has also been developed. (Fig. 9.2). This system enables a far wider range of formations to be drilled efficiently with substantially reduced air requirements compared to a single large-diameter hammer whilst at the same time using standardised components, which reduces the individual cost of spare parts. The technique can be modified to accommodate any specific diameter requirement.

The cluster drill is generally applied in the construction industry for shallow holes, such as in trench drilling, which are drilled at close proximity to form a complete elongated excavation, or for sockets into which power poles or pylons can be inserted. Where the depth is of such a limited nature, it is not essential to use a drilling rig since the cluster drill package can be attached to any suitable crane or excavator to which a small self-powered rotary table can be attached. The concept of the cluster drill is literally turned upside down in a development used for raise drilling application (Fig. 9.3).

In this application a pilot hole is drilled between two gallery levels in a mine. With the drill rig set at the top level the pilot bit is removed and replaced by a cluster drill in which two or three DTH hammers are mounted in an inverted position with a trailing stabiliser assembly. The unit is drawn back to the higher level by the drill rig, creating holes up to 30 inches in diameter.

The drill bits employed do not utilise the traditional spline shaft to drive them. Instead rotation of the drill bit is accomplished by friction against the side of the hole.

Because the cuttings are free to fall back down the hole, the air volume requirement is determined solely by the percussion component of the hammer.

Holes of high inclination, up to $87°$, are readily accommodated and hole depths to 600 ft in length are standard accomplishments.

In an adaptation of the design the DTH hammers are mounted on retractable arms with variable geometry which enables the bits to be extended to maintain the diameter of the hole even as they wear down in gauge size.

This concept has found a ready market in many mines which found small-diameter short pass raises uneconomic using the more conventional roller cutter raise heads with their attendant cost structures. Because the DTH percussive hammer is not as susceptible as the roller cutter in high-strength formations, the power and size of the rig required is considerably reduced, eliminating the need for expensive chambering as a prerequisite to setting up.

The concept has been expanded to create holes larger than the initial size of the head pass. In this technique the pilot hole and the subsequent large-diameter raise are first completed, followed by a series of peripheral 6-inch diameter holes which are charged. The rock spoil is evacuated into the previously drilled large-diameter hole to create a final diameter of up to 22 feet. Because this is drilled and charged from the top the procedure eliminates the need for operators to work in recently blasted areas and the need for climber raise drilling is totally dispensed with.

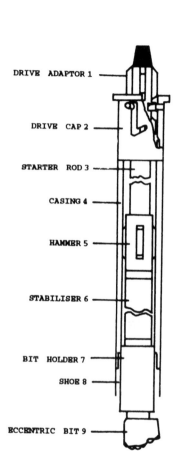

DRIVE ADAPTOR 1

DRIVE CAP 2

STARTER ROD 3

CASING 4

HAMMER 5

STABILISER 6

BIT HOLDER 7

SHOE 8

ECCENTRIC BIT 9

Fig. 9.1: Bulroc rotex hammer.

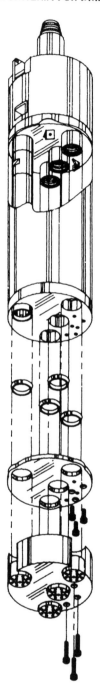

Fig. 9.2: Drillquip cluster drill.

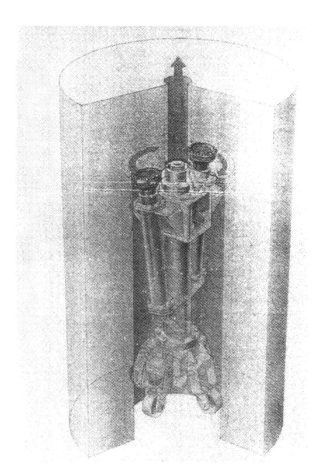

Fig. 9.3: SDS V30 Raise Drill

Possibly the most novel adaptation of the downhole hammer concept is embodied in the DTH water hammer commercially released in 1993 by SDS.

Drilling Tools of Australia.

In this application the principle of the cylinder with reciprocating piston striking the bit is retained but the motive power of the compressed air is replaced by high-pressure water, thus employing the greater efficiency of water hydraulics and dispensing with the need for large high-volume compressors.

The water hammer is run in conjunction with a dual-wall drill pipe, as used in reverse circulation DTH drilling with the water to power the piston being supplied through the outer annulus. The piston achieves a strike rate in the order of 50 blows per second. The need for high-velocity air to transport the debris is eliminated and the water returned to the surface can be recycled to continue the drilling operation.

This new system holds out some very exciting possibilities, especially in deep-hole applications more normally associated with the oil and gas exploration industry, without the need for introducing large-capacity boosters into the set-up.

CHAPTER 10

ROLE OF REVERSE CIRCULATION
IN EXPLORATION DRILLING

Alternative Sampling Methods

Mineral exploration has always been an expensive high-risk venture. In recent years technology has advanced in quantum leaps in the areas of geophysics with the advent of Landsat surveys and overfly surveys coupled with advanced interpretation of mass data for identifying areas of prospective interest.

Once the accumulated data has been assessed, an area identified and surface investigations completed by the geologist, the drilling operation may commence. This has traditionally been the domain of the diamond drill carrying out a coring programme.

Diamond drilling has always been regarded as the most accurate method of obtaining subsurface geological information. However, a substantial expense is generally associated with it.

The question of whether the speed of development in drilling has kept pace with developments in the area of geophysics is an interesting one. The answer might well depend on which side of the activity area one stands. Certainly there have been major breakthroughs and changes over the last twenty years.

The introduction of the wireline system in 1965 revolutionised the drilling operation overnight and is now universally accepted. Since its introduction, although the principles have remained unchanged, refinements have been introduced which have increased the system's efficiency and the depth rating of the equipment.

To the casual observer the changes in internal design and metallurgy now employed may not be apparent, but they differ radically from the initially introduced product.

Although the wireline system has been universally accepted, it is by no means universally employed and conventional drilling using equipment designs which would be familiar to drilling crews from another era are still in everyday use in various parts of the world.

In the Scandinavian countries 'thin-wall' systems have been developed to a degree which rivals wireline drilling in terms of output per shift. Although the barrel and core have to be removed from the borehole by evacuating rods one at a time, the area which requires to be cut by the the thin-wall diamond bit is greatly reduced, which substantially assists penetration.

Reduction in the ratio of outside diameter to inside diameter of the diamond bit creates a situation wherein the relative peripheral speeds are far more closely related than those of the wireline system, which of necessity must provide room to accommodate the engineering of a wireline barrel in the internals; hence a much higher rotational speed can be employed. With the reduced face cutting area, the high technology impregnated bits using synthetic diamonds find suitable application. The high ratio of core size to hole size permits the use of lightweight aluminium drill rods, aiding greater depths to be achieved with relatively smaller machines, more rapid rod breakout and handling mechanically, speedier round tripping and in many cases reduction in manpower requirements.

So we have yet another system which is fully developed and which has gained wide acceptance but not universal application. Herein lies the crux of the problem.

For any given exploration programme, when it comes down to the drilling operation the choice of method to be employed must be controlled by a knowledge of all the factors which will affect the ultimate performance, the geology, the topography, accessibility of the site and surrounding terrain. The intended depth of drilling, proximity to water, time-frame restrictions due to weather access and a myriad other factors come into play in making such a judgement. Central to this equation must be cost allocated to drilling and the time frame granted to achieve the desired objective.

It can be argued that no method can compare with a diamond drilled programme in terms of the information acquired for each foot drilled. True, a properly conducted investigation can reveal geotechnical, geological and geochemical information together with, in many cases, hydrological information if logged correctly, all of which is of considerable concern to the mining engineer. However, the exploration geologist has other parameters as his primary concern and unless the drilling activity is taking place within an existing site on which mining is currently underway or is to be shortly embarked upon, the opportunity to gather ancillary information is more often lost and requires a subsequent and separate programme.

It is these very factors which give rise to separate development and although drilling methodology traverses geographic boundaries, it does not always find the ready acceptance one might expect.

Numerous instances could be cited, such as the general application of long-stroke top drive hydraulic drilling rigs in Australia compared to the prevalence of the more traditional, albeit hydraulically advanced design of diamond drill in Canada. This does not indicate the superiority of one system over another but serves to indicate a difference in the approach to and perception of efficient application of two competing technologies related to the situation on the ground.

No matter what the geographic or cultural differences the geologists mandate to the drilling contractor or operator, they must essentially be along the lines of returning the maximum amount of information or sample in as uncontaminated and unaltered condition as possible within an acceptable scale of costs.

Accepting this as the goal, it becomes a matter of judgement as to how far the objective can be acceptably compromised and in which specific area the compromise is to be achieved. In the first instance cost and time are generally the major concerns and to achieve these objectives it is necessary to introduce some alternative form of drilling, which, although faster in operation, presents the evidence of the existing geology in some altered form, i.e., chip sample rather than core, but still with a correlation to specific quantity and the depth from which it was obtained.

Provided the party interpreting the results obtained understands the extent to which the integrity of the information has been potentially compromised, the exercise remains valid; but the level of interpretation remains open to question unless further drilling is conducted using a more precise and, by inference, expensive method. For example, in gold exploration, the concentration of the target objective is often measured in parts per million with an unpredictable distribution.

If one takes the Australian example, it is possible to highlight the steps which have taken place, particularly in mineral exploration, and which have seen the evolution of what is now both an accepted and proven method of exploration drilling in reverse circulation, displacing the need for some of the traditional diamond coring but supplementing rather than replacing it in totality and permitting the coring activity to be focussed into specific targets rather than blanketing the exploration area.

ALTERNATIVE SAMPLING METHODS

Rotary Air Blast

The cheapest alternative to coring is rotary air blast (RAB) drilling in which chip samples are obtained from the drilled debris or dust gathered at the surface collar and assessed for content (Fig. 10.1). Although cheap, the method is both basic and unreliable since any mineral show cannot be accurately related to the depth from which it was produced or the percentage content. At best it may show that something exists at some place in the area penetrated, at worst it may pass or penetrate something but fail to show its presence. If one is dealing with a bulk sample at depth with the requirement to ascertain if it exists and a general indication of its depth, the exercise may be acceptable.

The risk factors are that in this system using conventional circulation, the flushing medium (generally high-velocity air) is conducted to the face of the cutting tool (the drill bit) through the internals of the drill pipe and the cuttings are returned up the outer annulus formed between the drill pipe and the wall of the borehole.

In general terms the return velocity up the hole will be in the order of some 4000 ft/h and as such has the capacity to scour the sides of the hole, leading to hole enlargement and possible enhancement by volume of the content returned at the surface.

In stratified formations in which one particular stratum is more friable than a more competent or cohesive neighbour, the friable one is more likely to present itself at the surface in a content value which is disproportionate to its true content.

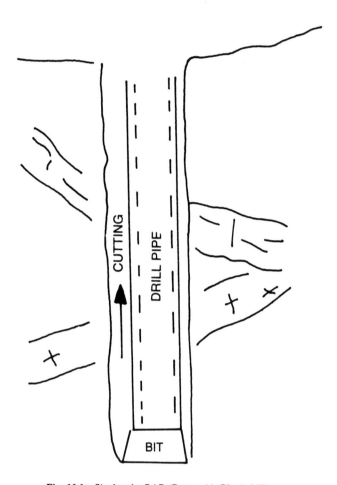

Fig. 10.1: Single tube RAB (Rotary Air Blast) drilling.

Open voids or massive hole enlargement lead to a drop in the uphole velocity of the circulation medium; this can lead to heavier particles dropping out of suspension into the void to be either lost or picked up at some later stage of the borehole's progress and thus recorded as having been produced from some greater depth.

The presence of water in the hole introduces a whole new dimension since, depending on the inflow rate, can either inhibit the flow of returns to it the surface or may carry with it a contaminating body leached from adjacent areas, to be logged at the surface as being produced from within the area of the borehole.

Despite the inadequacies of the system and its known potential for providing misinformation or missing information completely, RAB drilling is still widely practiced and is accepted by many as sufficiently accurate for scout drilling, to be followed up by a more accurate method at some later stage. The question has to be asked: What happens if the RAB method fails to provide a sample of a mineral revealed which it has penetrated? Is this the link between the increased technology of geosciences in the laboratory and the ability to detect lower levels of mineralisation due to advanced technology in the laboratory?

The answer lies firmly within the ambit of the geologist who commissions such a drilling exercise. He must carry the decision as to the extent he is prepared to compromise on the mandate outlined earlier, namely 'providing the maximum possible sample as uncontaminated and in as unaltered a condition as possible within an acceptable scale of costs'.

The number of mine potential properties which have been 'discovered' by a more reliable method of drilling, after being abandoned consequent to RAB drilling is by no means negligible!

RAB drilling has its place in the general scheme of exploration but the potential downside risks must be assessed and weighed against any potential cost savings in terms of reliability of detection.

Some in the geological profession have published papers defending the RAB method and pointing out minimum standards of field practice to improve sample recovery. These include drilling through stuffing boxes, using moistened air to flocculate the dust, blowing down rods between holes and collecting samples in the cyclone. Whilst such precautions may mitigate the deficiencies of the system, they cannot overcome the basic problems inherent in it. Hence the method should be used and the results interpreted with caution and with an appreciation of the risk factors. In the final analysis the question must be—does the cost saving of an RAB programme outweigh the cost of having to return with a more accurate method of drilling to obtain increased information at a later stage?.

The answer to such a question may be more influenced by market conditions than any scientific content since in 'boom' conditions in Australia the number of drilling projects often outnumber the available rig fleet and the RAB drilling set-up represents a relatively cheap entry level for the would-be drilling contractor.

What may not be so readily understood is the pace of development in the arena of reverse circulation and how in a relatively short period of time the advances in both speed and accuracy of production have overcome the problems associated with the industry's earliest venture into this technology.

Reverse Circulation

Reverse circulation as a concept has been with us for many years. The term was first developed commercially in the 1950s and used in large-hole drilling associated mainly with water-well and shaft drilling operations. The technology essentially infers that the flushing medium, generally water or mud, is circulated down the external annulus formed by the the drill pipe and the borehole wall and returned to the surface up the internal diameter of the drill pipe. The method was introduced to enable hole drilling through unconsolidated ground in areas of high water table. Since the water or fluid circulation is lagooned at the surface and flows into the borehole to supplement the head of the groundwater when air is injected into the inner diameter of the drill pipe, the differential in the water head promotes a flow in the pipe by air lifting, carrying the drilled cuttings back to the surface. The drill pipe has externally flanged connections and as the depth is increased air is injected at a lower point in the drill string through external airlines, which form an integral part of the drill pipe.

Because the external water has a low circulation velocity, being required simply to maintain the head of water in the hole, the stability of the unconsolidated ground is not disturbed. When rotation of the drill string is stopped, the lagooned water flowing into the borehole serves to provide a higher static water-head pressure acting on the borehole wall to keep the hole open.

A further adaptation of the method dispenses with the airlift and uses a pump to draw the circulating water back through the inner annulus carrying the cuttings in suspension. (Fig. 10.2).

When applied to mineral exploration the reverse circulation theory underwent several adaptations before attaining its current level of development.

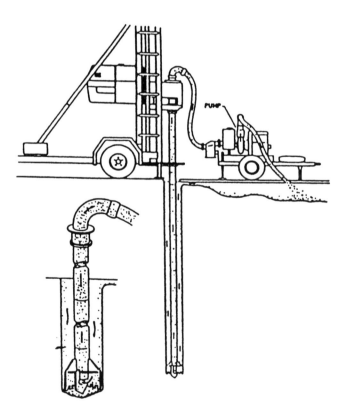

Fig. 10.2: Reverse circulation drilling with water pump.

In the initial stages it was used to indicate a reverse flow of a water circulation medium through a single-wall drill pipe with a rock bit acting as the cutting tool. In the late 1960s the American Drilco Co introduced a dual-wall drill pipe with a separate internal and external annulus, through which the circulation medium could be conducted and compressed air used as the medium. The cutting tool, generally a rock roller or blade bit, was connected through a sub which conducted the air circulation through the bit and was shrouded to just below the dimensions of the bit to promote the circulation flow to be directed back to the surface through the internal annulus (Fig. 10.3).

The system was acclaimed for its ability to bring uncontaminated sample back to and the surface through the internal annulus. With the high-velocity air circulation, cuttings were removed to the surface as fast as they were produced and could be related directly to the depth from which they originated. With this system lost circulation zones were not a problem and the sample was not affected by inflows of water.

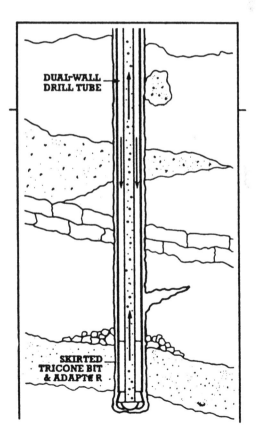

Fig. 10.3: RC drilling with shrouded roller bit and dual-wall pipe.

This technology was used in Australia but the drill-pipe dimensions reduced to enable its accommodation on the hydraulic top drive machines being introduced at the time.

The chief difficulty with this system was the inability of the roller bit or blade bit to handle the harder strata. In the sizes in use (6–4-1/2 inch dia) the bearings of the rock bit could not tolerate high loadings. Given the dual-wall design of the drill pipe, it was not possible to provide downhole loading with drill collars as would be the normal methodology. The system in this format was therefore not universal in application.

When a separate crossover/diverter sub was introduced into the system together with a downhole hammer, the method gained wider acceptance but again not universal application. The DTH hammer certainly helped to improve the penetration rate but suffered the usual difficulty of drilling through soft clay formations, which tend to block off the bit and may block off the diverter sub. The diverter sub itself is situated behind the hammer and connects it to the dual-wall drill pipe (Fig. 10.4). The cuttings produced by the bit travel up the outside of the DTH hammer casing and at the point of the crossover sub, the circulation is diverted into the inner annulus to travel back to the surface in the inside drill pipe.

To help overcome the softer formations moistened air or foam can be injected into the air stream as in drilling with a DTH hammer in conventional holes.

The 'new' system was very widely used but doubts arose as to the accuracy of sampling gathered in this manner.

Because the sample has to travel the length of the hammer casing before entering the protection of the inner annulus, the sample can be contaminated by the formation previously penetrated. Similarly, if drilling into a void, the sample can drop out of circulation and be lost. Inflows of water carrying samples from adjacent areas can contaminate the sample returned to the surface and create difficulties in interpretation. Although the diverter or crossover sub has an outside diameter which is set as close as possible to the bit diameter, it is not a packed hole condition and the sample travelling past the hammer casing in the outer hole is capable of passing the diverter—a condition known as 'blowby'. A sample which fails to enter the sub remains a 'flier' in the system and may fall to the bottom of the hole when the air supply is switched off, only to be collected and subsequently logged as having been obtained from some lower level in the hole.

There have been some high profile case studies documented in both Canada and the USA wherein the method as described was held to be directly responsible for misrepresentation of downhole conditions with embarrassing consequences.

The method was used, for example to obtain figures for estimating reserves in the Echo Bay mine in Canada. A subsequent diamond-drill programme failed to confirm the findings of the reverse circulation drilling and the reserves were downgraded from 4.5 million ounces to 0.9 million ounces. Subsequent investigations attributed the 'upgrading solely to water inflows contaminating the results of the RC drilling using DTH hammer and crossover sub'.

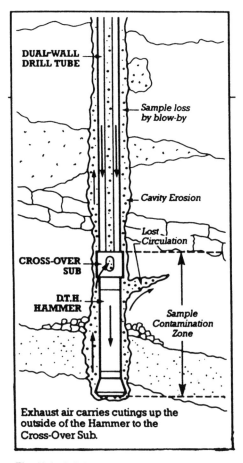

Fig. 10.4: RC drilling with DTH and diverter sub

Face Sampling Reverse Circulation DTH Hammer Drilling

In recognition of the potential problems associated with the diverter sub method of drilling, the industry set out to develop a face sampling reverse circulation hammer in which the diverter sub could be dispensed with and the sample collected at the face of the bit, could enter directly into the protected environment of the inner tube of the dual-wall pipe (Fig. 10.5).

To achieve this situation the basic concept of a downhole hammer was retained but the engineering and design underwent substantial redevelopment.

For the flow of cuttings to be reversed from the normal route of exiting the face of the bit externally and being collected through the internal face of the bit, the bit had to be redesigned.

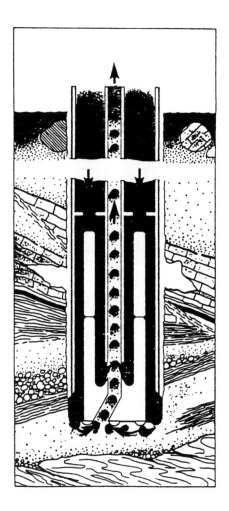

Fig. 10.5: RC air and sample flow.

Now the piston had necessarily to be hollowed out and a conducting tube introduced, which would collect the cuttings from the face of the bit and pass them through the internals of the hammer into the inner tube of the drill pipe. A system had to be evolved which would promote the flow of cuttings from the bit into the newly created inner tube passage back to the surface.

These design parameters had to be accomplished whilst retaining the DTH hammers high strike rate or blow count, without sacrificing blow energy and whilst still providing sufficient air to the gauge areas of the bit to both cool the inserts and clear the cuttings if the hammer were to retain the same penetration potential as the equivalent conventional hammer of the same dimension.

Clearly the mission was complex since the basic relationships previously established of piston diameter, weight and stroke still prevailed. If the piston were to sacrifice mass by being hollowed out, with consequent loss of surface area and only minimal variation permitted in the stroke length, certain compromises were inevitable (Fig. 10.6).

The objective of obtaining larger size chip samples was almost a by-product since if the bit could be made to drill correctly, the debris produced would in theory enter the central collecting point without further diminution.

However, whilst this might satisfy the geologist a slower rate of penetration would not be readily accepted by the drilling contractor, who is generally paid on the basis of footage produced; the dual burden of a more expensive piece of equipment with a lower performance potential and hence inferior earning capacity would not be attractive.

Two approaches were adopted to the question of how to induce air flow from the face of the bit into the central air tube. In the first method the air supply is

CONVENTIONAL DTH BIT

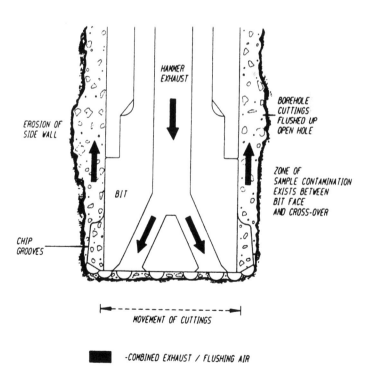

Fig. 10.6: Conventional air and sample flow.

channelled to the bottom of the hammer external to the central collector tube and at a point inside the bit assembly and just above the centre collector point, the flow diverted through small jets into the central tube. The high-velocity air injected in an upwards direction through the jets creates an envelope of air and the venturi effect creates suction, lifting the cuttings from the bit face through the central sample tube and back into the inner pipe of the dual-wall drill pipe for transportation back to the surface. (Fig. 10.7).

The alternative approach conducts high pressure to the face of the specially designed bit in which the air flow is conducted from the outside of the bit face to the collection holes in the face of the bit. The bottom of the hammer is shrouded to almost full hole size to promote flow back into the face of the bit rather than exiting back past the outside of the hammer.

Early versions of the hammers held out hope for improvement but whilst indicating that the concept could work, could not match a conventional hammer for penetration rate, mainly due to the compromises in design of the internals involving reduced piston area and piston wieght. The design of the early reverse

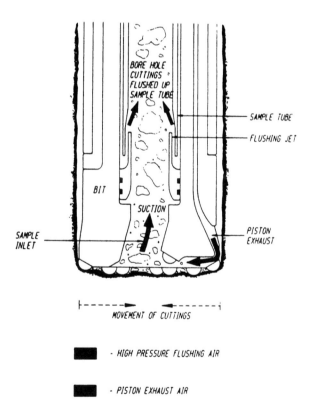

Fig. 10.7: Sample collection system.

circulation bits also proved to be a problem and substantially escalated the cost per foot drilled.

Subsequent experience and development accelerated the design of the DTH reverse circulation hammer to a point where it is now accepted as a field standard and the previous problems of extremes in soft formations, hard formations, broken ground and water inflows can be readily accommodated. Furthermore, production rates now rival conventional hammers and component prices have become more economically viable due to increased throughput in manufacture. Lastly, geologists and mining companies alike now have confidence in the results provided by the system.

To make the system fully acceptable, however, there had to be parallel development in the methods of handling the sample back at the surface through the cyclone and splitters to match the increased rate of sample production from the DTH reverse circulation hammers and bits. Specially developed cyclones and splitting systems now make it possible to draw samples even in areas of large water inflow.

The DTH reverse circulation hammer was first introduced by Bakerdrill in the USA in 1976. The method was sparingly used and improvement negligible for several years until the introduction of the Entec Samplex hammer in 1989, following some seven years of research. This was followed by several other entrants in the field which, in the exploration industry and prevailing conditions of Australia found only limited acceptance. In 1990 the local Australian manufacturing companies were approached to develop a competitive hammer. This activity has been so successful that the RC face sampling hammer system developed with a matching range of bits. A series of hammer sizes has enabled diversity of range and in a period of just three years the exploration drilling industry has undergone a complete transformation in methodology.

If we now examine the geologist's brief to the drilling contractor in the light of this development, the industry can presently offer almost complete samples within an acceptable scale of costs. It is for the geologist to determine whether he can accept the compromise of disturbance and alteration of sample condition caused by the drilling method and, having assessed the findings, determine the focus of the follow-up diamond drilling programme.

CHAPTER 11

DEVIATION IN DRILL HOLES

Most dictionaries define deviation in terms of 'departure away from a straight line or norm'. When considering deviation in a drill hole it is probably more appropriate to accept the generality that there is no such thing as a straight drill hole. The concept appears to be as abhorrent as a vacuum is to nature and a straight hole can be considered as the exception rather than the rule.

Any search of the literature will find innumerable references to deviated drill holes and very few to a naturally occurring straight one.

Reference to holes which have failed to penetrate known zones of minerali-sation, or have deviated to align themselves within the ore body to produce what would appear to be spectacular results, or have terminated tens and in many cases, hundreds of feet away from their inferred or intended path are readily found in the literature (Fig. 11.1)

What is surprising perhaps is that with the substantial monetary waste which accompanies decisions based on such erroneous information, the whole subject has not been the object of deeper study and investigation and that the lessons learned from such expensive mistakes have not resulted in a far more disciplined approach to the drilling operation.

There is, of course, an element of relativity in discussing this topic. A hole may be considered 'straight' if the deviation is measured in terms of thousandths of an inch per inch of advance. However, let us consider a hole which deviates at ten-thousandths of an inch per inch consistently over a 1000 foot total depth. The hole would terminate some 10 feet away from the projected target. This assumes the devi-ation to be consistently in one direction, an equally undetermined factor and, in truth, the termination could be anywhere within a radius of ten feet of the projected target zone. Such errors are not unusual and there are documented instances of vertical holes which have flattened by 30° in the first 300 feet of drilling and a 1500 ft hole which started vertical and finished horizontal having swung through 180° in the process.

Examples of such magnitudes of error are not hard to find and present a major problem in both the mining and civil engineering sectors of industry.

With the extreme depths of operation associated with oil and gas drilling oper-ations, deviation is a cause of major concern since at the simple level, a crooked

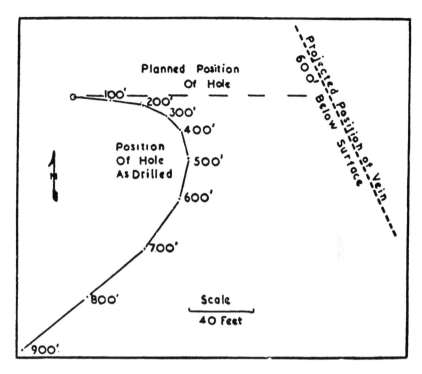

Fig. 11.1: Deviation of drill hole.

well is difficult to run casing in. Furthermore, if the well is allowed to drift uncontrolled, it can run outside the boundaries of the property. A well drilled to 7000 feet with a consistent 3° deviation could bottom out at any point within a circle of 732 feet in radius.

Many of the causes which lead to deviation in drill holes have been identified, even if the effects cannot always be quantified. By and large these causes are the reactive forces imposed by formation changes or bedding planes, the forces generated by a formation after penetration, as with schists and swelling clays etc., and the planes separating rocks of differing hardness.

The degree of deviation effect caused by these naturally occurring situations is reactive with the type of drilling technique employed, the equipment in use and the drilling parameters, i.e., rotational speed, feed pressures, pumping pressures and clearances involved. The type of drilling bit, its condition and design exert considerable influence on the situation, (Fig. 11.2).

The anisotropic formation theory was developed by researchers in the oil and gas industry wherein, given the extremes in hole depth, doglegs and deviation present a major problem.

The theory assumes that in uniform or isotropic formations a bit will drill in the direction of the resultant forces. In an anisotropic formation, one which

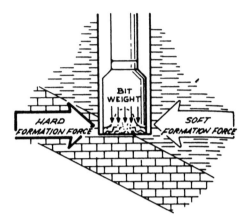

Fig. 11.2: Formation drillability theory.

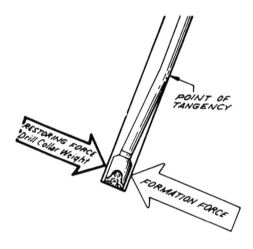

Fig. 11.3: Anisotropic drillability theory.

shows different properties in different directions, the bit will drill in a preferential direction. The stratified or anisotropic formation is assumed to possess differing drilling properties, both parallel and normal to the bedding plane, the result being that the bit does not drill in the direction of the resultant force (Fig. 11.3).

Each of the formations identified by the researchers was characterised by an anisotropic index and dip angle, unrelated to the specific geological properties of the rock. This was derived from observation of actual drilling measurements and allocated an empirical number related to a constant.

The anisotropic theory has been applied in various mines in which continual drilling data can be accumulated in a database with all information collated as to drilling technique and running regimes employed. As the database expands, the predictability factor for deviation in any drill hole planning can be refined.

This can be developed to a point where 'steering' of drill holes can be achieved, that is the tendency for deviation can be increased or reduced by the equipment and technique being employed.

As a general observation, it has been noted that holes drilled parallel to the planes of parting will drift to parallelism while those drilled beyond the range in which they drift to parallelism with the planes of parting will bend until they are normal with them.

The rates of deviation towards either parallelism or normal depend on the angle between the axis of the drill hole and the planes of parting in the rock.

Increases in bit pressure tend to increase the rate of deviation and higher rotational drill speeds tend to induce a spiralling deviation.

So far we have considered deviation as a naturally occurring phenomenon whose effects can be accentuated or reduced within certain restrictive limits, by changes in the operating conditions of the drilling equipment or technique employed. What must now be considered is a deliberately induced alteration to the path of the drill hole, which is termed a deflection.

Controlled drilling techniques have been evolved over a number a years which enable the path of a drill hole to be steered towards a target area.

The techniques have many areas of practical application. Apart from the obvious requirement of maintaining a specific alignment and overcoming the natural tendencies of a drill hole to deviate, it is often beneficial to be able to sidetrack a hole at differing levels to sample the adjacent formations.

When considering deep exploration holes in mining applications, the multi branch hole can be utilised to confirm areas of mineralisation which are predicted to exist based on the known formation and characteristics of previously mined areas on the property (Fig. 11.4).

In a more simple application, in the event of a 'twist off' occurring which cannot be fished, if the expenditure incurred prior to the 'twist off' is of such magnitude that total abandonment would be detrimental to the programme, rather than redrill the entire hole, the 'fish' can be bypassed by deviating the hole path from a level only slightly higher than the obstruction.

In many deep-hole programmes in which the top levels are of less concern to the geologist, deep precollaring can be accomplished with heavy-capacity machines which can produce such holes to depth faster and therefore more cheaply than diamond drilling. Invariably, as a trade-off in using higher penetration rates, there is a far higher tendency to deviation. The first operation prior to commencement of diamond drilling from the deep precollar must be a survey and introduction of a suitable technique to establish the subsequent diamond drill hole on the correct path.

There is some debate in geological circles as to the relative cost saving in precollar speed against the expense of deflecting the deeper hole onto a subsequent

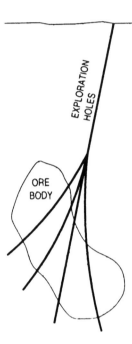

Fig. 11.4: Typical drill fan used in exploration drilling.

target zone. There is no universal answer to the question and it must remain a matter of subjective judgement. However, it is possible that a compromise in the precollar operation penetration rate with a suitably designed drill string bottomhole assembly, could minimise the deviation to within a more acceptable limit.

Similarly, if the conventions of the anisotropic theory hold true, by accumulating data from the initial branch holes with all the drilling data correctly recorded, it should be possible to set a heading for the deep precollar which would terminate at a point more approximate to the geologist's requirements. A great deal of research was carried out in the oil and gas industry in the 1950s to establish the basics of deviation control and to evolve methods to overcome deviation. Arthur Lubinsky is probably the name most associated with this research and thanks mainly to his efforts equipment and drilling practices were established which helped to bring a control factor into play.

Various centralising stabilisers had been used extensively in the oil and gas industry until 1953, the year in which Arthur Lubinsky, a research engineer with Amoco, produced a paper titled 'Factors Affecting the Angle of Inclination and Doglegging in Boreholes'. This paper set out with an element of mathematical precision the effects of bit weights, formation characteristics, drillcollar to hole clearance and stabiliser placement on deviation.

Prior to this publication the practice was to drill with a large-diameter drill pipe and no drill collars but with as much applied weight as could be contrived.

FACTORS CONSIDERED :

1. **STIFFNESS AND UNIT WEIGHT OF DRILL COLLAR**
2. **INCLINATION OF HOLE**
3. **CLEARANE**
4. **AXIAL LOAD**
5. **FORMATION CHARACTERISTICS**

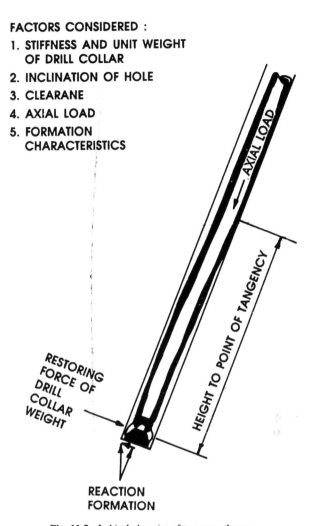

Fig. 11.5: Lubinsky's point of tangency theory.

Lubinsky advanced the theory that deviation was the result of the drill pipe loaded in compression deflecting off the low side of the hole. Based on this assertion he postulated that by using stiffened assemblies with drill collars of larger diameter than the drill pipe, running immediately behind the drill bit to provide an overweighted situation, and with stabilisers located at strategic points the drill string could be run in tension, thereby creating a pendulum effect wherein the bit would drill in a more centralised condition (Fig. 11.5).

Further research established optimum bottom hole assembly requirements to assist with concentric running and the results are now accepted oilfield practice with specific bottomhole assemblies producing predictable performance.

Much of the basic theory established by oilfield researchers has a general area of application in other drilling activities but the size of the holes and rigs involved impose severe limitations on how far the practices can be followed.

In the diamond drilling field, it is recognised that a larger diameter hole will deviate to a lesser extent than one of smaller diameter. Similarly, larger diameter core barrels tend to drift to a lesser extent since they replicate a packed bottomhole self-centralising assembly. Again, larger diameter core barrels of longer length produce the same effect.

To extend the theory a conventional barrel with a 'sludge pot' running behind and attached to the core barrel back head produces even better straightline control. The sludge pot must, of course, be threaded contra-rotation to the rest of the string.

Oversize core barrels can be produced with either ribbed stabilisers or with ribbed outer barrels to assist in centralising the cutting assembly.

The practice of spirally setting a reamer shell to provide better control is of use only in certain formations. A preferred method for centralised control and minimalisation of deviation might possibly be a taper set in which the diamonds are set into tungsten ribs that are then let into the reamer shell body in slots which have been machined to produce a 1 to 2° taper. This has the advantage of retaining a gauging stability throughout the life of the reamer shell.

The configuration of the face of the diamond drill bit exerts considerable influence on deviation and it is considered that a multistepped bit will tend to hold a better alignment than rounded profiles and that an internally stepped bit will offer the best alignment characteristics.

The above-outlined methods when used with good drilling practice will help contribute to the drilling of straighter holes but there are times when the desirable must be compromised to achieve the acceptable. The use of impregnated bits presents just such a situation. Due to the more limited sizes and hence potential rock penetration, it is, normally recommended that impregnated bits be run at far higher rotational speeds than their surface-set counterparts. If it is shown that the higher revolutions are proving detrimental to deviation control, some compromise must be made. Various devices have been developed which allow the path of a drill hole to be deliberately deflected in a known direction. An exhaustive review of all the equipment which has been developed to assist in deviation control and deflection drilling has not been attempted here, but rather an overview of some of the more diverse techniques which have been or are currently being used, illustrated with some typical case studies.

In considering the deflection of drill holes, two aspects are of equal importance. The first is the ability to survey the hole with sufficient accuracy to determine both inclination and azimuth bearing together with a device of sufficient reliability so that a deflection track can be set for continuation of the hole.

Two milestones were achieved in the diamond drilling industry in the area of identifying deviation as a problem in the closing years of the 19th century. In 1874, a German engineer patented the use of a hydrofluoric acid in a glass vial lowered into the hole with the drill string to etch a mark relating to the angle of inclination of the borehole. In 1885, a South African engineer named Mcgreger produced his 'clinostat', essentially a magnetic needle and a plumb bob immersed in a gelatin which was lowered into the hole and allowed to set. The assembly was retrieved and the settings read direct.

From such early beginnings the search for what has now become a highly developed range of downhole surveying techniques and sensing units which display the readings back on the surface in real time on a pc monitor as the drill hole progresses will mark the close of the 20th century.

The progress which has been made between the two points has been marked by a series of innovative ideas which have each pushed forward the frontiers whilst gradually transforming the techniques of deviation drilling from an art into a science.

The sensing and surveying techniques have been accompanied by development of a range of downhole tools that enable the information reported back on the surface to be transformed into a subsurface command to the drilling mechanism whereby the drill hole can be directed and steered in a given direction.

Although several of the techniques used in the diamond drilling field have been developed through the innovation and persistence of individuals specialising in that sector, it must be acknowledged that oilfield practitioners and suppliers have stimulated many of the 'high tech' performance tools now available. Such is inevitable what with the hourly rig costs, risk factors and returns in the oil and gas domain being so much greater than those in the mineral diamond drilling exploration industry. The cost of down-sizing the equipment from one area of activity to an other is often such that the potential applications do not always appear attractive. However, in the field of microelectronics for sensors and integrated computer programmes for disseminating and displaying information, the mechanics of such a scaling-down exercise are no longer of such magnitude as was the case with scaling down mechanical contrivances. Thus the prospect of transferring technology at a much greater pace looms brighter now.

If the technology transfer is to take place, it is equally essential that the design and instrumentation on the diamond rig be upgraded to provide more accurate assessment of the running parameters down the hole. This would also place great emphasis on the role of the driller, elevating him from merely an operator, no matter how skilled, to a technician with a much expanded range of skills. In the oilfield the driller is no longer required to be a master of all aspects of the operation since the contractors hire appropriately qualified technicians from the service companies to operate their own specialised equipment. Yet because of the different cost structures the hourly rate for the hire of such specialists is such that the mineral exploration sectors have difficulty in coming to terms with this factor. The result must either be a radical rethink in the way mineral exploration contracts are written or a considerable upgrading in the levels of skill available from the drill crews.

Several methods of deflecting the path of a drill hole by the introduction of mechanical wedges have been developed over the years and remain in use on a daily basis as the 'simplest' means of achieving this goal. The wedge has been progressively developed but essentially remains a long blade which is introduced into the borehole at the point from which the deflection is required. The blade has a long taper of semi-rounded profile which, once set into the hole, pushes the drill bit to one side as it tries to pass and forces it to cut a new path in the side wall of the existing borehole. A bull-nosed bit is generally used to drill off the wedge until a deflected track has been established (Fig. 11.6).

The mechanical wedge requires considerable skill for setting to the correct orientation in the hole. Each wedge set, is restricted to deviating the path by 1° only, generally has to be set on the low side of the borehole, is not always 100% successful and varies in performance depending on the formation in which it is set. The wedge is an expensive item to produce and although some designs are recoverable, whilst it is in the hole poses a potential hazard to all subsequent drilling operations in the borehole and can cause high rates of wear on the drill pipe and high build-up of rotational torque.

If deflections greater than 1° are required, multiple wedges must be set progressively in the hole to create a series of 'doglegs'.

Surveying the hole to both detect drift and to set the orientation of the wedge is accomplished with a single-shot camera which, when lowered to the appropriate depth and after a a preset time has elapsed, takes a photograph on a display card which, when developed back on the surface, indicates both the deviation from vertical in degrees and the compass heading or azimuth (Fig. 11.7).

Care must be taken to shield the compass from outside magnetic influences. The single-shot survey instrument is normally used in an open hole or placed inside a stainless steel, aluminum or similar, non-magnetic drill collar run in the string.

When interpreting the results it is essential that allowance be made to correct the magnetic compass display to a true bearing by applying an appropriate magnetic declination factor for the global location of operation if reference to surface mapping is required.

Mechanical wedging is not an operation to be embarked upon lightly and requires development of a range of skills through experience and training for effective use of equipment and correct interpretation of survey instruments.

The single-shot camera technique outlined above can be replaced by a multishot package which makes provision for up to 100 separate surveys to be made in one run whilst progressing down the drill hole. It can be further enhanced in areas of magnetic influence by the use of gyro compasses to replace the standard unit.

A photographic survey package which obviates the problems associated with gyro compasses and magnetic influence is the Reflex Maxibore unit (Fig. 11.8). This package comprises a probe assembly made up of four three-metre rods screwed together, one section holding a series of reflector rings and a liquid level sensor which definesthe vertical plane. The reflector rings are placed at 1.5 or 3-m intervals, depending on the curvature of the hole. The shorter system is used wherever the curvature is anticipated to be less than 100 m. The optical system is based on a light source directed at and reflected by the rings. Using video camera technology, a CCD area sensor recieves the reflected image and passes it to a real-time analyser.

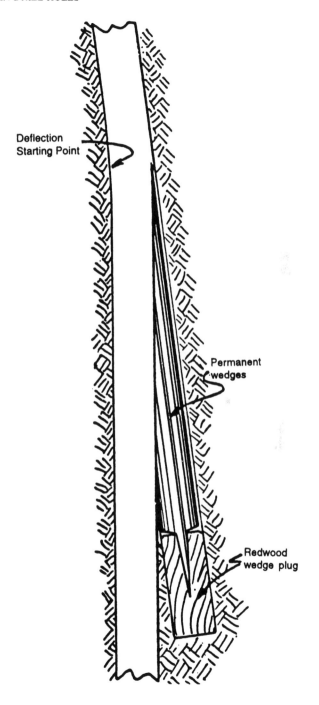

Fig. 11.6: Mechanical wedge.

This employs advanced pattern recognition techniques to extract deviation data and stores it directly in the onboard memory.

The image sensor records three reflector rings, each set in one of the probes and hence located 3, 6, and 9 metres from the lens. The rings are illuminated by small miniature lamps placed in a perspex ring in front of the main lens (Fig. 11.8). As the probe assembly bends to the contours of the hole, the camera images show the changes in the relative positions of the three rings in relation to each other. A ring-shaped bubble set between the plane of the film and the main lens enables the camera to photograph the bubble on each frame to provide a reference point for each survey image to define the vertical plane (Figs. 11.9 and 11.10).

When the survey has been completed, the Maxibore probe is retrieved by its wireline cable and the tool linked into a computer where the information is down loaded and converted by the software package to provide downhole regional and local co-ordinates, as well as dip and direction angles for each of the survey stations employed in the survey run. These are displayed either as a tabulated readout or plotted as a graphic display. All the information, including the downhole temperature readings measured by the probe, are filed for reference on a disc.

To accompany the advances in surveying techniques several 'inhole' deflection techniques have been developed. Probably the single most used method is the downhole motor, which was evolved for the oil and gas sector and then scaled down to suit the mineral exploration industry.

The downhole motor (DHM) uses the Moineau pump principle in which the stator is sleeved with a rubber obround cross-section inside which the solid steel helical rotor is fitted. When fluid is forced under pressure into the cavities formed between the rotor and the stator, the rotor develops both rotation and torque.

A universal joint is attached to the lower end of the rotor, which transforms the eccentric rotation of the stator into concentric rotation, which in turn is passed on to the drive sub attached to the other end of the universal joint. The drive sub runs inside a bearing housing and the drill bit is attached to the drive sub.

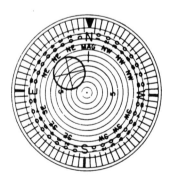

Fig. 11.7: Single-shot camera display.

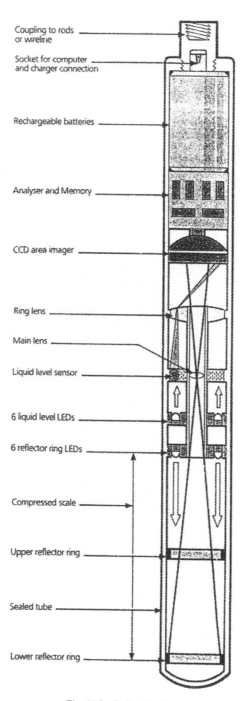

Coupling to rods
or wireline

Socket for computer
and charger connection

Rechargeable batteries

Analyser and Memory

CCD area imager

Ring lens

Main lens

Liquid level sensor

6 liquid level LEDs

6 reflector ring LEDs

Compressed scale

Upper reflector ring

Sealed tube

Lower reflector ring

Fig. 11.8: Reflex Maxibore.

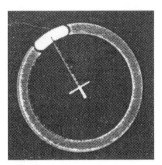

Fig. 11.9: Defining the vertical plane.

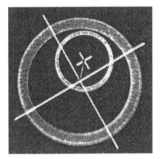

Fig. 11.10: Position of a ring relative to vertical plane.

The top sub of the DHM houses a bypass valve which allows the drill pipe to fill whilst tripping in and to drain whilst tripping out of the hole.

The rotational speed and torque developed by the downhole motor are a function of the volume of fluid pumped to the motor and the operating differential pressure, i.e., the difference between the pressure of the fluid measured at the stand-pipe without bit load and that with the bit loaded to drill. Being a positive displacement motor, the torque is directly proportional to the pressure differential across the motor. As the weight on the bit is increased, the surface fluid flow pressure increases. As the bit drills off, the pressure reading at the surface will fall away. The surface pressure gauge is used to indicate both weight on bit and torque generated.

Whilst drilling is taking place, with all rotational forces being generated by the DHM at the bottom of the hole, the drill string remains stationary. This enables three alternative methods of steering to be introduced into the system.

In one a bent sub is positioned between the bottom drill pipe and the downhole motor, which can introduce an angular displacement in the alignment of the drill pipe and the drill-off alignment of the drill bit (Fig. 11.11).

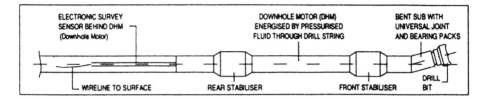

Fig. 11.11: Downhole motor with bent-sub assembly.

A second refinement to the bent-sub method uses a bent housing on the down-hole motor.

A third method of deflection is the deflection shoe, which is located at the lower bearing housing and constitutes the lowest non-rotating point in the system. The deflection shoe is a flat spring-loaded eccentric device which imposes an adjustable side load on the rotating components to produce a deflection in the path assumed by the drill bit (Fig. 11.12).

The DHM assembly is generally run with a non-magnetic drill collar immediately before the bent sub or top sub. This facilitates insertion of the inhole survey instruments if a single-shot camera is to be used and enables readings to be made with the drill string and downhole assembly left in place.

The downhole motor and its associated equipment, in particular the survey instrumentation packages, represent what is arguably the most sophisticated advances in drilling technology in the past two decades. This is particularly true in the area of MWD (measurement whilst drilling) systems. Here again the drilling industry owes a debt of gratitude to the oil and gas sector which developed the technology and the electronics industry which enabled the equipment to be miniaturised to suit the applications of both the mining exploration and construction industry, spawning a new drilling activity referred to as 'trenchless technology'.

The catalyst for these developments was essentially the MWD sensor package, which consists of a three-axis magnetometer and a three-axis inclinometer. The instrumentation is run behind the downhole motor in a non-magnetic collar located as a non-rotating drill string member immediately behind the rear stabiliser of the DHM and is run together with a data probe fitted to the sensor probe. The electrical signals developed by the devices are transmitted along a wireline run inside the drill string back to surface where they are decoded through dedicated software and displayed via laptop PC as a continual readout of the location attitude and bearing of the DHM drilling assembly whilst drilling proceeds. The course of the drill bit and drilling assembly is steered by rotation of the bent-sub assembly via the drill pipe to achieve targeted headings without the need to either withdraw the sensor package or the drill string (Fig. 11.13).

In the oil and gas sector development of MWD continues unabated with and transfer of data pertaining to temperature, lithology, gas detection etc. from the downhole sensor systems either by wire or electronic pulsation, as well as a variety

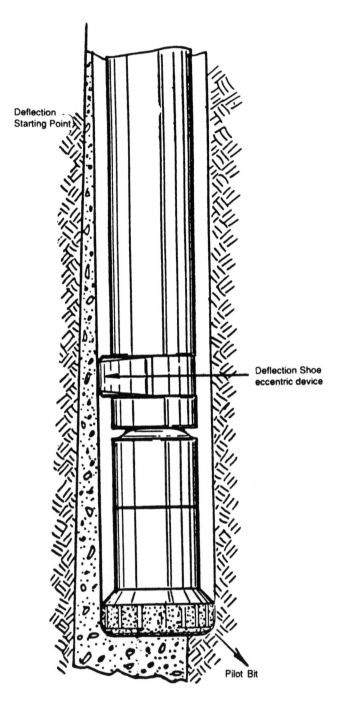

Fig. 11.12: Deflection shoe.

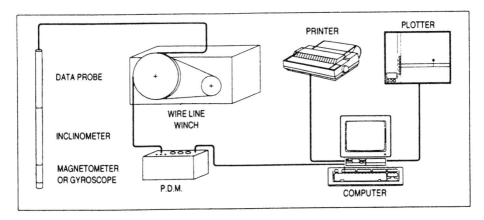

Fig. 11.13: Schematic layout of electronic borehole survey components.

of information acquired through some testing procedure in conjunction with another subsequent operation or separate procedure.

Where multiple wells are drilled from either one mother hole or adjacent well slots on a platform, the prespud planning and operational monitoring which can be provided whilst drilling commences is truly astounding, as can be judged by the plots illustrated in Figs. 11.14 and 11.15. These plots show both a vertical and horizontal plan for three adjacent wells. The major danger of one hole intersecting another is averted by a software programme in which a travelling cylinder analyses and calculates the relative position of each hole, which is then displayed as a polar plot, as illustrated in Fig. 11.16.

The achievements in directional control in both the mining and civil engineering sectors have been no less spectacular in pushing back frontiers that would have been considered insurmountable only a few short years ago.

Some of the operations which have been completed using the down hole motor and directional monitoring units are briefly described below.

At the Elura mine site in New South Wales Australia a pilot hole had to be drilled for a subsequent rise drilling operation in what was acknowledged to be a highly difficult formation noted for its deviation characteristics. The requirement was a 3.66″ H size hole between the 400 m and 800 m (1312 and 2624 ft) levels. The hole needed to remain within a 300 mm (12″) target cylinder over the 400 m (1312 ft) distance.

The parameters were set by the need to precision ream a 330 mm (13″) pilot hole to enable a 6-m (20-ft) rise to be pulled.

Although no actual core was required for the steeply dipping formations it was decided to core-drill the pilot hole since a long-core barrel and diamond bit were considered less prone to deviation. Despite the adoption of a formalised stabilisation program and good drilling practices, the first two holes had to be abandoned at 50 m and 60 m (164 and 197 ft) respectively, when they exited the target cylinder. The major difficulty other than the natural tendency for deviation in the formations

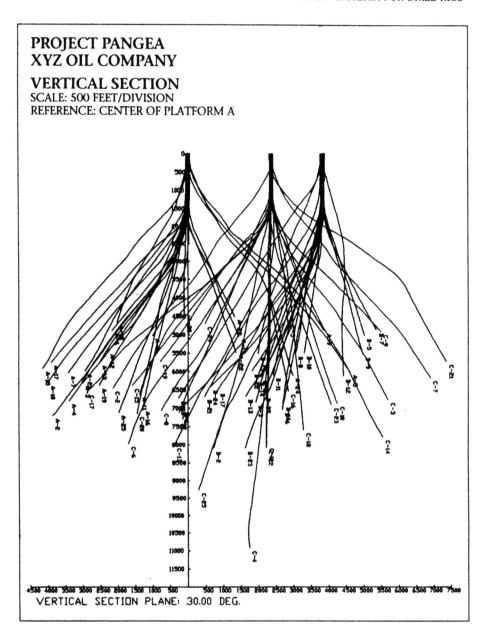

PROJECT PANGEA
XYZ OIL COMPANY

VERTICAL SECTION
SCALE: 500 FEET/DIVISION
REFERENCE: CENTER OF PLATFORM A

VERTICAL SECTION PLANE: 30.00 DEG.

Fig. 11.14: Multiple wells, vertical section.

was the use of a single-shot survey tool in the vertical hole and lack of prompt detection of deviation between survey runs.

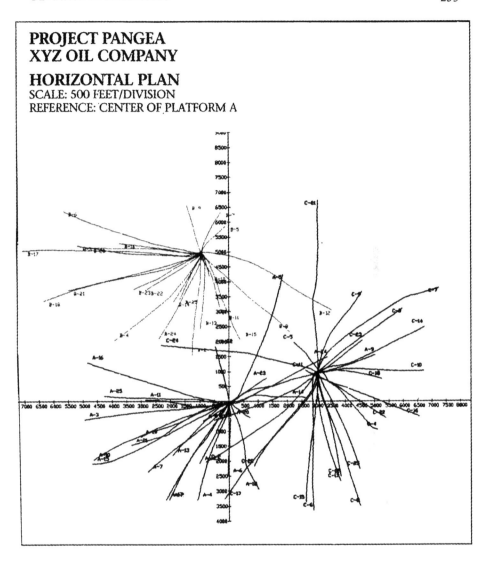

Fig. 11.15: Multiple wells, horizontal plan.

An alternative approach was adopted by introducing a DHM system with a deflecting shoe, together with a multishot instrument package which would run cluster shot hole surveys every 6 m. The instrumentation package was wirelined into the hole after every core run and the drill string rotated through 45° after each 8-shot cluster.

At the first sign of deviation the downhole motor assembly with deflection shoe and stabiliser assembly was run into the hole and the course adjusted to

PROJECT PANGEA
XYZ OIL COMPANY

TVD: 0 feet -> 4207 feet
MD: 0 feet -> 5000 feet   Interval: 20.00 feet
Separation Factor Plot   Method = Least Distance
Reference Well: A-21   Slot: A-21

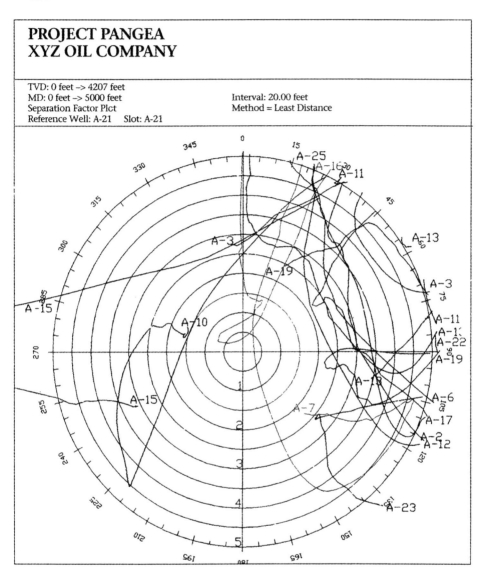

Fig. 11.16: Collision avoidance (travelling cylinder plot).

retain the alignment within the target cylinder. Once the corrected track had been established, the hole was reamed from 3.625″ (diameter of the side tracker bit) to 3.66″ to accommodate the stabilised core barrel.

At no time was the hole more than 0.5° off line; Fig. 11.17 shows the plot of the pilot hole at depth intervals within the 12″ cylinder target zone.

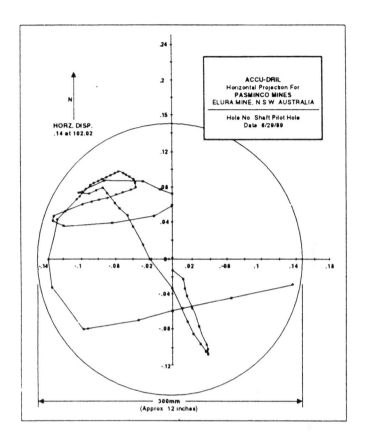

Fig. 11.17: Plot of pilot hole at the Elura mine.

As a result of this successful outcome, the contractors (Pontil Drilling Company) were given the contract to conduct a similar exercise at the Mount Iza Mine in Queensland where they drilled a pilot hole through 600 m (1969 ft) through a target cylinder 1-m dia (3.3 ft), which on completion was just 5.4″ out of alignment. It is reputed to be the straightest pilot hole in Australia.

In many countries of the world, inner city redevelopment has thrown up new challenges for the installation of commodity and service lines. The traditional methods of digging trenches for the installation of power lines, electric or gas, water and sewerage together with underground cabling are increasingly being questioned since they involve severe disruption of traffic and interference and disruption of essential services. Concomitantly, the proliferation of new improved communication links utilising fibre optic cables as transmission lines has created new opportunities to exploit the possibilities contained in directional drilling as an alternative.

In Australia, the establishment of a second communications carrier has required the installation of a network of some thousands of miles of fibre optic cables

across the continent. Where these have had to cross existing service roads, rivers
or building obstructions the new technology has been extensively used (Fig. 11.18).

Perhaps the most challenging project to date, however, is that recently undertaken in the USA. The project involved installation of a cable conduit but required
the drilling of two holes to accept a conduit with a minimum inside diameter of
5″. Hole number one was 519 ft long from the surface, an to a depth of 40 ft,
turning through 90° to run horizontally before being turned back through 90° to
emerge at a predetermined exit point (Figs. 11.19 and 11.20).

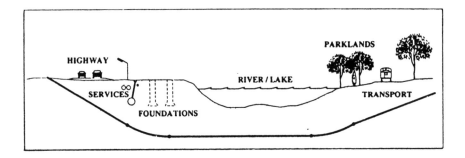

Fig. 11.18: Schematic illustration of directionally drilled service hole.

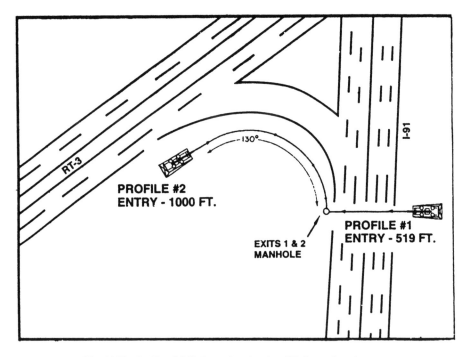

Fig. 11.19: Profile of drilled crossing showing 130 degree lateral turn.

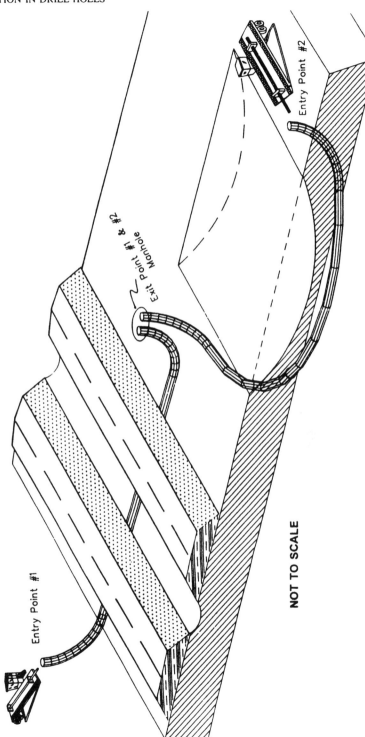

NOT TO SCALE

Fig. 11.20: Schematic layout of holes #1 & #2.

Hole number two was drilled some 1000 ft but in the process required a grade change from $-12°$ to $+10°$ with a $130°$ lateral turn throughout the 1000 ft to terminate in the same manhole as hole number one. The entire operation of drilling hole number two was achieved in just 5 days. The $130°$ lateral turn is believed to be a world record.

In the latter part of the millenium it is easy to become blasé about the pace of development, with each day seeming to bring about a new technological miracle. But if the rate of change in the drilling industry for the first 60 years of the 20th century is plotted against the technology changes achieved in the last 10 or 12 years, one is forced to conclude that here is an industry on the move. There appears to be no limit to the imagination of many of the people working in the industry or to what is achievable; yet, as an industry internationally, we are still struggling to establish the artistry and craftsmanship of the driller as a profession.

Is it that the electronics and computer technicians are to be the drillers of the future or is there still room for the solitary pioneering innovator with an inquisitive spirit for what might be, coupled with a blacksmith's ability to fashion a solution out of two pieces of metal which will set the pace of the industry in the coming century?

CHAPTER 12

WIRELINE CORING COUPLED WITH HIGH-FREQUENCY IMPACTOR*

INTRODUCTION

56-mm wireline coring is widely used in geological exploration in China. It gives fast drilling in solid intermediate to hard rocks and is easy to handle and transport. However, penetration rates drop sharply, even to zero, in extremely hard and non-abrasive rocks. In fractured strata core blocking happens very frequently with a marked reduction in length per core run. To overcome these problems a small-diameter hydraulic impactor is attached to the inner tube assembly which transmits impulse forces to the bit. The loads exerted on the rock are now not only the static thrust and torque, but also a dynamic force. This combination of three loads improves drilling efficiency and gives larger core runs.

POSSIBILITY OF FRACTURING ROCK WITH A SMALL DIAMETER HYDRAULIC IMPACTOR

To make it possible for the small-diameter hydraulic impactor to fracture rock, the amplitude of impulsive stress wave must be chosen properly according to rock properties. Only when the stress wave amplitude, σ_m, is equal to or larger than rock dynamic failure strength, σ_f, can the rock be fractured, i.e.:

$$\sigma_m \geq \sigma_f \qquad \ldots (12.1)$$

Using the Hopkinson rod, a high-speed impacting experimental set-up, rock dynamic failure strength was determined under impulsive forces (stress waves). During the experiment the wavelengths of the incident stress waves, τ, were 50,

---

*Excerpted from Yuan Gongyu. 1987. Wireline coring coupled with high-frequency impactor. *Drillex* 87. Inst. Mining and Metallurgy

100, 150 and 200 μs, amplitude of the stress waves from 39 to 290 MPa, impacting terminal velocity from 2 to 15 m/s, strain rate of the rod from 10 to 100 cycles per second, and the sample diameter 22 mm.

The results of the experiment indicated that the main factor related to rock dynamic failure strength is strain rate. Their relationship is

$$\sigma_f = A\dot{\varepsilon}^B \qquad \qquad \ldots (12.2)$$

where $\dot{\varepsilon}$ = strain rate; A and B = experimental constants.

If $\tau = 100\mu s$, then $\sigma_f = 485 \cdot \dot{\varepsilon}^{0.3}$ for skarn and $\sigma_f = 523 \cdot \dot{\varepsilon}^{0.25}$ for limestone. The dynamic failure strengths of skarn and limestone are 145 and 136 MPa respectively at the strain rate of 50 cycles per second. The hydraulic impactor for use with 54/43 mm drill rods has an impact hammer of 30 mm diameter whose cross-sectional area is 7.0686 cm². The anvil and the 54/44 mm outer core barrel also have a cross-sectional area of 7.0686 cm². Letting the hammer impact the anvil at a terminal velocity of 8 m/s (Fig. 12.1), the impacting force on the anvil is:

$$Pb = \mu_{a-b}(MaV_o)/2 \qquad \qquad \ldots (12.3)$$

where μ_{a-b} = transmissivity of stress wave,

μ_{a-b} = 2 Mb/(Ma+Mb);

   Ma = wave-drag of impact hammer, Ma = $\rho \cdot cA1$;

    c = velocity of stress-wave propagation in the hammer,

    c = 5130 m/s for steel;

    ρ = density of impact hammer,

    ρ = 7850 kg/m³ for steel;

   A1 = cross-section of impact hammer, here A1 = 7.0686cm²;

   Mb = wave-drag of anvil, Mb = ρ c · A2;

   A2 = equivalent cross-section of anvil and core barrel;

   V_o = terminal velocity of impact hammer, here V_o = 8 m/s.

If A1 = A2, then μ_{a-b} = 1. Substituting related data eqn. (12.3) yields Pb=113862 N. The stress amplitude on the impacted surface of the anvil is σ_m = Pb/A2 = 161 MPa. Apparently, σ_m is larger than σ_f, which means the rocks can be fractured.

Meanwhile, with the repetition of impacting, rock dynamic failure strength, σ_f, diminishes and approaches rock fatigue limit, σ_{-1}, (Fig. 12.2). For example, if σ_m is larger than 78.87 MPa and less than 145 MPa, fatigue failure will happen to skarn.

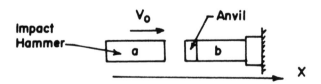

Fig. 12.1: Sketch showing impact system.

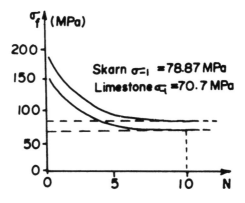

Fig. 12.2: Relationship between rock failure strength, σ_f, and repetition number of impacting, N.

From this point of view, the fracturing process of rock, based on static thrust and torsion is strengthened by applying an impulsive force.

STRUCTURE AND FUNDAMENT OF SMALL-DIAMETER WIRELINE CORING IMPACTOR

The drilling assembly includes a wireline core drill string, an impactor with its hanging positioner and force-transferring mechanism etc. (Fig. 12.3).

Hanging positioner:

Used to hand and position the inner tube assembly in the spring-lock chamber of the outer tube with the help of upper and lower spring-locks.

Impactor and force transfer mechanism

An impactor controlled by one valve has been developed, which strikes the anvil at a high terminal impact velocity. This is produced by the strong internal hydraulic pressure pulsation. The impactor has an outer diameter of 41.5 mm to pass through the wireline drill rods. The impact force is transferred to the bit via the bearing block, support ring, multiple keys and the outer core barrel. The bearing block can expand and contract to make the impactor work more reliably. The main parameters of the impactor are:

Borehole diameter	56.5 mm
Outer diameter of impactor	41.5 mm
Wireline drill rods	54 × 43 mm
Impact force per blow	Not more than 12.74 J
Pump volume	Not less than 70 1/min
Pressure drop	1.0–1.5 MPa

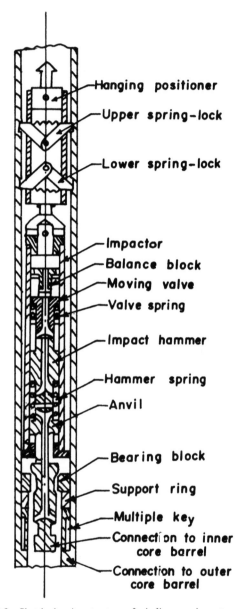

Fig. 12.3: Sketch showing structure of wireline core impactor assembly.

Working principles of the impactor

When the impactor is hung in the lower spring-lock chamber, drilling fluid flows through the annular space between the upper part of the impactor and the outer tube via the water slot in the anvil to the bottom of the hole. In this position the impactor will not operate. When the drill string is lowered to the bottom of the hole the impactor unit moves down along the multiple keys of the outer barrel; the upper spring-lock and its own weight force it down and applied thrust adds to this movement. This forces the bearing block to expand and seat itself on the support ring. This downward movement shuts off the slot in the anvil, causing a sharp increase in hydraulic pressure in the space between the top of the impactor and the outer barrel. This high pressure forces the moving valve and the impact hammer downwards together. The valve stops moving at its lower limit whilst the hammer continues downwards under its own inertia until it hits the anvil. This opens the central port on the hammer which evens out the pressure difference. The hammer and valve springs now return the floating valve and hammer and the whole cycle is repeated.

This whole operation is simple and reliable with few moving parts and a short multiple splined section. When pulling the tube assembly, the bearing block contracts and the whole assembly can pass through the drill rods. The float valve mechanism assures smooth travel to the moving valve which has a balance block on its upper face, which seals off the central port of the valve to restrict flow at the start of the downward stroke. As the hammer drops this port is opened which lets fluid flow rapidly to the upper space of the moving valve.

PROVEN PERFORMANCE OF WIRELINE CORE ROTARY-PERCUSSIVE DRILLING

Equipment, Accessories and Tools

Drilling machine	— Diamond drilling machine for 600–1000 m depth
Water pump	— Variable pump whose flow volume is not less than 70 l/min, and pressure not larger than 6 MPa
Accumulator	— Volume about 30 litres and pressure limit 8 MPa. Used to smoothen pressure fluctuation.
Pressure hose	— High-pressure hose of 1 to 2 in (25–38 mm) dia and length of nearly 30 m; pressure limit 8 MPa
Drilling bits	— Surface-set or impregnated diamond bits. Which are selected depends on rock properties. Outer dia 34 mm, hardness of impregnated bit matrix 40–45 HRC, diamond density 75% (in 400% scale) and diamond grain size 36–60 meshes. The dynamic strength of a single diamond must be over 60,000 J/m^2.

Drilling Operation Parameters

For surface-set diamond bits,

rotation speed —500–600 rpm
thrust —5000–6000 N.

For impregnated diamond bits,

  rotation speed —800–1000 rpm
  thrust   —6000–8000 N.

Clean water, non-solid fluid or low-solid mud may all be used as drilling fluids.

Proven Performance

This technique was tested in the field from 1985 to 1986. The results are given in the Table below, in which the tested depth of hole ranged from 167 to 493 metres, and the improved efficiency obtained by comparing the average values with that of conventional wireline core drilling. The tests show that this small-diameter hydraulic impactor designed on the basis of wave mechanics is successful. A small-diameter impactor attached to a wireline drill string can increase drilling rate and run footage appreciably. The impactor is characterised by its starting sensibility, simple structure, working reliability and convenient handling.

	Drilling Rate		Run footage	
Rocks	Average value	Improved efficiency	Average value	Improved efficiency
	(m/n)	(%)	(m)	(%)
Extremely hard and solid granite	2.60	47.50	2.00	50
Fractured dolomite	7.41	44.00	3.74	85

REFERENCES

[1] Xu Xiaohe, et al. (1984). *Fragmentation of Rock*. China Coal Industry Publishing House (in Chinese).
[2] Wan Buyan. (1985). The development of CS-56S hydraulic impactor for diamond wireline core drilling. M.Sc thesis, Central South Univ. of Tech., Changsha, China (in Chinese).
[3] Li Xibing. (1986). The dissipation of energy and fragmentation mechanics of rocks under percussive load. Central South Univ. of Tech., Changsha, China (in Chinese).

CHAPTER 13

RISK REDUCTION IN DEEP-LEVEL MINING VENTURES THROUGH APPLICATION OF UNDERGROUND DEFLECTION DRILLING*

Deviation and Deflection of a Borehole
Underground Deflection Drilling
Analysis of Deflection Drilling Configuration
Discussion and Conclusion

DEVIATION AND DEFLECTION OF A BOREHOLE

The accidental curved or spiral wandering common to most boreholes is known as borehole deviation and is distinguished here from deflection, which indicates a borehole trajectory which has been reorientated away from some former course by the drilling engineer using technological methods.

As is widely appreciated natural deviation of a borehole is induced both by rock formation conditions and by the drilling procedure employed[1]. For example a high-angle, oblique or cross-cutting borehole tends to deviate towards a trajectory normal to closely spaced rock foliation, jointing or bedding. Vertical boreholes almost invariably also spiral due to reactive torque on the drill string. However, boreholes inclined at a small angle to these structures often deviate around to a direction which is sub-parallel to them although Cottle[2] suggests that this effect is less common than presently thought especially for very thinly-bedded or foliated rocks. In addition, shallowly inclined boreholes commonly deviate to the right of the straight trajectory as a result of gravity and friction acting on the low side of the hole. Walsham[3] illustrates graphically how these difficulties can cause severe problems in an exploration programme directed at steeply dipping targets in West Cornwall. However, it has been found that it is possible to control these effects by

* Excerpted from Thorne, M.G., C.A. Jackson, J.K. Bawden and I.P.J. Doper, 1987. Risk reduction in deep-level mining ventures through application of underground deflection drilling. *Drillex 87.* Inst. Mining and Metallurgy

varying the drilling technology employed. Large diameter long-barrel core drilling equipment tends to deviate less than short-barrel, small diameter equipment. Packed hole assemblies with close tolerances between reamer and core barrel diameter also tend to produce a straighter hole. However, if the reamer is placed further back the bottom hole assembly then a fulcrum effect is created and downward deviation will be induced. Similarly, rigid thickwalled and internally stepped bits will generally minimize deviation but thinwalled bits exaggerate any natural tendencies and can be further enhanced by applying a greater weight or thrust to the bit. More costly remedial measures used with success include setting long casing strings[3] or mechanical wedges.

Thus, there is considerable scope for the drilling engineer to use his art to construct a borehole deflected back to a straight course from a deviated trajectory or to employ enhanced natural deviation to achieve an intersection with a specified target. Woods and Hopley[4] describe an excellent example from N.E. England where the Boulby Potash Seam is regularly evaluated up to 1000 metres ahead of the working face by means of a borehole initially drilled in the sub-seam halite bed and deflected upwards into the potash by means of a specially designed bottom hole assembly. The economic advantage of this system is further enhanced in that subsidiary boreholes or "daughter holes" can be deflected off from the original "mother hole" and thus the ground along the whole length of the hole can be explored without redrilling the initial straight sub-seam halite section. Similarly there are several mining companies in the Witwatersrand Basin, R.S.A. utilising controlled natural deviation to achieve multiple daughter hole intersections with the auriferous reef[3]. In this example boreholes tend to deviate strongly towards the normal to bedding which is of great advantage in obtaining accurate ore reserve intersections but as may be seen from Fig. 13.1 very careful control over initial mother hole orientation must be maintained or a premature intersection with the overlying reef is obtained. Steele[6] reports that at Mt. Isa Mine, Australia, sophisticated control techniques are used not to generate daughter hole intersections but to ensure that deep level targets are reached in a formation which induces severe deviation. A computer programme is employed to project the likely course of a borehole based on historical results in similar rocks. The programme incorporates factors to account for deviation induced by the core-rock bedding angle, proportion of non-foliated rock in the local succession and core size being recovered. It is employed first to set out the required trajectory before drilling begins and subsequently to update the trajectory achieved as survey data become available. As illustrated in Fig. 13.2 remarkably close agreement between predicted and actual trajectory can be obtained.

More robust deflection drilling schemes either override natural deviation tendencies or derive little benefit from them. Such schemes have long been common in the petroleum industry, in which deflection (or directional) drilling of production wells from a central platform is standard practice in many areas, e.g. the North Sea. McDonald, Rehm and Maurer[7] and Bakke[8] have presented useful reviews of these practices which, as reported by Beswick and Forrest[9], have recently been applied

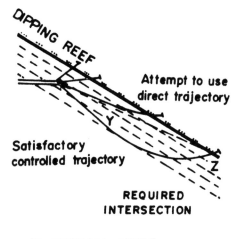

Y—CRITICAL ANGLE
Z—CORE-BEDDING ANGLE

Fig. 13.1: Controlling deviation for a planned down-dip intersection of a South African gold reef.
(after Dukas and Morkel[3]).

by the Camborne School of Mines in drilling large-diameter geothermal energy heat extraction boreholes in Cornish granite. Similarly, the soft-rock coal mining industry has produced effective schemes for exploration and degasification ahead of extraction panels similar to those described by Thakur and Dahl[10]. The system developed by Conoco (Mining Research Division) for its coal mines consists of a drill rig, drill bit guidance system and 43-mm borehole survey instruments. However, as most drilling sites are easily accessible on surface and low-cost small-diameter drilling technology is currently employed in most metalliferous mining operations, few applications of unassisted deflection drilling, apart from simple-wedge deflections on deep level gold and copper mines, have been attempted. Encouraging papers by authors such as Pritchard-Davies[11] and Steele[6] do not seem to have stimulated any new developments although the available technology has been constantly improved, e.g. Braithwaite[12]. It is commonly felt that both mechanical wedge and mud motor/bent-sub, deflection shoe or bent-housing deflection systems are still too inefficient and expensive for application to metalliferous mine exploration. The only widely published exception is the Kambalda nickel mine in Australia where, as illustrated in Fig. 13.3 from Steele[6], a combination of mud motor and mechanical wedges was used to spin off repeated daughter intersections of the nickel orebody from a near vertical mother hole. Steele has reported that the drillers at Kambalda claim that this technique avoids redrilling a difficult, highly weathered, near-surface section and provides much more accurate orebody intersections at a considerable cost saving. He also reports claims that slim-hole mud

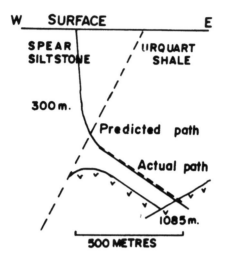

Fig. 13.2: Accurate computerised prediction of borehole trajectory at Mt. Isa Mine, Australia (after Steele[6]).

motor deflection drilling systems might be capable of angular orientation changes of up to 12° per 30 metres. McDonald, Rehm and Maurer[7] suggested 5° per 30 m as a more sustainable deflection rate (dogleg severity) and this is supported by unpublished reports of Boyles Bros. contract work in North America where mud motor hole deflections of 5–7° per 30 m were obtained.

In a typical operation in 'siliceous lava and volcanics, hardness 6–6.5' this company has employed a 2-3/8 in diameter (60.3 mm) Navi-Drill mud motor with a hard set spring-loaded deflection shoe, BXWL rods, a 70 gpm (5.3 l/s) pump operating at 40 gpm (3.0 l/s)/350 psi (24 bar) and a Boyles-designed full-face impregnated sidetracking bit rotating at 800 rpm to obtain a dogleg severity of 6° per 30 m. No difficulty in reverting to core drilling at the end of the deflected sections was reported and N-sized wireline core of good quality was recovered.

UNDERGROUND DEFLECTION DRILLING

The system of underground deflection drilling proposed in this chapter offers considerable potential for both reducing costs and increasing the quantity of information obtained from exploration core drilling in deep mines. It is primarily designed for use at the bottom of existing deep-level workings where there is limited strike development available for simple short-hole exploration programmes and thus few

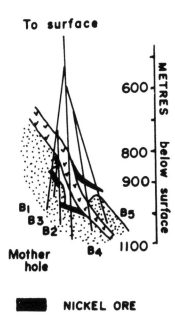

Fig. 13.3: Sophisticated deflection drilling scheme at Kambalda, Australia
(after Steele[6]).

opportunities for evaluating downdip extensions of the mineralisation. The proposed system is conceptually illustrated in Fig. 13.4. As can be seen there is a long declined mother hole giving the required strike extension and a series of daughter holes which are kicked off from it and obtain the downdip intersections. In this particular case the drill chamber is sited some way along an existing or purpose-built hanging wall cross-cut. Only eight daughter intersections are shown but any number may theoretically be obtained subject to the drilling and equipment limitations employed. Those constraints imposed by technical feasibility are discussed in greater detail below but in essence the factors which control borehole trajectory and thus intersection pattern are:

1) Dip and strike of target
2) Declination of motor hole (mother hole plunge)
3) Reach of the drill rig employed
4) Rate of curvature of deflected section (dogleg severity)
5) Length of cross-cut provided for the drill chamber

The interplay between these various factors is complex and a computer program BOREHOL has been developed to study the effect of varying parameters on a drilling situation. Thus in order to demonstrate the advantage of the proposed

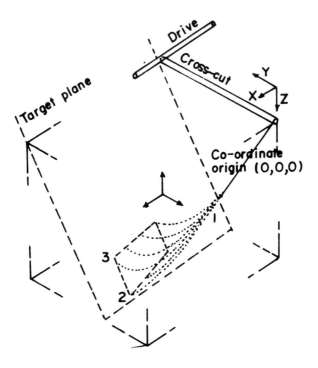

Fig. 13.4: Typical underground deflection drilling configuration as proposed in this paper. Note declined motherhole from origin to point "1", deepest and furthest intersection at point "2" and subsequent intersections ("3"...) to produce explored area as outlined on the target plane. (Output from GINO graphics routine within program BOREHOL).

system its technical constraints are discussed first, the operation of computer analysis demonstrated and the system then evaluated for a typical deep-level Cornish tin mine.

Drilling System Specifications

Two drilling systems are suitable for deflection drilling under the conditions outlined above: (i) a conventional core-drill/mechanical wedge system or (ii) alternatively a positive displacement (mud motor) drill/bent-sub system.

Conventional system

The conventional drilling system would employ a modern borehole deflection system such as that designed and developed by Encore Drilling Co. Ltd. In essence this system consists of a self-locking permanent wedge ('mole'), a wedge orientation survey tool equipped with SONAR telemetry to drill the chamber ('dolphin'), a retrievable steering partial wedge ('mule') and a wireline recoverable reaming spear ('ratty') which ensures the hole deflected by the 'mole' or 'mule' is effectively reamed to the required dimensions. This system[12] has the advantage of utilising standard core drilling and reaming equipment and also uses retrievable wedges ('mules') for the long deflected sections which can be lowered, oriented by means of the 'dolphin' and set in a single round trip. It is capable of producing good core recovery in deflected sections and is available in 46 mm and also B(60 mm) and N(75 mm) sizes.

A substantial amount of performance data is not yet available concerning the dogleg severity attainable with the Encore system. One of the authors, Mr. J. Bawden (Orebit Drilling Ltd.) feels that a severity of 4° per 30 m giving a total deflection (build) angle of 60° ought to be regularly attainable with the normal Craelius Diamec 260 drilling system his company operates. The drilling reach of this electrohydraulic rig is rated at 350 m with BQ steel rods but this has been exceeded by Orebit on many occasions in hard Cornish rocks including granite. One recent subhorizontal hole was drilled to 600 metres with BQ (56 mm) diameter rods. Thus it is suggested that a maximum build angle of 60°, dogleg severity of 4° per 30 m and reach of up to the rated capacity of the electrohydraulic rig used should be technically feasible. Several rigs and pumps are available. Some are more common in surface drilling modes but could be modified for underground use. Fifty horsepower rigs should be capable of a reach of 1000 m with a 60° build angle, provided a smooth trajectory has been maintained and effective flush-water drill-string lubricants are used.

In order to maintain close control of the borehole course it is necessary to supplement the 'dolphin' steering tool control with regular check surveys by a magnetic photographic survey tool.

Downhole Motor Drilling Assembly

Two basic types of downhole motors are available. The turbine drive or turbo-drill is not manufactured in small diameters suitable for this drilling system. The positive displacement type essentially consists of an elastomer stator and a solid-steel helical rotor. When fluid is forced between the rotor and the stator, a drive shaft connected to the rotor by means of a universal joint begins to rotate and a drill bit is turned without rotation of the drill string in the hole. The most commonly known positive displacement motors are the Dyna-Drill Delta 500 or the Christensen Navi-Drill Mach 3 series. However, Drilex of Aberdeen are developing a 2-3/8 in diameter multilobe positive displacement motor which may be suitable for metalliferous mining applications. Existing high-speed low-torque motors are compared in Fig. 13.5a, modified from Jackson[13]. Directional control over the motor is provided either by an oriented offset to the drive shaft at the universal joint housing or by a spring-loaded deflection shoe on the first non-rotating member above the bit (lower bearing housing). The latter method is more likely to be effective in hard rock environments and may need to be supplemented by a second deflection shoe fixed at 180° to the first at the upper end of the tool. A range of conventional bits can be employed with a motor drill but full-hole side-tracking bits are the most effective. For hard rocks an impregnated diamond bit should be tried.

Experience on the Geothermal Project[9] showed that in granite circulating drill cuttings inevitably cause motor wear and must be controlled by adopting a single-pass, open-circuit, freshwater flushing system. As operation of the $1\frac{3}{4}$ in Navi-Drill would require 37 gpm (170 l/min) at a pressure drop of 465 psi (32 bars) to provide a 1750 rpm operating speed, both water supply and pumping requirements are considerably in excess of those normally needed for exploration drilling. Fig. 13.5b illustrates the degree of variation in speed of rotation and torque which are available through pump pressure and capacity control. The drill rig is likely to be the same, however, as for conventional deflection drilling described above since at the completion of the deflected hole section a normal core barrel must be used to core the target.

To correctly orient the deflection assembly a carefully aligned orientation sub and survey collar are located above the motor's bypass valve and a single-shot survey camera is fastened to a mule shoe located in the survey collar. Above the tool 10 metres of non-magnetic drill collars shield the compass unit from the effects of the steel drill string. When the motor and deflection shoe have been surveyed into correct alignment, the drill chuck is locked and motor drilling can begin.

There is no doubt that existing commercially available motor drilling systems are capable of dogleg severities of 4° per 30 m, a total build angle of 60° and a

CHRISTENSENS NAVI-DRILL SERIES

TOOL SIZE O.D. INCH (mm)	MAX. HOLE SIZE INCH (mm)	STRING	PUMP RATE RANGE U.S GPM (l/min)	DIFF. PRESS PSI (bar)	BIT ROT. RANGE RPM	TORQUE FT/LB (Nm)	H.P. RANGE H.P (Km)
1 3/4 (44.5)	2 3/4 (69.9)	AW ROD AW ROD	20 - 45 (75 - 170)	465 (32)	720-1750 720-1750	25 (32)	4 - 8 (3 - 6)
2 3/8 (60.3)	3 1/2 (88.9)	BW ROD BW ROD	29 - 75 (110 - 275)	700 (48)	550-1370 550-1370	80 (110)	8 - 21 (6 - 16)
2 3/4 (69.9)	3 15/16 (98.4)	NW ROD NW ROD	37 - 100 (140 - 385)	700 (48)	485-1200 485-1200	135 (183)	12 - 31 (9 - 23)

DYNA-DRILL DELTA 500

1 3/4 (44.5)	2 3/4 (69.9)		18 (83)	(41)	810 810	25 (33)	4 (2.8)
2 3/8 (60.3)	3 1/2 (88.9)		22 (98)	(69)	1100 1100	(61)	(7)

Fig. 13.5a: Comparison of Positive Displacement Motor Operating Characteristics

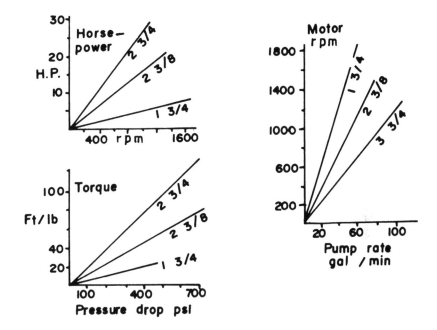

Fig. 13.5b: Operating characteristics· of the Christensen microslim series of Navi-Drill positive displacement mud motors (from Jackson 13 and Boyles Bros. operators).

reach of 1000 m. They have an advantage over the conventional wedging system described above in that the drilling method is proven for large build angles but also have a disadvantage in that the technology is new to the metal mining business and is expensive to introduce by either rental or purchase.

ANALYSIS OF DEFLECTION DRILLING CONFIGURATIONS

Analysis of drilling configurations is carried out initially by stereographic manipulation to establish geologically acceptable target intersection orientations. These are then input in program BOREHOL which calculates the collar site required to produce these intersections with specified equipment constraints. Output from BOREHOL consists of the co-ordinates of eight acceptable target intersections and a plane of reef plan showing both the location of these intersections and the area of reef investigated by them. It also produces a three-dimensional view of the drilling configuration, represents the location of the borehole collar, calculates the length of the hanging-wall cross-cut required to reach this location and computes the total length of drilling required for the eight idealised intersections. Fig. 13.6 illustrates the various stages in this analytical procedure. A detailed discussion of some aspects of the program and the stereographic analyses is given in Jackson[13].

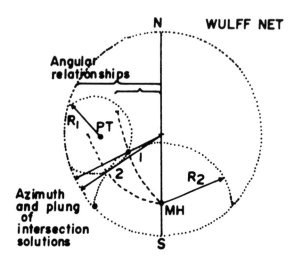

Fig. 13.6: Computational steps for evaluation of a deflection drilling scheme using program BOREHOL.

1. Plot circle radius angular deflection (R2) about mother hole (MH) and circle radius 90 minus permitted intersection angle (R1) about pole (PT) to target plane on Wulff stereonet. Intersection points (1,2) give two permitted daughter hole solutions. Azimuth, plunge and angular relationship with low side of mother hole are read off as shown.

2. Input to program BOREHOL: Mother-hole azimuth and plunge; Intersection One azimuth, plunge and angular relationship with low side of mother hole; Intersection Two azimuth, plunge and angular relationship with low side of mother hole; Dogleg severity (normally 4°/30 m); Drilling rig reach; Kick-off point spacing; Permitted maximum angular deflection (normally 60°).

3. Program BOREHOL calculates:

3A. Length of straight section before first kick-off point (K.O.P.) taken from drill rig reach, length of deflected section, K.O.P. spacing and final 15-m straight core run (point 1 on diagram).

3B. Co-ordinates of end of mother hole projected to end of furthest deflected section (XCOR, YCOR, ZCOR).

3C. Length of required hanging wall cross-cut for furthest intersection to be achieved.

3D. Lengths of straight coring sections for all other intersections to be achieved.

3E. Final co-ordinates of all intersections (XI (n), YI (n), ZI (n))

3F. Points of intersection in plane of lode plan and intersected area (points 2,3 etc.).

4. Program BOREHOL plots results and isometric diagram of drilling configuration.

In the present analysis certain drilling and target intersection constraints are fixed or must be input interactively. Firstly, it is assumed that meaningful intersections must make an angle of at least 60° with the target. More oblique intersections increase the viability of deflection drilling but risk biased sampling and uncontrollable deviation if the ore is soft. Secondly, the dogleg severity is assumed to be 4° per 30 m to produce a maximum build angle of 60°. Finally, the overall drill rig reach is entered interactively (in metres) with the option of specifying the length of the initial straight mother-hole section to the first kick-off point. At this point the distance in metres between subsequent kick-off points is requested; a minimum 15-m straight section after completion of the deflected section but before intersection of the target is automatically assumed. This final straight section is to prevent accidental drilling of the target with the mud motor and may need to be longer in areas of uncertain geology. However, as can be seen in Fig. 13.4 the near-collar daughter holes have much longer straight sections and thus an additional safeguard is built into the scheme. As mentioned above, the two variables which may be manipulated to obtain the desired intersections are mother-hole azimuth and plunge relative to strike and dip of the target. At present, these specifications must be optimised by trial and error but work is in progress to adapt program BOREHOL to perform this task.

Evaluation of deflection drilling system-South Crofty Mine, Cornwall

South Crofty tin mine is the largest underground tin mine in Europe and produces about 1500 tonnes of tin metal in concentrates each year. It is located on the margin of the Carn Brea granite cupola although most of the modern workings are entirely in the granite. Tin as cassiterite is won from a large number of narrow (2 m wide) vein-lode structures over a strike length of several miles and to depths of up to 730 m below the surface (Fig. 13.7). The lodes dip at more than 50° both north and south and have a subparallel NE - SW strike. In-depth extensions of these lodes have not been explored by surface drilling for many years and all evaluation drilling has been conducted by short-hole drilling from parallel development. However, the mine has now approached the bottom of its existing shaft system and a subincline is being constructed.

The current lower working level is the 400 fathom (730 metres below surface) and as can be seen from Fig. 13.8 the area of future resources which may be investigated by conventional methods is strictly limited unless extremely oblique vein intersections are to be tolerated. In contrast, even the simplest deflection drilling configurations employing the drilling constraints discussed above permit large areas of ground, up to 800 m along the strike and 500 m below the bottom level, to be investigated. Figure 13.8 is a plan of the 400 level and shows how current excavations can be utilised for drilling chambers to drill out the important No. 4, No. 8, and Roskear 'B' lodes along the strike and in depth. Further possible development is shown in broken lines and would enable additional intersections on these lodes and also on the major Tincroft lode. Obviously, on-lode development would enable additional intersections to be obtained further along the strike. Similarly, if the full reach of the drill rig is not employed, then some intersections closer to the drill bay could be obtained. A tabulated summary of the proposed deflection drilling schemes for the lodes and drill sites illustrated in Figs. 13.7 and 13.8 is given in Fig. 13.9. As can be seen these intersected areas are for a 60° minimum lode intersection angle, a drill rig reach of between 600 and 1000 m, mother-hole azimuths between 0 and 15° towards the strike and mother-hole plunges of 10 to 25°. However, in order to evaluate the versatility of this deflection drilling system the five major constraints on drilling configurations were varied:

1) Mother-hole azimuth and plunge
2) Available collar locations

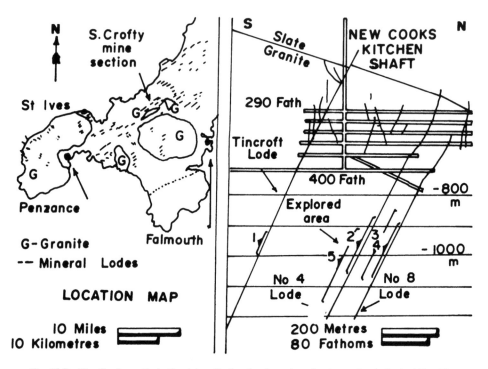

Fig. 13.7: The Camborne-Redruth mining district showing mineralisation at South Crofty Mine. The numerals 1-5 refer to the area of lode investigated by various deflection drilling schemes discussed in the text.

3) Drill rig reach
4) Downhole distance between kick-off points
5) Permitted intersection angle with lode.

For South Crofty lode targets dipping at 65° it was found that a drilling site in the hanging wall produced solutions which were computationally simple and required the shortest drilling distances. In these trials the mother-hole azimuth and plunge were varied together for a variety of drilling rig reaches. The required collar location thus became a dependent variable. Initially the permitted intersection angle drilling rig reach and kick-off point spacing were fixed at 60°, 1000 m and 100 m respectively. Computer analysis under these constraints produced the results graphed in Fig. 13.10 which may be summarised as follows:

a) As the bearing of the mother hole increases so as to converge with the target, the area of intersection of the eight daughter holes also increases.

b) As the bearing of the mother hole increases, the plunge of the mother hole to achieve intersection is reduced and shallower intersections can be obtained. These are also laterally farther away (along the strike) from the drill chamber.

c) For these steeply dipping targets there is an optimum plunge of mother hole which will produce a maximum target intersection area.

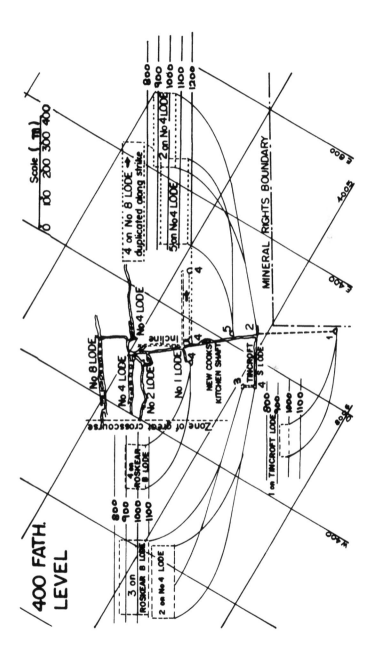

Fig. 13.8: Plan of South Crofty Mine 400 fathom level showing sample deflection drilling schemes and a selection of technically feasible exploration targets. Drilling schemes are denoted by numerals 1-5 and are the same as in Figs. 13.7 and 13.9.

NO.	MOTHER HOLE AZIMUTH/ PLUNGE	MAX. RIG REACH (m)	K.O.P. SPAC-ING (m)	MAX./MIN. X, Y, Z CO-ORDS	INTER-SECTED AREA (m)	CROSS-CUT LENGTH (m)	REQ'D DRILL LENGTH (m)	REQ'D DEVP'T LENGTH (m)
I	180/15	600	40	455,201,268 340,171,202	5293	295	4122	295
2	190/10	1000	100	849,379,324 580,298,151	43954	450	4598	0
3	195/10	1000	100	825,455,355 562,348,127	53797	514	4694	46
4	180/25	600	40	453,235,342 296,162,184	15558	321	4137	II
5	180/25	800	100	634,266,426 321,162,204	41993	361	4308	0

Fig. 13.9: Deflection drilling data for South Crofty Mine illustrations.

*Where the strike of the target mineralisation is standardised to 180° before beginning the analysis. An azimuth of 190° indicates a mother hole oriented 10° towards the dipping target on the hanging wall side.

d) To increase the mother-hole plunge means an inevitable shift in the required collar position, which may necessitate additional hanging-wall cross-cut excavation and a slight increase in permitted total drilled length.

It can be seen from Fig. 13.11 that the proposed drilling system is capable of producing a very wide range of intersections with extreme maximum and minimum Z co-ordinates ranging from 678 m and 428 m below drill bay elevation for a parallel mother-hole azimuth and 15° plunge, to 228 m and 63 m below drill bay elevation for a mother-hole azimuth 15° towards the strike and zero plunge. Use of Fig. 13.11 together with Fig. 13.10 allows a quick visual estimate of the likely drilling configuration required to intersect any particular underground section of the 65° dipping lode.

To obtain other areas of intersection of the lode it would be possible to change the drill rig reach with obvious effects, to move the drill bay along the strike, again with obvious effects, or to permit a more oblique intersection with the lode. Figure 13.10D illustrates the effect of varying the permitted intersection angle from 40° up to 60°. In this case permitting an angle of 50° has very considerable advantages and would certainly be worth further consideration. Similarly, by reference to Fig. 13.9 it can be seen that the maximum area of intersection is dependent on downhole distance between kick-off points. By changing the K.O.P. spacing from 75 m to 150 m the intersected area is increased from a mere 30,000 sq. m to 67,000 sq. m. At normal lode widths and densities the latter area could conceivably contain a quarter million tonne oreshoot which is precisely the type of target expectation this kind of exploration programme would be designed to evaluate.

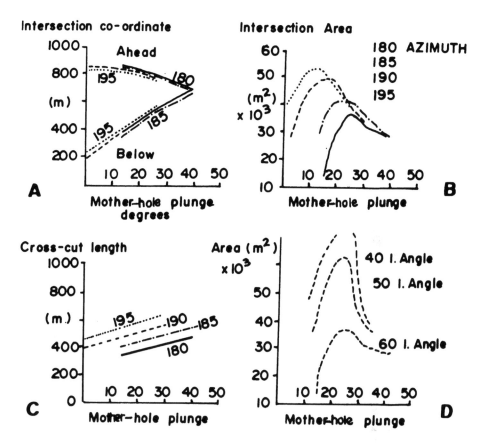

Fig. 13.10: Results of computerised simulation of alternative deflection drilling schemes. A—Variation of intersection position co-ordinates with mother-hole azimuth and plunge; B—Variation of explored area with similar azimuth and plunge variation; C—Required hanging-wall cross-cut length for each of these schemes; D—How explored areas vary with change in the permitted core-target intersection angle. All cases developed for a 65° dipping target which strikes at 180°.

With such a diversity of possible configurations the deflection drilling programme must be carefully designed in order to minimise the cost of exploration of a unit area.

Illustrative cost projections

The eventual application of deflection drilling methods to metalliferous mine exploration will inevitably be dependent on the capital and operating costs of the system. The drilling configurations discussed above have advantages but relatively complex technology is unlikely to be available cheaply. In order to arrive at illustrative costs for the deflection drilling configurations derived as above for South Crofty Mine the following assumptions are made:

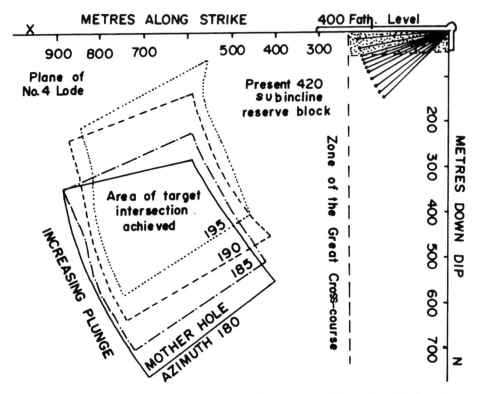

Fig. 13.11: Plan of lode diagram showing location of target areas which are obtainable from the collar position shown with mother-hole azimuth and plunge as illustrated. No. 4 lode, South Crofty Mine. (Note extent of currently known reserves which were evaluated from the subincline shaft being developed below the 400 fathom level. Further explored areas can be obtained by simply moving the borehole collar along the strike).

a) To drill a borehole in excess of 500 m with conventional drilling equipment requires a staged reduction in hole diameter from H(99 mm), N(76 mm), B(60 mm) to a final diameter of TT46 (46 mm). to simplify costing over this range an inclusive average of £36 per m was taken. (Underground, Diamec 260, granite, 1985-86, ex Orebit Drilling Ltd).

b) The cost of excavation of an approximately 3 m × 3 m tracked hanging-wall cross-cut development to the drill bay including forced ventilation and excavation of drill bay itself is taken to be £300 per m.

c) The costs of the wedging system manufactured by Encore Drilling (now under licence by Diamant Boart) as supplied by Orebit Drilling Ltd. are:-

Retrievable wedge = £800

Permanent wedge = £1200

For a retrievable wedge approximately six to eight runs could be made before loss of gauge. In hard granite six runs per wedge are assumed.

d) Cost data for drilling with small-diameter downhole motors is scant. However, experience from the C.S.M. Geothermal Project, admittedly for larger diameter motors, suggests that in granite actual drilling costs are certainly no more and possibly less than the cost per metre for conventional rotary drilling. The comparison with diamond drilling is not as straightforward but it is expected that rate of penetration will be no worse, wear on drill string will be substantially less and increased cost will arise only from bit cost. Power consumption at the pumps will be greater but will be less on the rig itself. Thus at present an actual operating cost of £36 per m is assumed, which is the same as for conventional diamond drilling. The major additional adverse factors are the hire or capital cost of the motor itself and its maintenance. These were estimated as follows for 1985/86:

Capital cost of Dyna-Drill microslim downhole motor 1-3/4 inch diameter from Dyna-Drill (Div. Smith Int.) = £6500.

Plus maintenance/refurbishment/replacement of bearing assembly after life of 125 drilling hours = £2400.

Hire cost for 1-3/4 inch Christensen Navi-Drill (ex Boyles Bros., N. American hard rock mines) additional to normal drilling rate = £35 per hour.

Drilling rate of 2.5 m per hour in granite.

Using these figures the projected cost of alternative schemes for exploration of South Crofty No. 4 Lode was evaluated. A drilling site is available on the existing lowest mining level (400 fathoms) which will produce eight favourable target intersections with a total drilling length of 4308 m using a drilling rig reach of 800 m and K.O.P. spacing of 100 m.

Alternative 1—Wedging

(a) Drilling 4308 m at £36/m   = £155,088

(b) Initial kick-off from mother hole, permanent wedge £1200 * 8   =   9,600

(c) Building angle at 1.5° per wedge, 60° = 40 wedges per hole,   =   42,880
life of retrievable wedge six runs = 6.7 wedges in each daughter
hole. Eight daughter holes = 8 * 6.7 * £800

TOTAL   =   £207,568

Alternative 2—Hired downhole motor

(a) Drilling 4308 m at £36/m   = £155,088

(b) Initial kick-off from mother hole, permanent wedge £1200 * 8   =   9,600

(c) Each daughter hole is 450 m, drilling rate 2.5 m/h, hire of   =   50,400
motor 450/2.5 * 8 hours at service cost of £35/h

TOTAL   = £215,088

Alternative 3—Purchased downhole motor

(a) Drilling 4308 m at £36/m   = £155,088
(b) Initial kick-off from mother hole, permanent wedge £1200 * 8   =   9,600
(c) Cost of two $1\frac{3}{4}$ inch downhole motors is 2 * £6500   =   13,000
(d) Total deflection drilling time is 450/2.5 * 8 = 1440 h.   =   27,648
    Refurbishment of motor costs £2400 after 125 h. Maintenance cost
    1440/125 * 2400

                                                    TOTAL = £205,336

Similar illustrative costs have been developed for the other drilling configurations discussed above and shown in Fig. 13.8. These costs are tabulated in
Fig. 13.12. Interestingly, they are all in the range £200,000 to £300,000 and as
expected are dominated by the cost of drilling itself and the additional cost of a
hanging-wall cross-cut when required.

DISCUSSION AND CONCLUSIONS

This study illustrates the potential usefulness of deflection drilling techniques for
exploration of a Cornish tin lode system. Geologically it is exciting in that each

TARGET	TOTAL DRILLED LENGTH (m)	CROSS-CUT DEVELOP-MENT (m, cost)	TOTAL COSTS		
			Wedges	Hire Motor	Buy Motor
TINROFT at 900-1000m. (500 fath) Scheme 1	4122	295, £88500	£289692	£286892	£275292
No. 4 LODE 900-1000m. (500 fath) Scheme 2	4598	0, £0	£218328	£225528	£203928
ROSKEAR "B" 800-1100m (560 fath) Scheme 3	4694	46, £13800	£235584	£242784	£221184
ROSKEAR "B" 800-1100m. (560 fath) Scheme 4	4137	11, £3300	£205032	£212232	£190632
No. 4 LODE 900-1200m (600 fath) Scheme 5	4308	0, £0	£207568	£215088	£205336

Fig. 13.12: Drilling cost projections for South Crofty Mine exploration schemes

eight-hole configuration is capable of probing an area of lode which comfortably contains the average Cornish ore shoot. However, there are objections to application of the method, not the least on grounds of cost. It is difficult to see how costs of even the most successful programme could be reduced below about £1 per tonne of ore discovered.

In addition, it is commonly argued that eight boreholes over an area of 66,000 sq. m are unlikely to provide an accurate evaluation of that section of the tin lode. This is true in the sense that a statistical reserve assessment of tin mineralisation must inevitably include a payability factor based on drivage and raising on lode along with more intensive diamond core drilling results. However, primary exploration for economic tin mineralisation must begin by proving sufficient continuity of structure and enough demonstrably payable ore grades for there to be a reasonable expectation of an ore shoot if further exploration is conducted. If this is possible from existing workings, then deflection drilling has few advantages. If the mineralisation continues below the current bottom level, however, then this conventional approach is impossible without further speculative development in depth. In this case deflection drilling provides the possibility of evaluating resources below the bottom level without a commitment to shaft deepening or subincline development and the concomitant investment risk is reduced. In one of the authors own experience in the 1970s at Wheal Jane Mine, near Truro, confidence in the future development of the mine was placed in less than eight intersections of payable grades in a structure or structures of unknown continuity. Since then and lately under new ownership this confidence has proved justified; the blind South Lode orebody has been in production for several years and until the recent tin price collapse a series of good profits were posted. However, as development has now progressed below these initial deep-level intersections it can be seen that they were in fact in several separate, parallel, discontinuous structures without significant between-hole continuity. As far as one of the authors is aware, the fact that other much better results have subsequently been obtained is largely due to successful corporate risk-taking by the previous and present owners, a general confidence in the style of mineralisation in the district, and the competence of the mine geological staff. Deflection drilling below the shaft bottom would have and still may permit very significant risk reduction in this particular enterprise.

Thus deflection drilling offers exciting new possibilities for deep-level exploration. The case has been developed in this paper for Cornish tin although work' is currently in progress for South African reef gold mines[13]. There are many outstanding technical questions such as, for example, what are the relative costs of downhole motors versus modern wedge drilling for the deflected sections, which are the most economical bits for motor drilling, which rig is most adaptable for the considerable thrust loading requirements and so on. These can only be answered by a demonstration project specifically designed to evaluate the alternative systems. Is there enough interest from manufacturers, operators and mining companies to consider a joint approach with the authors and Camborne School of Mines to the E.E.C.

Raw Materials Programme for funds under the heading of technology transfer from the soft-rock/oil industry to the rock-hard mining sector?

REFERENCES

[1] Gauthier, G. (1967). Basic theories of deflection in diamond drilling. Science and technology of industrial diamonds, *Proceedings Oxford Conference*, Vol. 2 Industrial Diamond Information Bureau, London.

[2] Cottle, V.M. (1958). Diamond drillhole deflection at Roseberry, Tasmania. The F.L. Stillwell Anniversary Volume, *Australasian Institution of Mining and Metallurgy*.

[3] Walsham, B.T. (1967). Exploration by diamond drilling for tin in West Cornwall. *Proceedings Institution of Mining and Metallurgy*, Section A, April 1967, pp. A49–A56.

[4] Woods, P.J.E. and Hopley, R.J. Horizontal long hole drilling underground at Boulby Mine, Cleveland Potash Ltd. *Mining Engineer*, Vol. 139.

[5] Dukas, B.A. and Morkel, H.C. (1983). Surface and underground drilling techniques used in exploration drilling. *Journal South African Institution of Mining and Metallurgy*, pp. 164–169.

[6] Steele, K. (1981). Deviation technique in diamond drilling. *Geodrilling* 13:8–12.

[7] McDonald, W.J., Rehm, W.A. and Maurer, W.C. (1979). Improved directional drilling will expand use. *Oil and Gas Journal*, February 1979.

[8] Bakke, S. (1986). Directional drilling: the state of the art. *Petroleum Review*, March 1986, pp. 48–56.

[9] Beswick, J. and Forrest, J. (1983). New low-speed, high-torque motor experience in Europe. *Society of Mining Engineers, September 1983 Conference, New Orleans*.

[10] Thakur, P.C. and Dahl, H.D. (1982). Horizontal drilling: A tool for increased productivity. *Mining Engineering*, March 1982, pp. 301–304.

[11] Pritchard-Davies, E.W.D. (1971). The future of directional control in diamond drilling. *Industrial Diamond Review*, February 1971, pp. 67–70.

[12] Braithwaite, R.W. (1984). Branching and steering small diameter cored holes. *Mining Magazine*, No. 2.

[13] Jackson, C.A. (1986). Deflection diamond drilling techniques and their application in underground exploration. Unpublished Final Year Project, Camborne School of Mines, Pool, Redruth, Cornwall (130 pages, appendices).

CHAPTER 14

LATERAL DRILLING SYSTEM

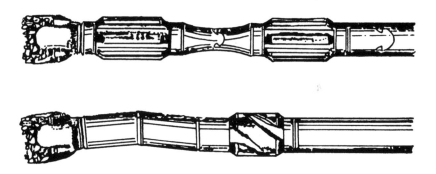

Fig. 14.0

Horizontal well technology can help oil companies solve completion problems and increase production from many types of reservoirs. In just the past few years application of horizontal wells has expanded rapidly to include new wells and recompletions on land and offshore, in oil- and gasfields throughout the world.

Proper planning is the most important factor in a horizontal drilling project. Horizontal wells have to be accurately placed in a three-dimensional target. Drilling fluid and hole-cleaning programmes are more critical than for conventional wells. Improper bit selection or inappropriate bottom-hole assembly design can result in low penetration rates and stuck pipes.

To avoid these problems, Eastman Christensen's horizontal well-planning process includes in-depth drilling engineering and advanced computer software. In addition, each horizontal well is planned based on thorough knowledge of drill bits and horizontal drilling equipment and how they should be applied in local drilling conditions.

Horizontal drilling experts can determine whether short-, medium- or long-radius techniques (or a combination) will meet the objectives (Fig. 14.1.) They can also recommend the well profile and drilling procedures that will most efficiently place the required length of horizontal in the target zone. Since 1979, Eastman Christensen—the world's leader in horizontal drilling services—has provided its proven Lateral Drilling Systems to the industry. These systems include specialised short-and medium-radius equipment, as well as Nor Trak Navigation Drilling System for long-radius profile wells.

Eastman Christensen has the industry's most experienced staff of horizontal drilling experts. Their engineers and co-ordinators have planned and supervised over 500 horizontal well bores.

COMPLETE RANGE OF BUILD RATES

Lateral drilling differs from conventional directional drilling in several ways. Build rates in lateral drilling are much higher than conventional drilling. Angle build rates

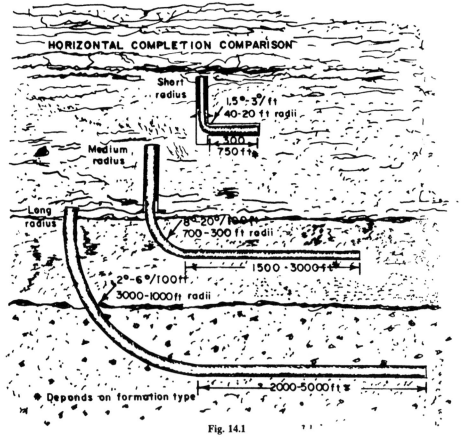

Fig. 14.1

(Courtesy: Baker Hughes Inteq)

for short-radius lateral drilling run from 1.5°–3°/ft (0.31 m) and medium-radius lateral drilling from 8°–50°/100 ft (31 m). Conventional directional drilling builds angle at a rate of 1°–5°/100 (31 m) (See Fig. 14.2).

Lateral drilling also allows a large portion of the horizontal borehole to be contained in a single zone of interest, so a large area of the pay-zone is exposed. In conventional directional drilling only a very small percentage of the well bore (typically only the termination point) makes contact with the producing zone, resulting in very limited exposure.

In addition, multiple lateral holes can be drilled from a single well bore allowing completion of more than one formation. This is rarely accomplished using more conventional techniques.

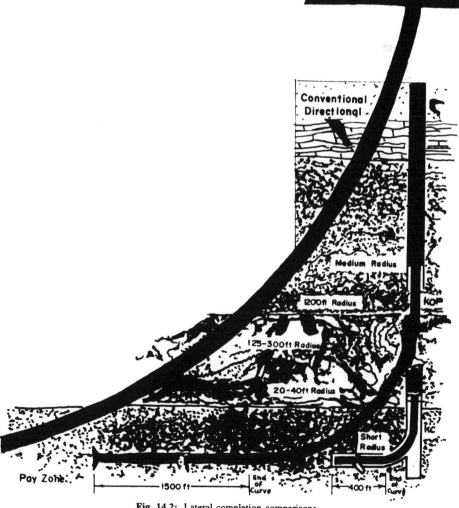

Fig. 14.2: Lateral completion comparisons.

SHORT-RADIUS SYSTEM

The short-radius system uses specialised rotary assemblies and a flexible drive pipe and is well suited for:

— Re-entering existing wells to improve production.
— Solving coning problems in existing wells; applications in which lease limitations or lithology require the curve and hori :ontal well bore to reside within the target formation.

A more in-depth description of Eastman Christensen's short-radius lateral drilling technology is as follows:

Major applications of lateral drilling can be used to reach irregular reservoirs, to limit invasion of unwanted formation fluid (coning), to cross natural vertical fractures and to maximise production of low permeability and low energy reservoirs. The technology can also be used in coal seams for methane drainage and *in-situ* combustion.

Water Coning: Whenever productive zones have a water drive, and especially when the viscosity of the oil is significantly higher than that of water, the potential for a vertical well to have a water-coning problem arises. In this situation the well produces water as well as oil. A laterally drilled well can be kicked off in the oil-leasing sand and drilled at a fairly precise depth across the productive formation, thus reducing the possibility of water coning (see Fig. 14.3a).

Gas coning: Since gas is much less viscous than oil, gas coning normally is more severe than water coning. If gas coning cannot be controlled, the gas must be reinjected or reservoir pressure will be drawn down prematurely. A laterally drilled well can help avoid gas-coning problems by assuring that the well is completed in the oil-leasing sand and by allowing economic production rates without excessive pressure gradients, which induce coning (see Fig. 14.3b).

Vertically fractured reservoirs: Such reservoir productive zones lie in vertical planes and typically are separated by 20–200 ft (6–61 m) of non-productive formations. A vertical well may hit a single productive zone or may miss the petroleum entirely. A horizontal hole drilled perpendicular to the productive bedding planes could intersect a number of producing fractures and yield significantly more oil than a vertical well (see Fig. 14.4a).

Irregular Formations: Lateral drilling has been used successfully to develop irregular reservoirs such as Niagaram rees in Michigan. In this application oil-bearing formations are located in discrete pockets and may be difficult to pinpoint with seismic surveys. These formations are easily missed by vertical wells or conventional directional drilling techniques.

Short-radius lateral drilling kicks off the existing well from a point very close to the productive depth while avoiding possible water- or gas-coning problems (Fig. 14.4b).

Low permeability formations: In some formations with low permeability, drilling a horizontal well through the pay-zone can produce better results than

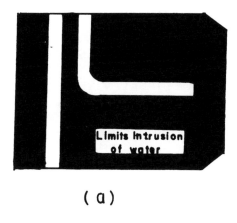

(a)

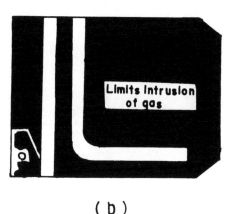

(b)

Fig. 14.3: a—Water Coning; b—Gas coning.

a hydraulic frac job. By drilling a lateral hole several hundred feet through the pay-zone the operator can increase the well's total millidarcy feet and improve production (see Fig. 14.5a)

Solution mining: Many minerals are currently extracted using solution mining. Lateral drilling can be used to improve the economics of producing these minerals found in bedded deposits (Fig. 14.5b).

Enhanced oil recovery: Lateral drilling techniques have the potential for improving production in heavy oil formations by placing horizontal holes through the reservoir to be used for producing and injection wells (Fig. 14.6a).

Coal degasification/*in-situ* gasification: Lateral drilling has been used to remove gas from coal seams prior to their being mined. Also lateral drilling can be used to drill into coal deposits to facilitate *in-situ* gasification (see Fig. 14.6b).

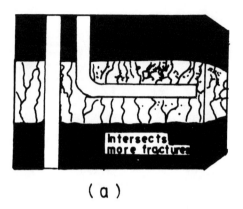

(a)

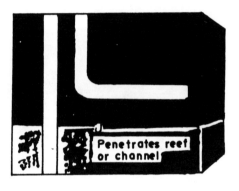

(b)

Fig. 14.4: a—Vertically fractured reservoirs; b—Irregular formations.

Short-Radius Rotary System

The short-radius rotary system has been used to drill over 350 horizontal well bores, more than any other comparable system.

The system includes an orientation assembly, non-rotating curve drilling (angle building) assembly (Fig. 14.7), flexible drive pipe and a stabilised drilling assembly (Fig. 14.8). Two sizes of tools accommodate the most common hole sizes from 4.5 in up to 6.5 in/114 up to 175 mm.

The Orienting Guide (OG) is a specially designed whipstock which is installed with an inflatable packer. Once the OG is oriented and set, the curve drilling assembly deflects the well bore from vertical to horizontal. This bottom-hole assembly consists of the flexible drive pipe and the Curved Drill Guide (CDG). The CDG's flexible shell remains stationary, defining the curve, while its internal drive shaft

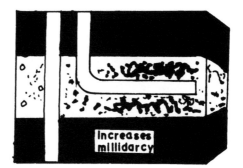

(a)

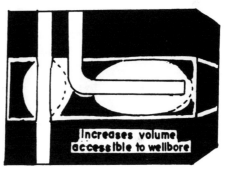

(b)

Fig. 14.5: a—Low permeability formations; b—Solution mining.

conveys the drill string's rotation to the bit. Hole direction is normally controlled within ±20° at the end of the curve.

The angle-hold assembly has two near-bit, undergauged stabilisers which are varied in diameter to build, drop or hold the planned well path. Inclination can be controlled within ±2°.

The short-radius rotary system's orienting guide is normally oriented with the Seeka Rate Gyro System or other gyro instrument. The drilling process is monitored using magnetic multishot and single-shot instruments.

Short-Radius Motor System

Eastman Christensen's new short-radius motor system employs advanced downhole motor technology to build well bore inclination to horizontal along a radius of only 40 ft/13 m. The steerable system affords precise directional control over inclination and hole azimuth and can drill horizontal sections of up to 1000 ft/310 m.

Powered by a short, articulated Positive Displacement mud Motor (PDM), the short-radius system puts more power at the bit for better ROP than rotary

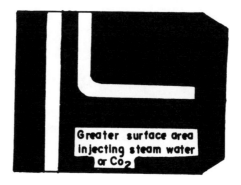

(a)

(b)

Fig. 14.6: a—Enhanced oil recovery; b—Coal degasification/in-situ Gasification.

techniques. The system also requires less string rotation, thus reducing the potential for hole damage in the short-radius curve.

The system can kick off the well in either open or cased intervals and can drill multiple horizontal lengths from a single vertical well bore. In addition, the steerable downhole motor's precise drilling capability keeps the well bore within very narrow target boundaries. Two bottom-hole assembly configurations are used—one to drill the curve, the other to drill the horizontal section. Both assemblies are steerable. Either tubing or articulated drive pipe, which is required to permit rotation through the curve, can be run through the curve.

The short-radius motor system uses a custom, flexible Directional Orientation Tool (DOT) to determine real-time tool face orientation and hole direction during drilling. Magnetic single and multishot surveys are taken at strategic intervals to monitor inclination and direction as the well progresses.

Angle- hold assembly

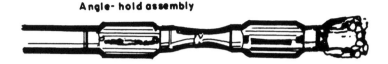

Fig. 14.7: Orientation assembly.

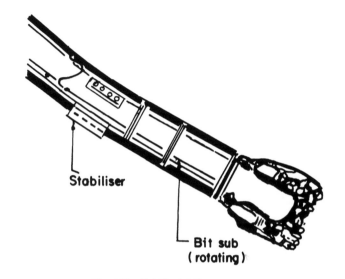

Stabiliser

**Bit sub
(rotating)**

Fig. 14.8: Stabilised drilling assembly.

Some examples of use of the Orienting Guide (OG) are illustrated in Figs. 14.9, a, b, c, d and e.

Surveying System Monitors Progress

The short-radius lateral drilling system uses a variety of Eastman Christensen's survey technology to orient, monitor and map the lateral well bore. The orienting guide used to initially deflect the curved drill guide is typically oriented using the Seeker Rate Gyro System or other gyro instrument.

The curve drilling process is monitored using magnetic multishot and special inclinometer single-shot instruments. Completion surveys are conducted using magnetic multishot instruments run in special lateral survey bottom-hole assemblies.

Medium-Radius System

Eastman Christensen's medium-radius lateral drilling system uses techniques and equipment that are logical extensions of conventional oilfield technology (see Fig. 14.10). The system can build hole angle at rates between 8° and 20°/100

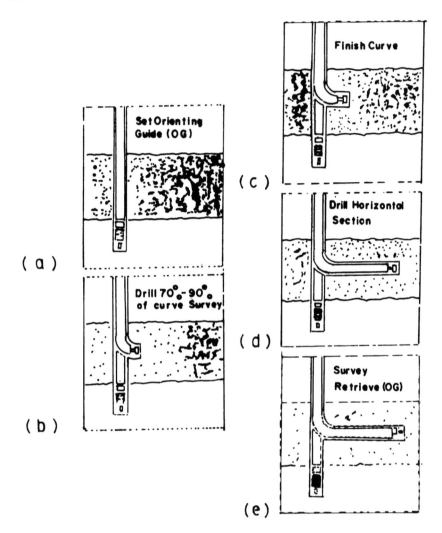

Fig. 14.9: a—Trip in hole with OG. Orient Set retrievable packer, Uniatch and Pooh; b—Run-in-curve drilling BHA Establish kick-off drill pull assembly from hole survey curve; c—Trip in hole with curve drilling BHA finish drilling curve Trip BHA; d—Trip in angle hold assembly drill lateral section Trip BHA; e—Survey well with special BHA Trip in pick-up BHA Release OG packer POOH.

ft (31 m) producing radii of 286–700 ft/87–213 m. Well bores can be turned from vertical to horizontal within 450–1000 drilled feet/137–305 m. The medium-radius system is driven by low-speed, high-torque positive displacement motors, customised with special stabilisers and deflection subs.

The angle-build assembly, utilising a fixed position or steerable angle-build motor, is used to kick off the well and drill-oriented, curved section to establish inclination. If necessary, straight or 'tangent' sections can be drilled within the

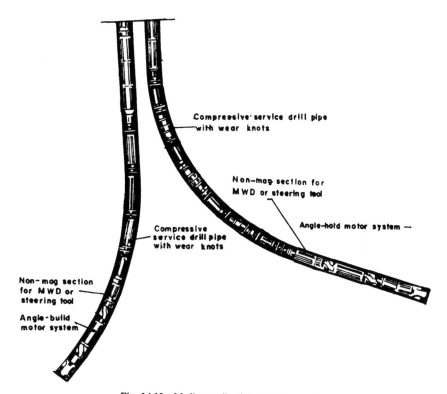

Fig. 14.10: Medium-radius lateral drilling system.

curve to assure that the well reaches horizontal at the target depth. Such straight sections can be drilled with either the steerable angle-build motor or with an angle-hold assembly.

The angle-hold assembly incorporates a customised Double Titled U-joint housing (DTU) and special stabilisers. When rotated, this assembly drills straight ahead; when held stationary, it drills on an oriented path to permit course corrections.

Medium-radius systems with build rates of 15°/100 ft (31 m) or greater typically use compressive service drill pipe with integral wear knots. Otherwise, standard heavy-weight drill pipe may be used.

Directional control is maintained using measurement-whilst-drilling (MWD) or steering tools. Wells drilled with the medium-radius system can be logged with tools conveyed by the drill string or coiled tubing.

Navi-Drill Mach 1

The Navi-Drill Mach 1 is a positive displacement motor that develops high torque at the bit at a relatively low speed range (80-340 rpm). This makes it ideal for directional applications and drilling with high weight on the bit. Navigation Drilling with roller cone or King Cutter PDC bits and coring operations are its other applications.

The Mach 1 motor has a multilobe (5/6) rotor/stator configuration (see Fig. 14.11). This generates more torque than other Eastman Christensen motors, permitting more weight on the bit and increasing ROP. The Mach 1 is specifically recommended for use with roller cone bits. Because the motor develops its power at low speeds, it can improve the performance of these bits without accelerating wear on bit bearings or cones.

A unique bearing assembly and improved elastomer compounds in the stator have increased the Mach 1 hydraulic horsepower and extended operating life. The Mach 1 also has a new motor nozzling system that allows the motor to be run at 50-100% over its maximum recommended motor speed. The additional mud passes through the motor's rotor and flow rate can be adjusted with interchangeable nozzles. The higher rates offer improved hole clearing and bit hydraulics.

Although primarily a directional performance drilling motor, the Mach 1 can also be used for straight-hole drilling and for coring operations.

Navi-Drill Mach 2

The Navi Drill Mach 2 (Fig. 14.12) is a positive displacement motor that can improve ROP in both straight-hole and directional applications.

The Mach 2 has a multistage, 1/2 rotor/stator configuration, which generates high torque at medium speeds (155-2100 rpm) for higher penetration rates with less weight on the bit. This makes it a good choice for drilling straight and directional holes in difficult formations.

The motor is particularly suited for doing interval performance drilling with an EC natural diamond Balla set, or PDC bit. The motor is available with integral blade stabilisers for optimum stabilisation in performance drilling applications.

Mach 2 motors also come in 1.75, 2.375, 3.75 and 4.75 in/45, 60, 95 and 120 mm ODs for slim-hole applications.

Navi-Drill Mach 3

The Mach 3 is a positive displacement motor with a 1/2 rotor stator configuration. Its low reactive torque makes it a good choice for directional drilling applications—kick-offs and correction runs (see Fig. 14.13).

The Mach 3 has relatively low mud-flow rate requirements, making it ideal for applications in which pump output is a limiting factor. Despite its low flow requirements, the Mach 3 has a wide range of speeds (125-800 rpm) and generates ample torque at the bit for increased ROP and precise directional control. It also offers good performance in short-interval applications such as rathole and mousehole drilling, controlled straight-hole drilling, and production/workover operations.

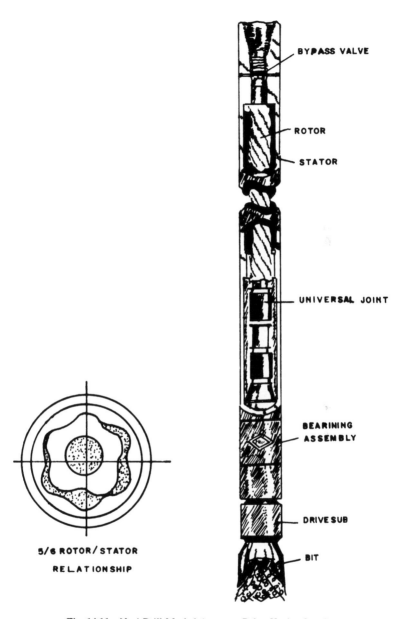

Fig. 14.11: Navi-Drill Mach 1 (courtesy Baker Hughes Inteq).

SEEKER RATE -GYRO SURVEYING SYSTEM

The Seeker rate-gyro survey system uses a unique method of measurement to provide surveys and tool orientations faster and more accurately than conventional surface-readout gyros.

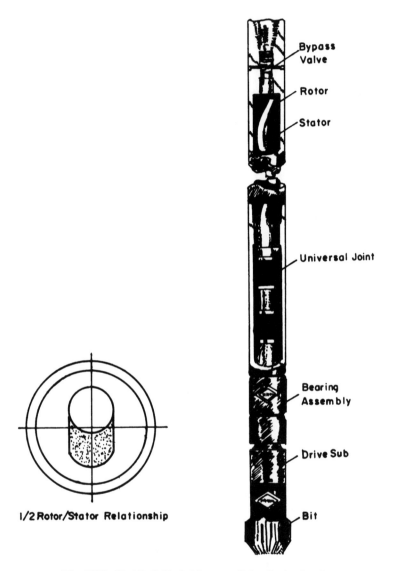

Fig. 14.12: Navi-Drill Mach 2 (courtesy Baker Hughes Inteq).

Unlike conventional gyros, which must be carefully aligned to a surface reference before being run downhole, Seeker's rate-integrating gyro requires no surface orientation. Instead of maintaining fixed heading, Seeker's sensor package—a rate-gyro and an accelerometer—is mounted in a revolving gimbal. The rate-gyro measures the earth's rate of rotation at each survey station and the accelerometer measures the force of gravity. Seeker uses this data to calculate hole inclination and direction independently for each station.

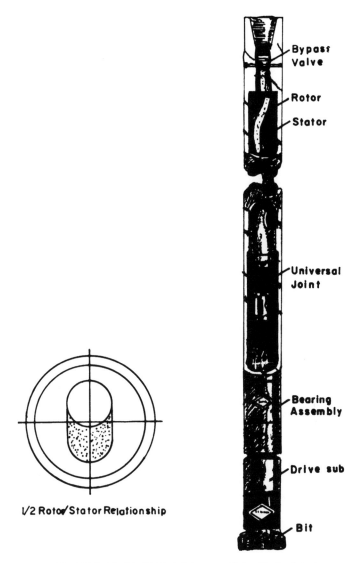

1/2 Rotor/Stator Relationship

Fig. 14.13: Navi-Drill Mach 3 (courtesy Baker Hughes Inteq).

This method of measuring eliminates errors that might be introduced by improper surface alignment. It also eliminates conventional gyro drift and draft checks, and minimises systematic survey errors.

Seeker is recommended for surveys in drill pipe, deep holes, cased holes, and production tubing. It is particularly effective when a high degree of accuracy is required in congested fields, offshore drilling, and relief-well applications.

The Seeker system includes a downhole probe, a power supply/computer interface, and a surface computer and printer.

The instrument probe is run downhole on a single conductor wireline. Besides the sensor package, the probe also contains an electronics module, which regulates power to the sensor, controls the gimbal rotation and transmits sensor data to the surface.

ELECTRONIC MAGNETIC SURVEY INSTRUMENT (EMS)

Eastman Christensen now offers a proven, solid-state electronic instrument for both multi- and singleshot magnetic surveys. Compared to conventional, camera-based magnetic tools, the electronic system is faster, more accurate and more reliable (see Fig. 14.14).

EC's electronic multishot uses triaxial accelerometers and magnetometers to obtain accurate readings of hole inclination, direction, and tool-face orientation. The downhole package also features a memory chip, which stores the survey data for each station, and a battery pack.

Surface gear includes a rugged portable computer, which processes survey results right at the site, and a small-system printer.

EC's electronic survey system is armed on the surface, then run like a standard magnetic multishot. Following the survey the total is reconnected to the system computer which processes the data and generates a survey report at the rig site.

Because it uses an electronic memory to store survey data, the system eliminates many of the error sources associated with camera-based systems, such as film problems and film reading errors. The system also offers enhanced accuracy over conventional film-based tools with inclination and azimuth resolution of 0.01°.

EC's electronic instrument measures a variety of downhole parameters to ensure a good survey. In addition to hole inclination and direction, it also calculates magnetic dip angle and field strength at each survey station. These values are used to determine the level of downhole magnetic interference on the instrument, providing a good measure of survey validity. (The magnetic parameters also indicate when a gyro survey should be run.) The instrument also measures downhole temperature and is modelled for a range from 0 to 125°C.

DIRECTIONAL ORIENTATION TOOL (DOT)

The DOT provides continuous real-time measurements of hole inclination, direction and tool-face orientation during drilling with a downhole motor. The DOT system comprises a solid-state electronic probe, a surface computer, a hand copy-printer and a driller's readout.

The probe is run down on a single conductor wireline and seats in a mule-shoe assembly in the UBHO sub immediately below a non-magnetic collar (see Figs. 14.15a, b and c). The probe is 1.75 in/44 mm in diameter and the mule-shoe stinger 1.875 in/47 mm, allowing the tool to be used in drill-string IDs as small as 2.125 in/55 mm.

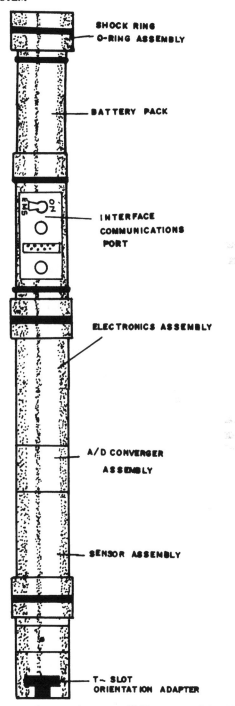

Fig. 14.14: Electro magnetic survey instrument (EMS). (courtesy Baker Hughes Inteq).

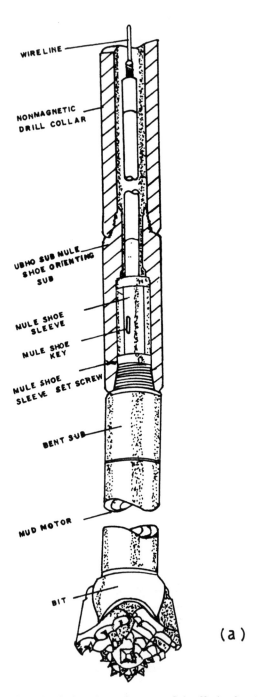

Fig. 14.15: a) Directional orientation tool (courtesy Baker Hughes Inteq).

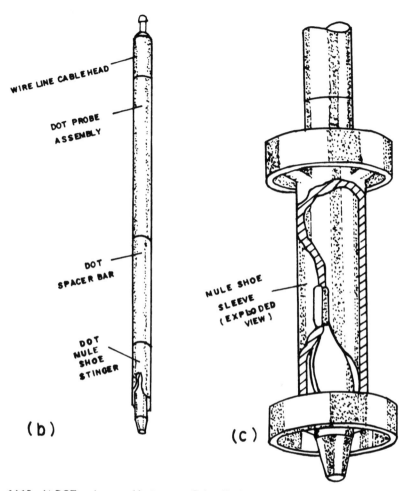

WIRE LINE CABLEHEAD

DOT PROBE
ASSEMBLY

DOT
SPACER BAR

DOT
MULE
SHOE
STINGER

MULE SHOE
SLEEVE
(EXPLODED
VIEW)

(b)   **(c)**

Fig. 14.15: b) DOT probe assembly (courtesy Baker Hughes Inteq). c) Mule-shoe sleeve (courtesy Baker Hughes Inteq).

Optional electronics and calibration procedures for the DOT system are available to increase its accuracy and stability, especially in higher angle holes (over 45°).

The system computer provides a continuous digital readout, revised every three seconds, and changes from magnetic to high side mole at the flip of a switch, without changing probes. The computer monitors tool circuitry and wireline conditions and localises any problems, allowing instant analysis of probe integrity.

The system computer relays information to the driller's readout, in which hole direction and tool-face orientation are registered on a compass and hole inclination is displayed digitally.

The Eastman Christensen wireline side entry sub permits drill pipe to be added as required without tripping the probe. (Use of a split Kelly bushing is essential for safe operation of the wireline entry system.)

GYROSCOPIC SURVEYING INSTRUMENTS

Eastman Christensen Gyroscopic instruments perform accurate, reliable surveys in cased holes, production tubing, or any area where magnetic interference precludes the use of magnetic instruments (see Fig. 14.16).

EC gyroscopes are built to extremely close tolerances using only the highest quality materials. The gyro spin motor is hermetically sealed and filled with helium to prevent contamination and lubricant evaporation. A 1.5 in/38 mm air-cooled gyro is also available for hotter hole applications.

Unique circuitry in the control sub provides precisely regulated power from the gyro battery pack, ensuring reliable performance. A shock-absorbing mount assembly compensates for vertical impacts during survey runs.

Gyroscopic multishot units are available in ODs of 1.75, 2 and 3 in/44, 51 and 76 mm.

Angle units are available for the 1.75 and 2 in/44 and 51 mm systems in ranges of 0–5°, 0–12° and 5–90° and for the 3 in/76 mm system in ranges of 0–2°, 0–5° and 5–90°.

The gryoscopic multishot systems use the standard Eastman Christensen Drip Multishot camera and electric timers.

The gyro heat shield rated at 600°F/315°C for 5 hours, with a 5.25 in/132 mm OD, permits gyro runs in deep, hot holes.

The gyro single-shot orientation system is available with 1.75 and 2 in/44 and 51 mm ODs with angle unit ranges of 0–12°.

Sigma Gyroscopic Surveying System

Sigma is a real-time, surface readout/recording gyroscopic instrument that provides fast, accurate tool orientation and multiple-shot surveys. The continuously transmitted data is unaffected by magnetic interference from the drill string, casing, or nearby wells. The sigma system comprises a gyroscopic probe, a surface computer and a hand copy-printer.

The probe, run on a single conductor wireline, continuously transmits measurements of hole azimuth, inclination and tool-face orientation. A precise optical encoder reads the gyro gimbal position and solid-state electronics relate the reading of hole inclination data gathered by accelerometers. The probe also monitors gyro system performance and hole temperature. The Sigma operates in downhole temperatures up to 250°F/121°C.

Probes are available in two sizes. The Sigma 175 incorporates the EC 1.5 in/38 mm gyro and has 1.75 in/44 mm OD. The Sigma 300 uses the EC 2.5 in/63 mm gyro and has a maximum OD of 3 in/76 mm. The Eastman Christensen 2.5 in/63 m gyroscope has long been the most reliable multiple-shot gyro in the oil field.

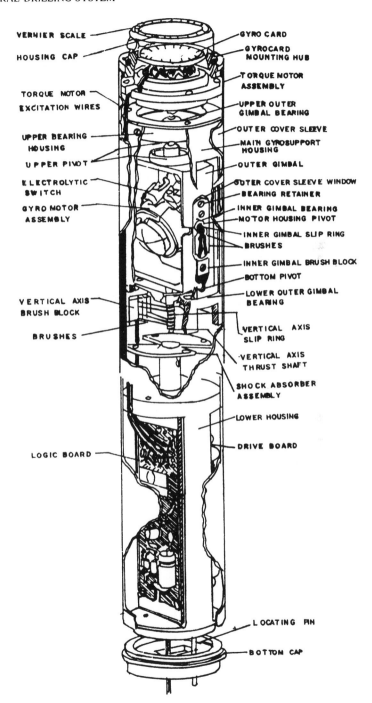

VERNIER SCALE

HOUSING CAP

TORQUE MOTOR
EXCITATION WIRES

UPPER BEARING
HOUSING

UPPER PIVOT

ELECTROLYTIC
SWITCH

GYRO MOTOR
ASSEMBLY

VERTICAL AXIS
BRUSH BLOCK

BRUSHES

LOGIC BOARD

GYRO CARD

GYROCARD
MOUNTING HUB

TORQUE MOTOR
ASSEMBLY

UPPER OUTER
GIMBAL BEARING

OUTER COVER SLEEVE

MAIN GYROSUPPORT
HOUSING

OUTER GIMBAL

OUTER COVER SLEEVE WINDOW

BEARING RETAINER

INNER GIMBAL BEARING

MOTOR HOUSING PIVOT

INNER GIMBAL SLIP RING

BRUSHES

INNER GIMBAL BRUSH BLOCK

BOTTOM PIVOT

LOWER OUTER GIMBAL
BEARING

VERTICAL AXIS
SLIP RING

VERTICAL AXIS
THRUST SHAFT

SHOCK ABSORBER
ASSEMBLY

LOWER HOUSING

DRIVE BOARD

LOCATING PIN

BOTTOM CAP

Fig. 14.16: Gyroscopic surveying innstrument (courtesy Baker Hughes Inteq).

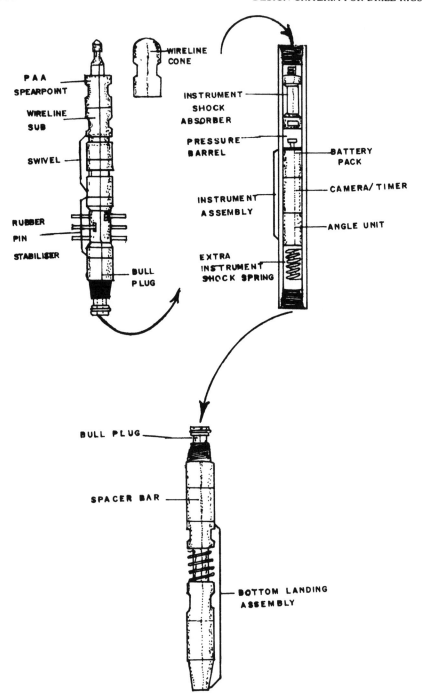

Fig. 14.17: Magnetic multiple-shot surveying instrument (courtesy Baker Hughes Inteq).

The Sigma system provides real-time orientation on the rig floor using the Eastman Christensen Driller's Readout (DRO). The Driller's Readout continuously displays the borehole inclination, azimuth and gyro tool-face orientation.

Magnetic Multi Shot Surveying Instruments

Eastman Christensen's multiple-shot surveying instruments perform accurate, reliable surveys of long-hole sections and of entire uncased well bores. The instruments permanently record hole inclination and magnetic direction at regular intervals along the well bore.

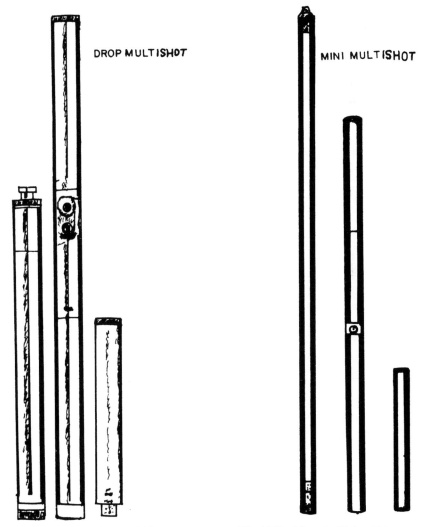

Fig. 14.18: Magnetic Drop multishot surveying instrument.

Fig. 14.19: Magnetic Mini multishot surveying instrument.

Magnetic multiple-shot units are available in two sizes: the Drop Multishot (DMS) with a barrel OD 1.75 in/44 mm Fig. 14.18 and the Mini Multishot (MMS) with a barrel OD of 1.375 in/35 mm (see Fig. 14.19).

These instruments typically are dropped through the drill pipe into a non-magnetic collar and the survey is performed as the drill string is pulled from the hole. Because the instrument barrels are designed to withstand great pressure (up to 26,000 psi/1820 kg sq. cm for the DMS and 245,000 psi/168 kg sq. cm for the

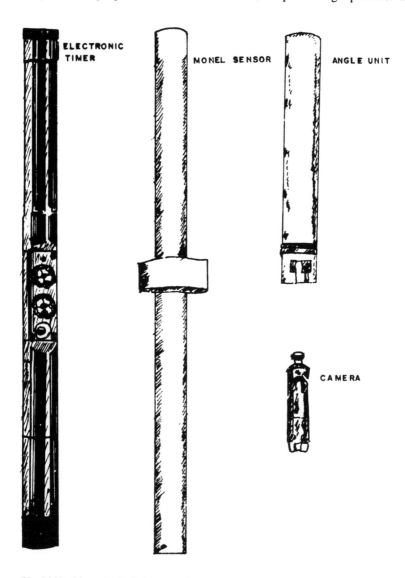

Fig. 14.20: Magnetic singleshot surveying instrument (courtesy Baker Hughes Inteq).

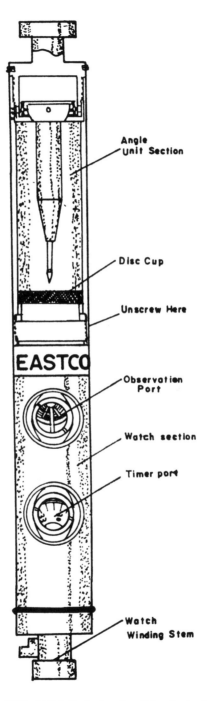

Fig. 14.21: Drift indicator (courtesy Baker Hughes Inteq).

MMS), the instruments produce accurate surveys even in very deep holes with heavy drilling fluids.

The multishot cameras have large film capacities, permitting extended surveys. The DMS camera can take 350 pictures and the MMS Camera 485 pictures. The solid-state electronic timers can be programmed for starting delays and variable pictures.

Magnetic Single-Shot Survey Instruments

Magnetic single-shot surveying instruments monitor the course of a well bore during drilling. The instruments simultaneously record the inclination and magnetic direction of the borehole at a single measured depth and provide a permanent record in the form of a film disc.

The complete assembly consists of a camera disc trap, angle unit, timing device, battery section and a protective outer sleeve see Fig. 14.20. The instruments require a non-magnetic drill collar.

The film discs can be daylight loaded and unloaded using a film injector and developed in 4 minutes without a darkroom or camera bag.

Drift Indicator

In vertical holes for which directional drilling data is not required, the Eastman Christensen drift indicator (Fig. 14.21) allows the driller to track borehole drift simply, quickly and accurately.

The simple mechanical tool uses no fluids, film or electrical power, providing reliable service in deep and hot holes. The snap-lock barrel is hand operated and requires no wrenches during make-up.

The patented self-checking punch records the well bore inclination from vertical twice on a paper disc, verifying survey accuracy and providing a permanent record.

Dual observation ports on the timer permit a visual check of watch operation, assuring proper timing and avoiding misruns.

Barrel assemblies and angle unit sizes are available to accommodate all hole sizes.

CHAPTER 15

HORIZONTAL AND DIRECTIONAL DRILLING IN SOILS

Horizontal Auger Boring
Pipe Jacking
Microtunnelling Technology
Directional Drilling Technology
Drill Stem Rock Boring System
DTH Air Hammer Drilling System
Remote Controlled Micro Tunneler
Directional Drills
TBM Systems
Pipe-Jacking System

To achieve optimisation and targets in construction it is essential to adopt innovative and modern construction techniques. This chapter describes the state-of-the-art horizontal and directional drilling techniques adopted for laying of water and sewer pipelines, telephone cables, gas pipelines under highways, railtracks, canals and built-up areas without disturbing normal life. The targets can be achieved cost effectively within the stipulated time schedule. The various methodologies currently in practice can be grouped under the following headings:

a) Horizontal earth boring and pipe jacking
b) Microtunnelling technology
c) Directional drilling technology.

Horizontal earth boring and pipe jacking is still more of an art than an exact science. Success with such projects calls for experience, ability and flexibility, i.e., innovativeness in adopting to various boring and jacking conditions. What is the best method? When should you auger bore, slurry bore or jack pipe? Specific soil conditions may in some cases restrict the method used. However, for most jobs, any of these methods is equally successful. Compared with open-cut and other methods, boring and jacking often provide the best protection to the surrounding environment and traffic, both natural and social. This gives a great advantage for road crossings, river crossings and intercity underground works.

In deciding the location of the crossing, several requirements should be taken into consideration, for example:

  i) Adequate space for the pit.
 ii) Overhead traffic conditions with reference to location of the pilots.
iii) Overhead obstructions to be cleared.
 iv) Accessibility of the work area.
  v) Depth requirement of the pit and bore.
 vi) Approaches to the actual bore.
vii) Underground obstructions that might be encountered, such as utilities.
viii) Length of the crossing.
 ix) Height of the water table that might be encountered in relation to height of
     the bore.

HORIZONTAL AUGER BORING

The horizontal boring method is a process of simultaneously jacking pipe through
the earth while removing the soil inside the encasement by means of a rotating auger
(Fig. 15.1). In unstable soil conditions the end of the auger is kept retracted inside
the encasement so as not to cause voids. The more unstable the soil, the farther
back in the encasement the auger should be retracted. In stable soil conditions the
auger can be successfully extended beyond the end of the encasement to facilitate
the breaking down of soil for removal.

While auger boring methods can successfully be used in most kinds of soil
conditions, a general rule of thumb should be kept in mind, concerning obstructions.
The encasement size in relation to the obstruction should be small enough to
bore through it or large enough to permit removal of the obstruction through the
encasement. This is also a good rule whenever the soil contains boulders: the size
of the bore depends on the size of the boulders.

Pit Preparation: The boring operation should proceed from a work pit that
adequately and safely accommodates the auger boring equipment, materials and
workmen.

Dewatering: When groundwater is present, known or anticipated, a dewater-
ing/sump system of sufficient capacity to handle the flow should be installed and
maintained at the site until the operation can be safely halted or the job is completed.

Preparation of the Casing Pipe: Casing lengths are critical when trying to
keep either the auger retracted inside the casing or extended beyond the end of
the casing. Soil conditions usually determine which choice is preferable but length
becomes critical to maintain that condition.

Skin Friction: Experienced operators are finding more and more that successful
crossings are more easily achieved by reducing the thrust required to push the pipe
through the strata. This friction may be reduced in a number of ways. One method
may be more desirable than another depending on the conditions encountered and
the individual in charge of the job. Two of the most common methods for reducing
skin friction are overcutting the bore or lubrication of the casing. These methods
can be used separately or in combination.

Use of Casing Lubricants: Suitable lubricants, such as bentonite, become a gel
when mixed with water and act as an excellent lubricant when applied to the outer

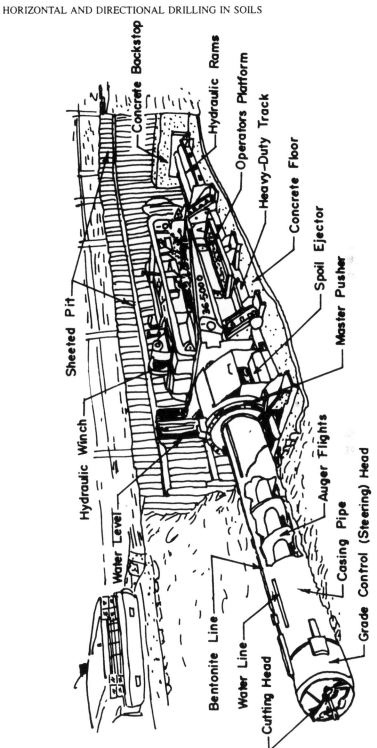

Fig. 15.1: Horizontal earth boring.

surface of the pipe. Bentonite can be applied to the outer casing surface in the space created by overcutting. It can be pressure pumped to and around the lead section of the pipe and when conditions require, to other locations along the surface of the casing.

Line and Grade Tolerance: In the boring process the line and grade of the pipe or casing will tend to continue through the bore as originally set up on the machine in the pit unless the course is altered by some external force.

Steering devices, such as steerable cutting heads, have been built and successfully used to mechanically adjust the direction of auger-bored road crossings and therefore the line and grade. On using a hinged lead section of casing at the front of the bore, resistance forces are encountered and the cutting head and cashing can be brought back into the desired location. Using this type of device, along with constant monitoring of the line and grade through sensing and checking devices, such as a level, an operator can make the desired and necessary changes and in many cases 'steer' the bore with accuracy in most soil conditions. Water can be used successfully to facilitate soil removal when it becomes necessary. Water is introduced into the front casing pipe behind the cutting head by means of a pumping system. This method has proven to be extremely useful in sticky ground conditions where the soil tends to adhere to the auger flighting.

Horizontal earth-boring machines are available in 16 different models, varying in boring capacity from 4 to 60 in (102–1524 mm) and advance thrust capacity from 20,000 pounds (88 kN) to 1,200,000 pounds (5538 kN).

PIPE JACKING

Pipe Jacking (Fig. 15.2) is another method introduced to facilitate construction of underground pipelines in circumstances where open cutting and the resultant disruption to traffic services and the local environment should be minimised. Reinforced concrete pipe and welded steel casings are jacked through the ground with high-performance hydraulic equipment with minimum disturbance to the subsoil, while the soil is removed from the front of the casing and transported through the casing to and out of the work pit. Any specific system can be set as a standard.

Since conventional jacking procedures require access by workmen through the pipe to the heading, a 36 in/900 mm diameter pipe is the smallest practical size used for most jacking operations. Concrete pipes as large as 132 in/3350 mm in diameter and steel casing pipes as large as 96 in/2440 mm OD have been installed by the jacking method.

The contractor should use jacking equipment designed to provide the forces necessary for installation of the pipes. The thrust load should be imparted to the pipe through a suitable thrust ring, which should be sufficiently rigid to ensure distribution of the load without creating point loading. The jacking should proceed from a work pit provided for the jacking equipment and workmen.

As in horizontal auger boring, when groundwater is known or anticipated a dewatering system of sufficient capacity to handle the flow should be maintained at the site until its operation can be safely halted. The dewatering system should be equipped with screens or filter media sufficient to prevent the displacement of fines.

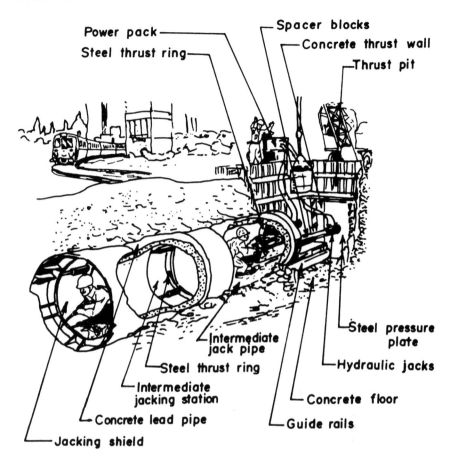

Fig. 15.2: Pipe jacking.

Likewise, skin friction is reduced by using suitable lubricants, such as bentonite, applied to the outer surface of the pipe. To check the line and grade of the pipe the self-levelling laser should be in use at all times.

MICROTUNNELLING TECHNOLOGY

Microtunnelling technology installs pipelines and underground water and sewer lines whilst maintaining free traffic above ground (Fig. 15.3).

Machines of various capacities for installing pipe from 100 mm diameter to 800 mm diameter with casing up to 120 m in different kinds of soils are available.

This system has numerous advantages in busy urban areas for the installation of utilities, especially where free traffic flow is an important factor. Conventional sewer and house connections can be installed from difficult and tight site locations. Jacking shafts, either circular or square, can be as small as 2 m diameter. Gas or water pipes and electrical cables can be installed under buildings and across main

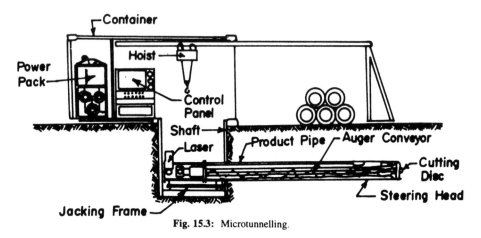

Fig. 15.3: Microtunnelling.

road junctions. These machines are well built and robust, ensuring high performance in many different applications. A house connection sewer up to 20 m long can be completely installed by a 2 to 3 men team in a normal working day. When this is compared with the time required for open-cut construction it becomes clear that installation of pipes using this method is commercially competitive. The capital cost of the machine is quickly recovered since most of the costly aspects, such as traffic control, service diversion and reinstatement, are reduced or eliminated altogether.

This system is easy to use and understand. With little cost or effort it can be adapted to install different pipe diameters. This machine has multidirectional steering control. Direction control of the pilot tube is maintained with the help of theodolite sighted onto a target mounted in the control head. A newly developed video-optical surveying system displays the direction of advancement on a monitor mounted in the jacking shaft. The direction can therefore be accurately monitored and controlled. Steering adjustments are made by turning the pilot tube whilst jacking.

DIRECTIONAL DRILLING TECHNOLOGY

The directionally controlled horizontal drilling process is used as a horizontal boring technique for crossing under natural or man-made obstacles. It is an outgrowth of the technology and methods developed for the directional drilling of oil wells. This method was developed in the United States and has revolutionised drilling on complicated pipeline river crossings through alluvial soils. The horizontal directional drilling method is a two-stage process in most cases. The first stage consists of drilling a small-diameter pilot hole along the desired centre line of the proposed pipeline. The second stage involves enlarging the pilot hole by reaming to the diameter requisite for accommodating the pipeline of desired diameter.

The pilot hole is drilled with a specially built drill rig that allows the drill string to enter the ground at an angle of entry which can vary from 5 to 30°; the optimum entry angle is 12° however. The drill rig pushes the drill string into the ground while bentonite drilling mud is pumped through the drill stem to a downhole drill motor located just behind the bit. The drill mud operates the downhole motor, functions as a coolant, and facilitates spoil removal by washing the cuttings to the surface where they settle out in a reception pit.

The drill stem is approximately 3 in/76 mm in diameter, non-rotating, and contains a slightly bent section which is called a bent housing. The bent housing (typically from 0.5 to 15°) is used to create a steering bias. A curved or straight profile is achieved by steering the drill rod as it is being pushed into the ground. Steering is controlled by the positioning of the bent housing.

The pilot hole path is monitored by a downhole survey system located behind the bent housing and provides data on the inclination, orientation and azimuth of the leading end. This data is transmitted to the surface where it is then interpreted and plotted. Normally, position readings are taken about every 30 ft/9 m. Should the pilot hole get out of alignment, the drill stem is pulled back and a new course is cut.

During the drilling operation, a 5 in/127 mm diameter steel washover pipe is rotated over the pilot drill stem. The washover pipe relieves the friction and resisting pressure caused by cuttings mixed with the drill mud. In addition, the washover pipe provides rigidity to the pilot drill stem. Bentonite slurry is pumped between the washover pipe and the pilot drill stem. Rotation of the washover pipe allows the diameter of the borehole to be increased to approximately 11 in/280 mm.

After the pilot hole has been constructed the pilot drill stem is withdrawn through the washover pipe. Reaming devices are then attached to the washover pipe and pulled back through the pilot hole, enlarging it to the desired diameter suitable to accept the designed pipeline. For a project in which the designed pipeline is equal to or less than 20 in/500 mm, the pipeline can be attached directly behind the reamer with a swivel device so that the total assembly can be pulled through in one pass. However, if the designed pipeline is greater than 20 in/500 mm, the borehole will probably require prereaming in order to obtain the desired diameter. The pull sections of the designed pipe are fabricated in one continuous length to avoid shutdown periods during pullback.

Figure 15.4 illustrates a typical drilling procedure and a typical pullback operation. Details of the bottom-hole assembly are given in Fig. 15.5.

The type of pipe installed must be such that it can accept sufficient axial tensile force to permit being pulled through the borehole. A steel pipe is the most commonly used at the present time; however, high-density polyethylene pipes may well be used in future.

Diameters range from 3 to 50 in/76–1270 mm. However, it is not uncommon for multiple lines to be installed in a single pull. The most significant multiple line crossing to date is 2800 ft/852 m length of five separate lines that range in size from 6 to 16 in/152–406 mm.

Disturbance to the ground must be taken into consideration in the design and construction of a directional drilled crossing. Significant forces are created by the

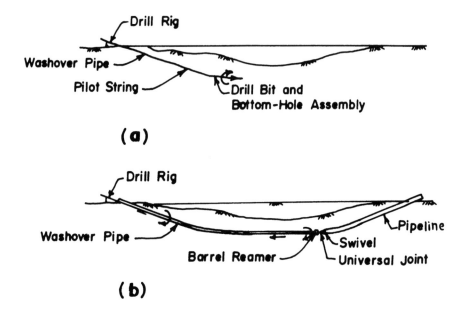

(a)

(b)

Fig. 15.4: Typical directional drilling and pullback procedure:
a) Typical drilling procedure;
b) Typical pullback operation.

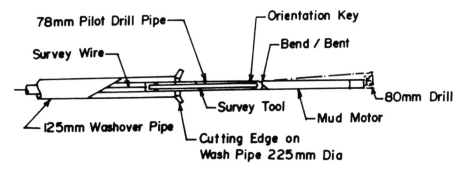

Fig. 15.5: Details of typical bottom-hole assembly for directional drilling method.

flow rate and pressures at which the bentonite slurry drilling mud is circulated through the drill stem to operate the downhole mud motors and then to wash the cuttings from the borehole. The flows and pressures must be monitored to prevent mud migration problems. Much of this problem is eliminated in the design by making sure that adequate depth is obtained and that a compatible stratum exists for the process to function properly. Therefore, subsurface conditions are critical to the success of directional drilled crossing.

Directional drilling equipment is sophisticated and requires a skilled operator. The operator must be knowledgeable about downhole drilling, impact of drilling in various geological formations, sensing and recording instrumentation, and interpretation of computer printout data.

Clay is considered the ideal material for the directional drilling method. Drillability of rock is dependent on its strength and hardness. Cohesionless silt and sand will generally behave in a fluid manner and remain in suspension in the drill fluid for a sufficient period to ensure being washed out of the borehole.

Directional drilling can accomplish long and complicated crossings quickly and economically with minimum environmental impact. This process requires neither a bore or receiving pits.

Cutting Heads

Generally five types of cutting heads, ranging in diameter from 4 to 60 in/102–1524 mm for durable service in dirt, clay, cobble or medium-grade rock are manufactured by the suppliers.

The cutting heads manufactured by American Augers of America are of the following types (see Figs. 15.6a–e)

(a) ADD-Dirt: Fitted with tapered lead screw and hardened cutting blades and routers.
(b) BHD-Clay: Tapered lead screw with backhoe style mud teeth.
(c) HTD-Multipurpose: Combination head for sand, clay, dirt and small gravel. Carbide pilot screw with backhoe teeth.
(d) AAR-Shale: Carbide cutting blade inserts in forged head ring cutters with bullet bits.
(e) HTR-Rock: Tungsten carbide-tipped bullet bits arranged in aggressive cutting angle. All bits engage uniformly for fast penetration and maximum wear. Useful in soft solid rock up to 6000 psi/421 sq. cm. Recommended for hard pan, shale, gravel, small boulders, landfill or rail road embankment. Also available with pilot screw.

Augers (Fig. 15.7)

The leading boring machines require equally dependable, high-quality auger sections. These should be manufactured in suitable sections to rigid specifications for guaranteed service. Each step from broaching the hex bushings to the fillet welds should be performed with proper quality control.

The following auger features are required for dependable service:

1) High grade steel plate
2) Stronger 1/2 in/13 mm wall tube
3) Heavy hellicoids
4) Heat-treated hex shanks
5) Cold-drawn flights
6) Tough 3/8 in/10 mm welds

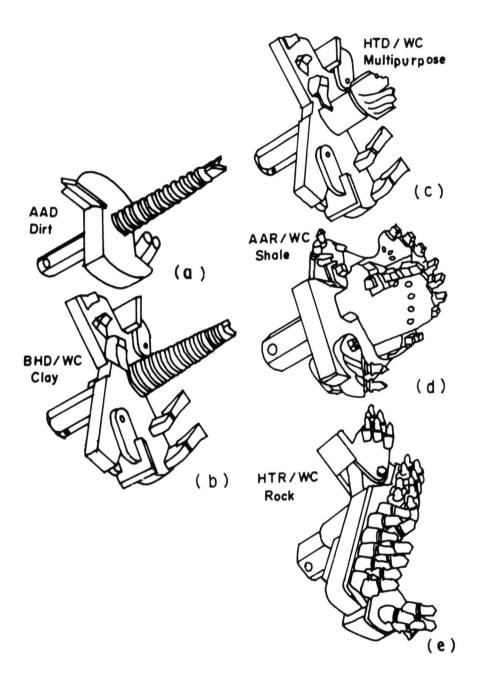

Fig. 15.6: Cutting heads.

Fig. 15.7: Auger.

7) Lap joint flights
8) Full year warranty.

DRILL STEM ROCK BORING SYSTEM

Hard rock no longer poses an impossible job for boring contractors. There are two means of crossing through solid rock—the Drill Stem Rock Boring System or a Downhole Air Hammer Drilling System.

Coupled with a larger boring machine, the Drill Stem Rock Boring System operates in a two-stage process. An initial 9.875 in/251 mm pilot hole is drilled through the full bore length; it requires a special roller core bit head mounted on a hollow drill stem (Fig. 15.8). Water is pumped through the drill stem by means of a water swivel to wash away cuttings from the drill face.

In the second stage the drill string is replaced with an auger and a rock roller bit reaming head. The reamer follows the pilot hole to cut the finished borehole diameter. Water is pumped back from the exit pit through the pilot hole to again flush cuttings from the face. The auger casing may be left in place in the finished hole or removed for a clean borehole through the solid rock.

Components needed for the Drill Stem Rock Boring System:

1) Water swivel
2) Roller core bit pilot drill bit 9.875 in/251 mm outer diameter
3) Drill stem stabilisers
4) Hollow drill stem, 10 or 20 ft/3 m or 6 m lengths, 7 in/170 mm outer diameter
5) Centraliser break-out plate assembly
6) Break-out wrench

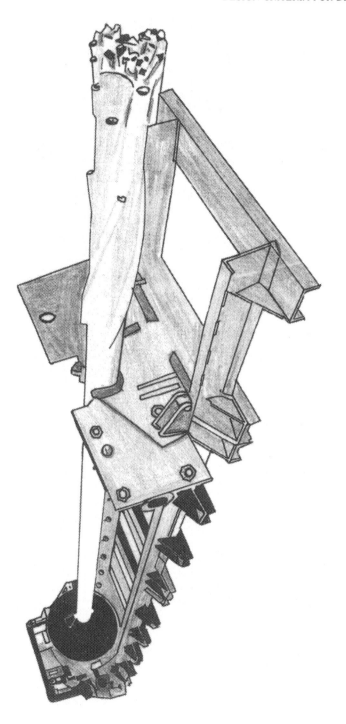

Fig. 15.8: Drill stem rock-boring system.

7) Safety plate
8) Roller core bit reaming cutter head
9) Auger
10) Casing

DTH AIR HAMMER DRILLING SYSTEM

An innovative method for drilling hard rock utilises a down-the-hole air hammer (Fig. 15.9) for horizontal boring. It not only drills solid rock formations, but also performs well in cobble, boulders or overburden.

The drilling design mounts a specialised drill bit and the DTH air hammer and is welded to a drive shoe just behind the drill bit. The air hammer then drives the drill bit through the rock and pulls the attached casing to line the finished hole.

A specialised air swivel conducts a steady air supply to the hammer. Water and lubricant can be injected through the swivel by means of an auger bentonite pump to control dust and reduce operating temperatures.

The boring machine gives the advance thrust and rotation needed to set the hammer and casing cuttings are flushed back through the casing by an air stream.

DTH hammer rock drill systems have hole sizes of 4 in/102 mm, 5 in/127 mm, 6 in/152 mm, 8 in/203 mm, 10 in/251 mm, 12 in/305 mm and 16 in/406 mm.

Horizontal earth-boring equipment for soil conditions ranging from sand through soil rock are discussed below.

Horizontal boring machines (Fig. 15.10) have a boring capacity that ranges from 4 to 60 in/102–1524 mm. These machines are designed around a unitised box-frame construction which assures lasting alignment of the power train.

Larger models use efficient Deutz Air-cooled Diesel Engines as the foundation of dependable power trains. Turbo options offer up to 40% more horsepower and torque for extra tough jobs.

Bore heads (Fig. 15.11) are made in sizes from 42 to 96 in/1067–2438 mm diameter. The bore heads make it possible to tackle large-diameter boring projects without enormous capital outlay.

The cutting head is independently powered and has its own auger spoil removal system.

A standard model incorporates a centre drive unit with an electric motor, four-speed transmission and a planetary gear box driving the cutting head. It gives great performance in dirt, clay, shale or mixed conditions.

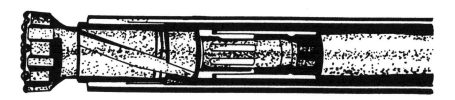

Fig. 15.9: DTH air hammer drilling system.

Fig. 15.10: Horizontal boring machine.

The bore head is a manned tunnelling operation. An operator monitors the cutting action and controls steering of the articulated head from the bore head. He works in voice-linked communication with the boring machine operator to co-ordinate cutting and jacking operations.

Laser guidance is used to control steering. The operator adjusts steering jacks in the articulated front section to maintain line and grade as shown on the laser target. In contrast to auger boring, control of both dimensions is possible. Accuracy of steering is dependent on the performance of the laser set-up and the operator.

— The bore head can be fitted with whichever spoil removal system is preferred.
— Augers, conveyors or muck cars.
— The cutting head can be driven by electric motor from an independent generator or by hydraulic power.
— Clearances within the shield allow easy access to the cutting head for inspection, alterations or repairs.
— The boring machine provides thrust and spoil removal power.

REMOTE CONTROLLED MICROTUNNELLER

Trenchless excavation techniques have become rational, cost-effective alternatives to conventional open-cut methods of installing pipes and sewers underground.

There is little more to see of the construction site than the container which is positioned over or next to the jacking shaft. The control mechanisms and all the power equipment needed for the job are housed in the container.

With the help of a hoist, the jacking frame is lowered into the shaft and set in place. Then the laser-controlled cutting head drills ahead exactly on target.

Spoil is crushed by the cone crushers removed by the auger system (see Fig. 15.12) and then lifted out of the jacking shaft by the hoist. For higher levels

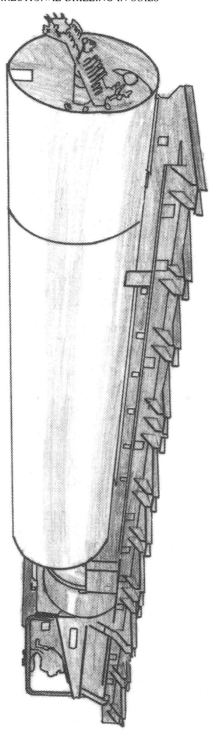

Fig. 15.11: Bore head.

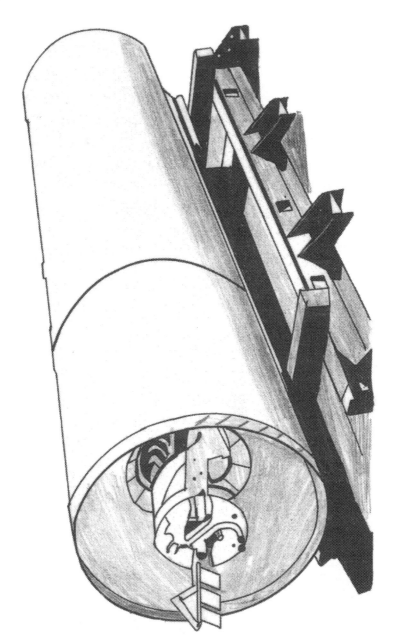

Fig. 15.12: Microtunnelling machine.

of ground-water, the microtunnelling head can be converted to the slurry system for spoil removal.

Step by step, the pipes are lowered into the shaft and jacked following line and grade. In this way, an installation up to 500 ft/152 m can be completed through the most diverse types of ground without disrupting traffic.

Using a special head for milling and relining (pipe eating) operations, damaged sewer pipes can be removed and simultaneously replaced by new ones.

Directional control is maintained by means of a laser beam sighted on a control device in the control head from the jacking shaft. Even deviations of $\frac{1}{10}$ of a millimetre are shown on the control panel and automatically adjusted.

One technique that is particularly economical involves adapting an auger-boring machine for microtunnelling. The auger-boring machine can be used as a combination jacking frame and power unit. It can also be used to remove the spoil if an auger system is present. The configuration works best with a large jacking shaft.

The auger system is often the most economical solution for spoil removal because of the simpler technical equipment required, the lower costs for installation and removal of the system, and the minimal space requirements for set-up.

Rock crushing heads with their integrated eccentric cone crushers ensure optimal performance in almost every type of soil.

In the slurry spoil removal system the ground is broken by a hydraulically driven cutting head and deposited in an eccentric core crusher in which the stones are broken down into removable particles. A strong rotary/centrifugal/turbine pump pumps the spoil through a feedpipe into a settling tank. A second pump transports the clean water back into the circulation cycle. The pressure necessary to support the face is constantly monitored and regulated by the control equipment.

Fig. 15.13: Typical directional drill.

DIRECTIONAL DRILLS

Medium-size directional drills are mounted on track carriers (excavator under carriage) (Fig. 15.13) and are diesel powered. The engine is mounted on the drilling itself so there are no hoses or lines to drag around. The control console is rig mounted or can be placed in an optimal heated/air-conditioned driller's cab.

Larger River Crossing Rigs: American Augers Inc. have produced their model DD-140. This smallest of the 'big rigs' is designed to handle most crossings.

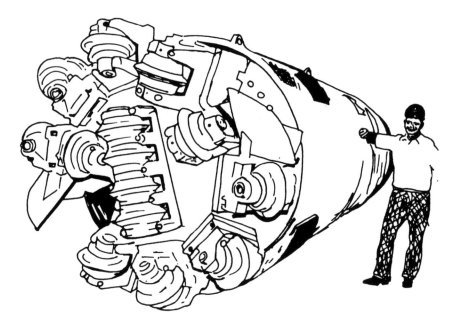

Fig. 15.14: Tunnel-boring machine.

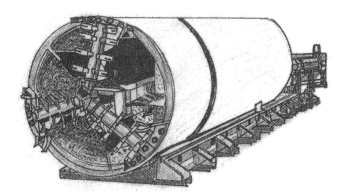

Fig. 15.15: Soft-ground TBM

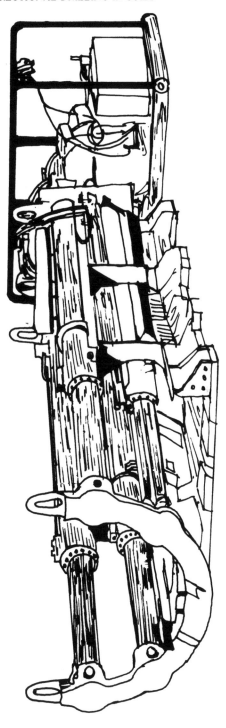

Fig. 15.16: Pipe-jacking system.

The DD-140 has drilled out over 3000 ft/900 m for conduit installation and has pulled in a 36 in/914 mm gasline under the Trans-Canada highway.

TBM SYSTEMS

Many kinds of machines are built for small-to medium-size rock tunnelling. As illustrated in Fig. 15.14, this TBM is equipped with fourteen 12 in/305 mm diameter single-disc cutters. The design of these cutters includes offset eccentricity to provide over cut above the TBM and to wipe the invert clean with each revolution. The cutting head never has to support weight when the thrust jacks are reset. A variable speed feature makes it possible to match cutting speed to rock consistency. The articulated tail unit enhances course correction ability. A trailing 'Sled' furnishes hydraulic power plus a conveyor extension. The machines design offers the options of thrust-off of excavated tunnel wall, ribs and lagging, and concrete pipe or liner plate providing lining behind the machine. The main bearing is of the durable crossed roller type. Eight steering stabilisers exert 50 tons/45,000 kg pressure each.

A soft-ground TBM of 63.5 in/1613 mm is illustrated. In Fig. 15.15.

PIPE-JACKING SYSTEM

Pipe-jacking systems (Fig. 15.16) employ the tremendous power of hydraulics to offer an economical method of trenchless excavation. Using a tunnel shield for excavation, the pipe-jacking station provides forward thrust to jack pipe and shield for continuous tunnelling operations.

Pipe-jacking stations provide uniform hydraulic thrust for smooth jacking power. The jacking stations produce 1 to 4 million pounds of thrust (4500–17,800 kN), as may be specified. A hydraulic power unit delivers 15-20 gpm (57–76 lit/m) in a psi 5000 (34,375 kPa) system to drive the jacking unit.

Powered by either diesel or electricity the hydraulic power unit is remotely controlled in the jacking pit.

Intermediate jacking stations are an effective solution for the extra thrust needed for long or difficult drives. They may be used with tunnel liners of RCP concrete, FRP fibreg lass or steel casing.

The intermediate station consists of sets of hydraulic jacks mounted in a steel casing which fits between the sections of jacked pipe.

When activated, the intermediate station advances the pipe, thus distributing the overall thrust required for any length of pipe. Each station can provide 300 thousand to one million pounds (1330–4450 kN) of additional thrust. The cylinders may be reused in other stations for pipes of various size.

CHAPTER 16

MODERN SURVEY SUPPORT
TO DRILLING BOREHOLES

Surveying
Diamond Drilling
Plane Tabling
Theodolite Traverse/Triangulation Methods
Total Station-(Combination of Precise Theodolite and EDM)
Coordinates through Aerial Photography
Global Positioning System (GPS)
Conclusion

SURVEYING

Surveying has been meeting the needs of other professionals who need to know the position (co-ordinates) of terrain points for planning or execution of engineering products. The survey profession started with the advent of cartography, which needed ground positions for simple map making. The very primitive yet useful technology of plane-tabling remained the workhorse for the profession for almost 100–150 years. The profession grew in technological content as the other professions of mechanical, optical and electronic technologies grew over time. Various techniques of surveying which can support drilling boreholes most of the time are described below.

DIAMOND DRILLING

Important applications of diamond drilling are related to its ability to produce core samples. Hence, uses of diamond drilling include reconnaissance and mineral prospecting and proving of reserves, gas and oil-well drilling, proving the quality of structural concrete, hydrogeological and geophysical investigations, and sampling in site-geological and quarry investigations.

The least thrust is provided by diamond during drilling. Hence falling or weathered sculptures or monuments can be preserved by drilling small-diameter holes through their bodies and inserting mild steel bars. This technique is also applied on laminated rock formations overhanging roads, power-houses and various structures. Steel bars of 31 mm, 38 mm and 51 mm are inserted in diamond-drilled holes to ensure safety below in 48 mm, 60 mm, 75.7 mm Ax, Bx, and Nx sizes respectively. Other applications of production drilling with a diamond drill are certain forms of blast-hole drilling, grout-hole drilling and making holes for drain pipes and cables in mines and structures.

Diamond drills are deployed on the surface and underground by quarrying, mining, civil engineering, and the oil industries. The holes vary in depth from a few centimetres to many thousands of metres, both in mineral exploration and in oil-well drilling.

To encounter lodes at depths and to decipher contours of subsurface structures, the holes are generally drilled either vertically or at an inclined angle, depending on the geological formations and the behaviour of mineralisation. The purpose is to encounter lodes and to decipher structures up to a predetermined depth.

The inclination and bearing fixed at the collar of the drill hole do not ensure the assigned orientation it is supposed to follow. No hole is found ideally straight. Adherence to all preventive means in this regard minimises the tendency of a drill hole to follow a crooked path. But even with the best technique and precautions, some curvature due to deviation does take place.

PLANE TABLING

A very basic technique indeed, plane tabling is described in all survey textbooks. In simple words, an unknown point is determined by means of other known points (known co-ordinates plotted on a sheet of paper on the desired scale). An intelligent joining of such determined points on a plane-table will provide a manuscript map.

The technique is still valid and eminently applicable in case position is required for one or two boreholes. The accuracy of determination of position of borehole(s) will depend on:

— accuracy of the co-ordinates of (known) control points,
— scale of plane-table section (paper),
— working skill of the plane-tabler (surveyor).

To give some idea of accuracy and scale, a box is constituted below on the assumption that maximum graphic accuracy can be assumed as 0.2 mm on paper.

Box	0.2 mm on		
		1:1000 scale is 0.2 × 1000 mm	= 0.2 m
		1:5000	= 1.0 m
		1:10,000	= 2.0 m
	Scale of topo-maps ⎱	1:25,000	= 5.0 m
	of Survey of India ⎰	1:50,000	= 10.0 m

Note: In the above box errors due to wrong plotting of controlled points, orientation or centring of the plane table are neglected. For more details, refer to survey books.

It is readily apparent that a borehole can be located within an accuracy of 2 metres on a 1:10,000 scale.

THEODOLITE TRAVERSE/TRIANGULATION METHODS

In the traverse method (see Fig. 16.1) angles L1, L2, L3 etc. and corresponding distances d1, d2, d3, are measured. Angles are measured by means of theodolite and distance by tapes (steel or metallic). The co-ordinates of station A are transferred to an unknown station, which may be a borehole.

Triangulation Method or Determining Co-ordinates

The propagation of errors in the method of traversing depends on the accuracy of angles and distances. Minor errors in measurements of these parameters may throw the final position in to larger error as the methodology is not self-checking. Besides, one needs cleared ways to measure distances; the terrain has to be comparatively flat.

The method of triangulation (Fig. 16.2) obviates some of the disadvantages of the traverse method. The method is self-checking because the sum of the internal angles of a triangle should total 180 degrees.

A and B are starting stations whose co-ordinates are known. This means we can calculate the distance AB. Point C is observed with the help of theodolite from A and B. Angles ABC and BAC are therefore observed. Theodolite is then taken to C and stations A and B are observed. We thus observe all the three internal angles of triangle ABC. The sum of these angles must total 180° If distance AB is known and angles are observed, the co-ordinates of C are known by adopting the sine law of co-ordinate geometry. Point D is then added likewise from triangle ACD. Points E,F,G and H are similarly taken forward through this method.

The co-ordinates of borehole I are determined from triangle GHI. The modern version of theodolites can measure 0.6 seconds as compared to 20 seconds in the theodolites of the sixties. The accuracy of angle measurement through theodolites has therefore taken a quantum jump.

The co-ordinates of the borehole can be determined to the level of submetre accuracy.

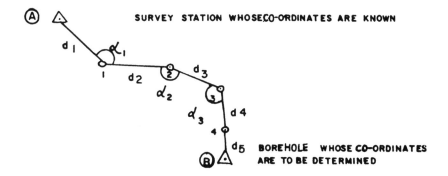

Fig. 16.1: Traverse Method.

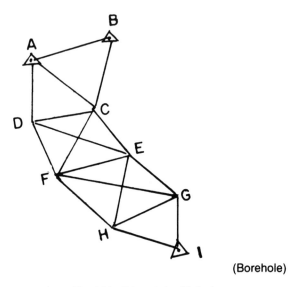

(Borehole)

Fig. 16.2: Triangulation Method.

Electronic Distance Measuring Equipment (EDM)

Just as in angle measurements, there has been a tremendous, indeed mind-boggling improvement in the accuracy of measurements of distance.

EDM equipment is capable of giving an accuracy of one centimetre in one kilometre, thereby reaching an accuracy of 1 in 100,000 in measurement of distance.

TOTAL STATION (COMBINATION OF PRECISE THEODOLITE AND EDM)

Through mechanical improvement in miniaturisation, EDM equipment and theodolite have been combined into one instrument called the 'Total Station' (Fig. 16.3). The total station is thus capable of measuring angles and distances and even recording the observations electronically in one instrument. Here again the accuracies achieved are within a few centimetres.

CO-ORDINATES THROUGH AERIAL PHOTOGRAPHY

Boreholes can be marked on aerial photographs of the terrain. A suitable scale for marking (termed post-pointing in survey terminology) is 1:25,000 or larger. In case such large-scale photographs are not available (or accessible) due to security reasons, a 1:50,000 scale may be used although post-pointing is comparatively more difficult. Post-pointing is done in the field using adjacent sharp features on the ground around the borehole. Aerial photography of 1:50,000 scale is available for the whole of India.

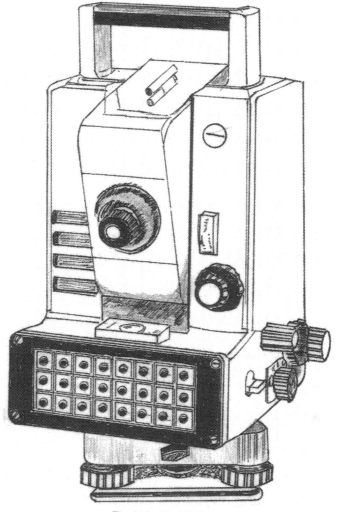

Fig. 16.3: Total Station.

After the borehole is marked on a photograph, its co-ordinates can be system-atically determined by means of the known co-ordinated points, termed 'control points', on the ground. The distortions in an aerial photograph ('tilt distortion' and 'height distortion') are removed through well-known and practised procedures.

This method is ideally suitable when the number of boreholes in a small area is large, e.g. seismic work. All such boreholes can be post-pointed and co-ordinates determined. The aerial photograph also provides a permanent record of the terrain while the boreholes are in the process of being dug/drilled.

Aerial photographs are taken with the help of camera fitted on the air-craft usually flown with 60% forward overlap. When these overlapping pictures (photographs) are viewed through an instrument called a 'stereoscope', a three-dimensional optical model is obtained. The boreholes can even be planned in the comfort of an office on this three-dimensional reduced model of the ground.

The only snag in this technique pertains to security restrictions; aerial photographs in India are 'classified documents'. For permission to use them and to ascertain their availability, contact the Survey of India office in New Delhi (Oc No. 73 party, West Block, New Delhi). The efforts invested in procurement of aerial photographs are well rewarded through the rich information such a photograph provides the user.

GLOBAL POSITIONING SYSTEM (GPS)

The Global Positioning System (GPS) is a new and revolutionary technology in the surveying discipline. Its main utilisation is in providing co-ordinates of a ground point in plan and elevation (X, Y and Z). Since each control point on the ground is determined independently (individually), the intervisibility conditions, so very essential but irksome in conventional surveying, are not operative. By virtue of this characteristic, GPS is ideally suited for providing control of the borehole.

The accuracy of results of GPS are unbelievably high, as much as 5 mm $\pm$ 1 part per million of measurement. Indeed, accuracy is of the 1st order. In a less accurate version, location of a ground point is determined within 1 to 10 metres, which is good enough for many novel applications, such as position of a vehicle or location of a boat.

In futuristic terms and because of its ultrahigh accuracy, the potential of GPS is being investigated for monitoring deformation of dams and other millimetre movements.

TECHNOLOGY OF THE GLOBAL POSITIONING SYSTEM

The basic concept of GPS, in its absolute simple form, can be compared to a basic survey operation called the 'resection method' in plane tabling, in which three or more ground control points are used for ascertaining the position of unknown station. Analogically, instead of three ground control points, in GPS three satellites are observed and resection is done automatically through an instrument. A check of the position is done by a fourth satellite.

In the American GPS there are 24 satellites orbiting the earth in differently inclined orbits. Please see Fig. 16.4.

The orbital parameters, namely distance from centre of the earth and its X, Y and Z positions are precisely known and disseminated to users by the U.S. Navy. In many situations the Navy may distort the information for security reasons.

A large number of GPS receiving instruments are available from different manufacturers. Some are represented in India: Wild-Leitz Delhi, Rolta of Bombay,

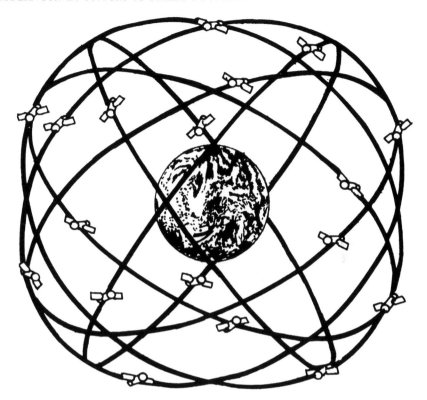

Fig. 16.4: Global Positioning System.

KLG Consultants, Delhi and Oceonics of Bombay are some of the firms providing hardware information. The following trade names are available in India:

TRIMBLE 4000 ST Field Surveyor	—	Oceonics, Bombay
ASHTECHZ XII GPS Receiver	—	Zeiss, Germany
WM 102 Satellite Receiver	—	WILD/LEICA (Cost about 17 lakhs)
GPS from MOTOROLA (USA)	—	KLG Consultants, New Delhi
GPS from Magellan (USA)	—	Rolta Corporation, Bombay

For the higher accuracy (geodetic) desirable for a tunnel project, the GPS is used in a 'differential mode' in which one GPS instrument is kept on a ground control point of known co-ordinates (say, India's Great Trigonometric Station) while the other is kept on the point whose co-ordinates are to be determined. Observations are taken simultaneously to eliminate many inherent errors, including those which may be artificially introduced by the American Navy.

Higher accuracies obtained in the differential mode are geodetic (first order level). The accuracy is represented as:

$$5 \text{ mm } \pm 1 \text{ ppm of measurement;}$$

where ppm is part per million.

GPS instruments which are capable of giving such high accuracy use both frequencies (dual frequency) of the satellites and thus remove ionospheric errors.

Incidentally, the GPS can be used in a coarser mode and provides accuracy from 1 to 5 metres for location of objects and vehicles. The cost of these coarser instruments is almost one-third that of higher accuracy instruments.

It is strongly recommended that for location of boreholes GPS is ideally suited from the point of view of convenience, cost and accuracy.

CONCLUSION

Various technical methodologies have been described in this chapter to make the drilling professional aware of how surveying technology which can be useful in furtherance of his objective, namely locating a borehole on the ground.

CHAPTER 17

GEOTECHNICAL INVESTIGATIONS, TYPES AND THEIR ECONOMICAL ASPECTS

Introduction
Types of Investigations
Choice of Methods and Costing
Selection of the Agency

INTRODUCTION

Geotechnical investigations provide a base for any construction work, large or small. It is important to know the existing topography of the area, position of drains or streams, any water logged areas, hillocks, rock outcrops, description of exposed cuts, vegetation, position regarding water table and its likely variations, geological features, available details regarding existing constructions and any other information typical for the area, such as occurrences of slides, subsidences, large-scale cracking in buildings, indicative of expansive soils, soft and marshy grounds etc.

Equally important is subsurface data—how the soil has behaved under existing constructions or will behave under changed conditions, during construction work and after its completion. Such subsurface investigations may be of two types—preliminary or detailed. However, the type and size of investigation will generally depend on the type and size of construction envisaged, nature of the terrain and time at one's disposal. For any project, timely decisions, proper liaison and control can result in considerable overall savings.

Example: A medium-size petrochemical complex was approved in Manali, Madras. Time of completion, one year with countdown effective immediately. Ground marshy, with very soft to soft soil up to about 3 m followed by medium compact soils up to 6 to 10 m overlying firm strata at depths ranging from 10 to 20 m.

Construction included roads, a large boundary wall, one- and two-storey buildings, a multistorey office complex, tanks and reservoirs, factory portion consisting of light, medium and heavily loaded structures, machine foundations, heavy flooring etc.

A preliminary layout plan of the complex was prepared and work begun on several fronts simultaneously, with well-defined targets and jobs allotted to various task forces. Starting with earth filling for roads and preliminary investigations, hectic activities in the design section provided details one after another. Several meetings were held with successful contractors—at the local office, at the site, at Bombay—and specific works were awarded to each at the earliest. It was decided to use (1) auger-bored stone columns for larger loaded areas so that work could commence on them immediately in specified portions, (2) normal underreamed piles for heavier loads after carrying out part filling, (3) underreamed compaction piles for heavy loads where permissible settlements were rather low, and (4) bored piles of larger depths, as required for supporting only very heavy loads. Work on superstructures and machinery installations followed and the complex was completed before the time limit with enormous benefits for all concerned.

TYPES OF INVESTIGATIONS

Subsoil strata can be very complex. One method alone cannot provide all the information required. Hence, depending on the likely nature of the subsoil strata and purpose of exploration, details of investigative procedure must be worked out.

Geophysical Investigations

Of the four systems making use of differences in the physical properties of geological formations in a given area, such as electrical conductivity, elastic moduli, density and magnetic susceptibilities, two are more widely used: a) electrical resistivity method and b) seismic method.

The electrical resistivity method, in which resistance to the flow of an electric current through the subsurface materials is measured at specified intervals on the ground surface, is useful for a general study of foundation problems and in particular for locating rocky strata under deep soil cover.

The seismic method makes use of the variation in elastic properties of the strata which affect the velocity of shock waves travelling through them, thus providing a useful tool for both dynamic elastic moduli determinations and mapping of the subsurface horizons. The requisite shock waves can be generated by hammer blows on the ground or by detonating a small charge of explosives. This method is quite useful in delineating the bedrock configuration and geological structures in the subsurface.

Subsurface Soundings

These provide primarily information regarding soil resistance but can also supply other details with depth, depending on the method of drivage:

  i) Dynamic cone penetration tests
 ii) Static cone penetration tests
iii) Static-cum-dynamic cone penetration tests

In dynamic cone penetration tests resistance to penetration of the cone in terms of number of blows per 300 mm of penetration may be correlated with the bearing capacity of cohesionless soils and possibly with the load-carrying capacity of piles. The correlations are qualitative rather than quantitative in nature and are influenced by the character of the soils, such as grain-size distribution, surcharge pressure, permeability and degree of saturation. The extra work required to determine the penetration resistance is small compared to the value of the data obtained. Resistance provides a rough indication of the consistency or relative density of the soil. However, these tests provide a simple and effective device for probing the soil strata and have an advantage over the standard penetration test, namely making of a borehole is avoided. Moreover, the data obtained by the cone test provides a continuous record of soil resistance. Records of the test should include: a) date of probing, b) location, c) elevation of ground surface, d) depth of water table and its likely variation (from locally available information), e) resistance offered at the required levels, f) any interruption in probing together with probable reasons, g) any other information available, for example type of soil, and h) diameter of the cone used in the test.

Dynamic Cone Tests

Dynamic cone tests can be of two types. In the first a 50 mm cone is used without bentonite slurry. This method provides resistance of different soil strata to dynamic penetration of a 50 mm dia cone and thus gives an indication of their relative strengths or density or both. It aids reconnaissance survey of wide areas in a shorter period of time, which enables selective *in-situ* testing or sampling for typical profile. It can provide useful data for local conditions where reliable correlations have been established. The maximum depth to which the cone should be driven depends on the type of soil, position of the water table and purpose of the test. If correlations between cone penetration values and values obtained by other methods are desired, then in cohesionless soils the depth may be limited to 5 m and in mixed soil with some binding material, to 10 m. If the test is used for obtaining a general qualitative idea of the strata, the cone may be driven to any convenient depth. The number of blows (N cd) as a continuous record for every 300 mm of penetration may be shown in a tabular statement or as a graph between N cd and depth.

In the second type of dynamic cone test a 62.5 cm dia cone is driven by dropping a 65 kg hammer and bentonite slurry simultaneously pumped through. Effective circulation of the slurry is essential for eliminating friction on the rods. A typical set-up for the dynamic cone penetration test using slurry is shown in Fig. 17.1.

Correlation between cone penetration values obtained using a 62.5 mm cone (N cbr) and penetration values obtained by other methods may be developed for a given site by conducting the latter tests adjacent (about 3 to 5 m) to the location of the cone test. However, for medium to fine sands the following relationships between the standard penetration value (N) and the cone penetration value (N cbr) exist. These relationships should be used with caution.

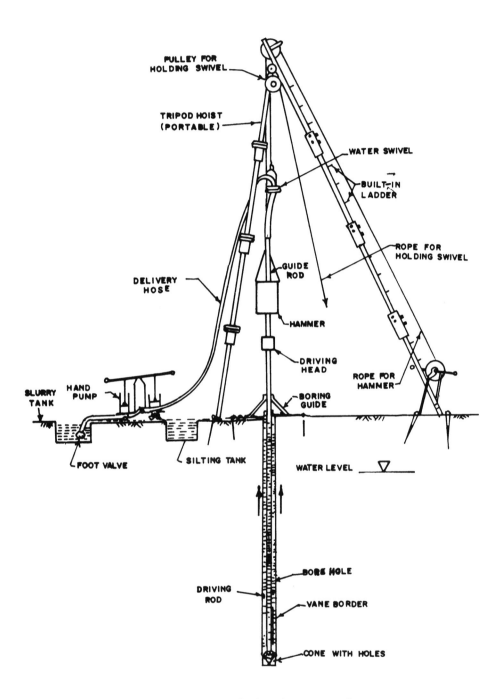

Fig. 17.1: Typical set-up for dynamic cone penetration test.

a) When the 62.5 mm cone is driven dry up to 9 m (without bentonite slurry):
   N cbr = 1.5 N ... up to a depth of 4 m;
   N cbr = 1.75 N ... for depths of 4 to 9 m.
b) When the 62.5 mm cone is penetrated by circulating bentonite slurry: N cbr = N.

Static Cone Penetration Test

Among the field sounding tests, the static cone test is a valuable method of recording variation in the *in-situ* penetration resistance of soils. It is particularly useful wherever *in-situ* density is disturbed by boring operations. The results of the test are also useful in determining the bearing capacity of the soil at various depths below the ground level. In addition to bearing capacity values, this test also enables determination of skin friction values, which are needed for determination of the required lengths of piles in a given situation. The static cone test is most successful in soft or loose soils, such as silty sands, loose sands, layered deposits of sands, silts and clays as well as in clayey deposits.

Experience has shown that a static cone penetration test up to depths of 15 to 20 m can be completed in a day with manually operated equipment. Thus it is an inexpensive and quick sounding method. A typical set-up for the static cone penetrometer (manually operated) is shown in Fig. 17.2.

In advanced countries the penetrometer is invariably used in the exploratory stage of investigations, when both time and money are at a premium. In areas where some information regarding the foundation strata is already available, the use of test piles and load tests can be avoided by conducting static code penetration tests only.

The set-up basically consists of the following:

1) A 60° steel cone of 10 sq. cm cross-sectional area
2) A friction jacket
3) Sounding rods
4) Mantle tubes
5) Driving mechanism (manual or manual-cum-power)
6) Measuring set-up
7) Anchoring device.

The rate of penetration is specified around 1 cm/s. The set-up can measure point resistance, total resistance, frictional resistance on the friction jacket and total frictional resistance on the mantle tubes.

The results should also be presented graphically, in two graphs: one showing the cone resistance with depth and the other showing frictional resistance with depth together with a borehole log.

The test is unsuitable for gravelly soils and for soils with a standard penetration value 'N' greater than 50. Also, in dense sands the anchorage becomes too cumbersome and expensive and for such cases dynamic cone penetration tests may be carried out. The test is also unsuitable for made-up or filled-up earth since erroneous values may be obtained due to the presence of loose stones, brickbats etc.

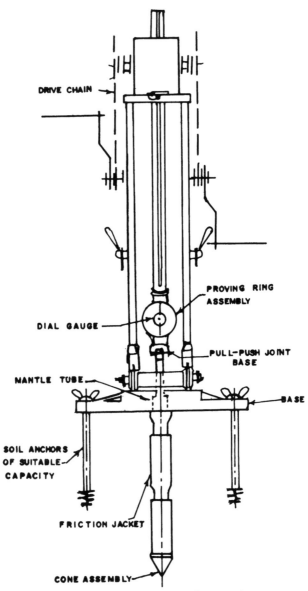

Fig. 17.2: Typical set-up for static cone, manually operated penetrometer.

In such places either the made-up soil shall be completely removed to expose the virgin soil layer, or readings in the filled-up depth shall be ignored.

Static-cum-Dynamic Cone Penetration Test

Sometimes thin layers of hard strata are encountered. These cannot be penetrated by static driving due to limitation of the driving capacity of the system.

In such situations the static driving mechanism may be moved aside, leaving the driving shaft with cone assembly free to be driven with a drop weight, i.e., by dynamic penetration. After the hard strata has been passed by means of dynamic penetration, normal static driving may be resumed.

Load Tests

These tests can be made on soils, gravelly and bouldery strata or rocks. Normally, three sizes of plates are used on soils, to assess the size affect. In gravelly and bouldery strata, cast *in-situ* footings provide better results but require marked reaction in the form of a loaded platform. Flat jacks are commonly used in rocks.

Load tests in soils are not considered very reliable. Changes in moisture content can very considerably effect the results as can changes in the water table. If at all these tests are required to be carried out in soils, the results should be interpreted along with borehole studies.

In-situ Shear Tests

These tests are best suited for determining *in-situ* strength parameters of gravelly and bouldery strata. These are two types, the boulder-boulder test and the concrete-boulder test.

The boulder-boulder test requires carving a block of bouldery soil and slowly confining it in a rigid steel former. This bouldery soil block is sheared under a normal load. The shear stress is plotted against displacement.

The concrete-boulder test requires casting a reinforced concrete block on the bouldery strata and pushing it laterally under a given normal load.

For shear stress, the corresponding displacement is noted and a shear-stress versus displacement curve plotted. This test has an advantage in that it eliminates the need for a steel former and carving an undisturbed bouldery soil block. Precast concrete blocks may be used after they are seated with mortar on the foundation bed.

Exploratory Borings and Drillings

Preliminary borings by augers, either power or manually driven, are quick and economical up to a depth of about 10 m in alluvial deposits. When detailed information is not required, wash boring with chopping and jetting may be utilised in cohesive as well as non-cohesive soils up to even larger depths. When a casing is not used, particularly in non-cohesive soils, the sides of the holes should be stabilised where required with drilling fluid consisting of bentonite slurry in water. In the wash-boring method changes in stratification can be ascertained only by the rate of progress of the drill or change in colour of the wash water or both. As the formation hardens, rotary drills, using churning bits, may be utilised. In gravelly materials percussion drilling with simultaneous advance of the casing is the only easy method.

In hard and cemented formations such as rock, the hole is advanced with cutting edges using steel shots, hardened metal bits, tungsten carbide or diamond bits.

a) *Auger Boring:* An auger may be used for boring holes to depths of up to about 15 m in soft to medium compact soils which can stand unsupported. It may also be used with lining tubes, if required, or sides held by pouring bentonite slurry.

b) *Shell and Auger Boring:* A manually operated rig may be used for vertical boring up to 200 mm in dia and 30 m in depth. In alluvial deposits the depth of the borehole may be extended up to 50 m with a semi-mechanised rig. The tools consist of augers for soft to stiff clay or shells for silty, clayey and sandy strata. These are attached to strings or sectional boring rods. Small boulders and thin strata of rock may be broken by a chisel bit attached to the boring rods. The boring rods are raised or lowered by means of a tripod rig and winch and are turned manually. The casing, where essential, is advanced using a drop weight.

c) *Percussion Boring:* This method consists of breaking the formation by repeated blows from a bit or a chisel. Water should be added to the hole at the time of boring and the debris baled out at intervals. The chisel may be suspended by a cable or rods.

Where boring is in soil or into soft rocks and provided that a sampler can be driven into them, cores may be obtained at intervals using suitable tools. However, in soils the material tends to become disturbed by this method of boring and samples may not be as reliable as obtained in the shell-and-auger method. As these machines are devised for rapid drilling by pulverizing the material, they are not suitable for careful investigation. However, this is the only method fast enough for drilling boreholes in rocky and bouldery strata.

d) *Wash Boring:* In this method the pipe or drill rod assembly with a swivel at the top and a cutting edge at its far end is rotated and also moved up and down, either manually or with a mechanised rig, and water is pumped in continuously. The bore may be with or without a casing as required. The cut soil mixes with the cutting fluid and floats to the surface. The slurry flowing out gives an indication of the soil type. In this method heavier particles of different soil layers remain under suspension in the bore and intermix. Thus this method is not suitable for obtaining samples for classification. Whenever a change in strata is indicated by the slurry flowing out, washing should be stopped. A suitable tube sampler can be attached to the end of the drill rod and fairly undisturbed samples of soil can be obtained by suitably driving the sampler into the soil.

Rotary Boring

a) *Simplified Mud Boring Method:* In this method boring is advanced by a suitable cutter fixed to drill rods which are rotated by means of pipe wrenches manually. Bentonite is pushed simultaneously by a double-piston pump. The slurry flowing out of the cutter bottom mixes with the cut soil and flows to the borehole surface, the settling tank and back to the slurry tank. The process is continuous and the same slurry can be used continuously. The cutting tool is slowly lowered down by means of a manually operated winch fixed on a tripod. After boring has been advanced to the desired depth, pumping of the slurry should be continued for 10 to 15 minutes. The set-up is similar to that shown in Fig. 17.1. The cone is replaced by the cutting tool. The hammer and driving head are not required until sampling is carried out using suitable sampling tubes. Dynamic cone or standard penetration tests can also be carried out by the same set-up.

In case gravel and kankar are encountered, a gravel trap fitted around the drill rod, a little above the cutter, may be used. The trap consists of an 80 to 100 cm long hollow cylinder that is conically shaped at the bottom. A few holes of 3 mm diameter are also drilled in the drill rod within the trap as well as in the conical portion of the trap. During boring, gravel and kankar rise a little and then settle in the trap. Given the provision of holes, finer materials cannot settle in the trap.

Small silt-sandstone or hard beds may be broken using conical or chisel-ended bits connected to the drill rod. The broken pieces can subsequently be removed by means of the gravel trap.

b) *Calyx Rig Boring and Drilling:* In this system, boring is effected by the cutting action of a rotating bit which should be kept in firm contact with the bottom of the hole. The bit is carried at the far end of jointed drill rods which are rotated by a suitable chuck. A mud-laden fluid is pumped continuously down the hollow drill rods; the fluid returns to the surface in the annular space between the rods and the side of the hole and so a protective casing is generally not necessary. In this method soil samples and rock cores may be obtained by the use of suitable samplers/coring tools.

c) *Shot Drilling:* This system is used in larger diameter holes, that is over 150 mm. Due to the necessity of maintaining the shots in adequate contact with the cutting bit and the formation, inclined holes cannot be drilled satisfactorily.

d) *Diamond Drilling Rigs:* These high-speed drills are meant for coring in rocks. In sound rock up to 90% recovery of cores is possible by using suitable equipment. Water is circulated down the hollow rods, which returns outside them carrying the rock cuttings to the surface as sludge. These are retained as samples in traversing friable rock from which cores cannot be recovered. It is important to ensure that boulders or layers of cemented soils are not mistaken for bedrock. This necessitates core drilling to a suitable depth in bedrock in areas where boulders are known to occur. For shear strength determination, a core with a diameter to height ratio of 1:1 is required. Rock pieces may be used for determination of specific gravity and classification.

Field Tests (Through Boreholes)

a) *Standard Penetration Tests:* In this test a standard spoon sampler (Fig. 17.3) is driven through a borehole by blows from a 65-kg hammer falling freely from a height of 75 cm. The number of blows required for the first 15 cm of penetration is taken as the resting load; the split spoon is then driven another 30 cm in 15-cm stages, total number of blows not exceeding 50. The total number of blows for the latter 30 cm is termed the standard penetration resistance 'N'.

These tests are widely used and provide a very simple means of comparing the results of different boreholes on the same site and obtaining an indication of the bearing value of non-cohesive soils which cannot be easily assessed by other means.

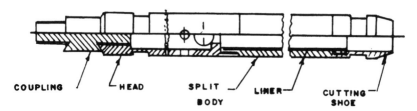

COUPLING   HEAD   SPLIT   LINER   CUTTING
                                 BODY   SHOE

Fig. 17.3: Split-spoon sampler assembly.

b) *Vane Shear Tests:* The vane test has proven to be a promising non-empirical method for measuring the shear strength of soft clay *in situ* at depths up to 30 m. It is particularly useful in the measurement of strength in deep beds of soft sensitive clays.

The shear strength is measured by pushing a small four-bladed vane attached to the end of a rod into the clay, and then measuring the maximum torque necessary to cause rotation. This torque is approximately equal to the moment developed by the shear strength of the clay acting over the surface of the cylinder with a radius and height equal to that of the vanes.

c) *Pressure-meter Tests:* A pressure meter applies a uniform radial stress to the borehole at any desired depth and measures consequent deformation. The test involves lowering an inflatable cylindrical probe to the test depth in a borehole. The probe is inflated by applying water pressure from a reservoir. Under pressure, it presses against the unlined wall of the borehole and causes volumetric deformation. The stress on the borehole wall is the pressure of the water applied. Deformation of the borehole is read in terms of volume corresponding to fall in water level of the reservoir.

Trial Pits

These pits are preferable for shallow depths since they enable sand and shingle strata to be seen in their undisturbed state and give a more accurate idea of shoring and pumping. Trial pits in stiff fissured clays also give a fairly accurate idea of the depth to which open excavations or vertical cuts can be carried out without shoring. They also give a better picture of the patchy ground wherever the soil lies in pockets. In the case of gravels and sandy soils, fines tend to be washed out and the various layers are apt to become mixed as a result of 'piping'.

Hence, it is difficult to obtain representative samples in such cases and unless proper precautions are taken, a misleading impression may be obtained. The best procedure is to obtain samples from trial pits dug after the groundwater has been lowered by suitable means.

CHOICE OF METHODS AND COSTING

Both depend on whether a preliminary investigation is being taken up or a detailed one, purpose of the investigations, its magnitude, time and funds at one's disposal and availability of required facilities, equipment and expertise. Certain norms for some of the more common tests are suggested here.

For a 30-m deep borehole in alluvial deposits, including boring/sampling, *in-situ* testing in the field, the time and manpower required for different processes may be as under (for guidance only).

a) Shell and auger: Four to six days with an 8-hour shift and 4 to 6 persons per rig. (Overall cost approx. Rs. 120.00 per hour + mobilisation charges.)

b) Mud boring: Four to five days with an 8-hour shift and 6 to 7 persons per rig. (Overall cost approx. Rs. 120.00 per hour + mobilisation charges.)

c) Rotary boring using calyx drilling rig: Four to five days with 8-hour shift and 4 to 8 persons per rig. (Overall cost approx. Rs. 150.00 per hour + mobilisation charges.)

(NOTE: Boring through gravelly/bouldery strata can be very time consuming, cumbersome and rather expensive.)

Continuous dynamic cone test up to refusal: Depending on depth penetrated, say 20 m, 2 days with an 8-hour shift and 3 to 5 persons per rig. (Overall cost for dry probing approx. Rs. 50.00 per hour, and with bentonite slurry Rs. 70.00 per hour + mobilisation charges.)

Static cone test for point bearing and total friction up to refusal: Depending on depth penetrated, 2 to 3 days with an 8-hour shift and 4 to 6 persons. (Overall cost for 5-t machine approx. Rs. 100.00 per hour + mobilisation and for 10-t machine approx. Rs. 150.00 per hour + mobilisation.)

Plate load test up to 20 t: With the Kentledge system, 4 to 5 days with an 8-hour shift and 6 to 8 persons. (Overall cost approx. Rs. 200.00 per hour + mobilisation.) With the truss anchoring system 3 to 4 days with an 8-hour shift and 4 to 6 persons. (Overall cost approx. Rs. 150.00 per hour + mobilisation.)

For lab investigations it is important to know the minimum number of various equipment required for a certain job. Norms for some of the common tests are suggested below.

For laboratory investigations, considering one operator to be on each test, the following operator hours may be projected.

a) LL/PL   — 1 to 2 hours per test
b) Unconfined compressive strength — 2 to 3 hours per test
c) Triaxial (quick)   — 6 to 8 hours per test of 3 specimens.
d) Triaxial consolidated (quick)   — 2 to 3 days including consolidation of clay samples and considering separate cells for consolidation under different pressures
e) Triaxial tests with pore pressure — 3 to 4 days
f) Consolidation tests   — 8 days per test (3 days for quick test) NOTE: One operator can handle up to 8 consolidometers
g) Box shear (quick)   — 6 to 8 hours per test for 3 specimens

Note: Lab investigation charges, based on operator hours, work out to approx. Rs. 15.00 to 30.00 per hour per operator.

SELECTION OF THE AGENCY

Selection of the agency depends on its capability to execute works of different magnitudes so as to be in a position to complete the specified field and lab work in time, and of course the requisite expertise. It also depends on whether the agency can take up special works, such as dynamic testing, electrical resistivity tests, field permeability etc., if required.

It is also essential to establish the availability of field testing equipment, lab test facilities, required personnel and experts for analysis of data and designs.

It is equally important to ascertain what works the agency has in hand and the spare capacity available for execution of the job in question. Clients, on the other hand, need to be more specific regarding period of award. Very often for a two-month job the validity of the offer is anything from 3 to 6 months, in which case any exercise to ascertain the spare capacity of the agency is almost futile.

Very often unworkable rates are cited by an agency just to obtain a job. Such unworkable rates either indicate that the party may not do the full job efficiently or has some vested interest. In either case the ultimate loser is the client. It should be appreciated that the party once selected becomes a consultant and should be associated throughout the project.

REFERENCES

CBRI Handbooks on Underreamed and Bored Compaction Piles. G.S. Jain & Associates Pvt. Ltd.,
Digest on Dynamic Cone Penetration Tests. Central Building Research Institute, Roorkee.
Kumar, A. and V. Kumar. An Improved Method and Device for Penetration Tests. Licensee M/s. G.S. Jain & Associates Pvt. Ltd.
IS:1892:1987. Code of Practice for Subsurface Investigation for Foundations.
IS:2131:1987. Method for Standard Penetration Test for Soils.
IS:4968 (Part I):1987. Dynamic Method Using 50 mm Cone without Bentonite slurry
IS:4968 (Part II):1987. Dynamic Method Using 50 mm Cone without Bentonite slurry

CHAPTER 18

DESIGN OF FOUNDATION OF DAMS*

Introduction
Foundation Treatment for Concrete Dams
Foundation and Abutment Treatment for High Embankment Dams at Rock

INTRODUCTION

The foundation for a dam is designed to provide:

i) Acceptable deformation under load
ii) Impermeability within acceptable limits
iii) Adequate resistance to shearing/sliding
iv) Protection against erosion
v) Stability of abutments.

The treatment can broadly be classified into two categories.

FOUNDATION TREATMENT FOR CONCRETE DAMS

Excavation

The entire area to be occupied by the base of the concrete dam should be excavated up to firm material capable of withstanding the loads imposed by the dam, reservoir, and appurtenant structures. Considerable attention must be given to blasting operations to assure that excessive blasting does not shatter, loosen or otherwise adversely affect the suitability of the foundation rock.

Foundations such as shales, chalks, mudstones and siltstones may require protection against air and water slaking and or in some environments, against freezing. Such excavations can be protected by leaving a temporary cover of several feet of unexcavated material, by immediately applying pneumatically a minimum thickness layer of mortar to the exposed surfaces, or by any other method that will prevent damage to the foundation.

---

* Adapted from V.M. Sharma. 1991 "Design of Foundation of Dams." National Workshop on Rock Mechanics, New Delhi, organised by the Central Board of Irrigation and Power.

Shaping

If the canyon profile for a dam site is relatively narrow with steep sloping walls, each vertical section of the dam from the centre towards the abutments is shorter in height than the preceding one. Consequently, sections closer to the abutments will be deflected less by the reservoir load and sections closer to the centre of the canyon will be deflected more. Since most gravity dams are keyed at the contraction joints, the result is a torsional effect in the dam that is transmitted to the foundation rock.

A sharp break in the excavated profile of the canyon will result in an abrupt change in the height of the dam. The effect of the irregularity of the foundation rock causes a marked change in stresses in both the dam and foundation, and in stability factors. For this reason, the foundation should be so shaped that a uniformly varying profile is obtained, free of sharp offsets or breaks.

Generally, a foundation surface will appear as horizontal in the transverse (upstream/downstream) direction. However, where an increased resistance to sliding is desired, particularly for structures founded on sedimentary rock foundations, the surface can be sloped upward from heel to toe of the dam.

Dental Treatment

Very often the exploratory drilling or final excavation uncovers faults, seams or shattered or inferior rock extending to such depths that it is impracticable to attempt to clear such areas out entirely. These conditions require special treatment in the form of removing the weak material and backfilling the resultant excavations with concrete. This procedure of reinforcing and stabilising such weak zones is frequently called 'dental treatment'.

Theoretical studies have been done to develop general rules for guidance as to how deep transverse seams should be excavated. These studies, based on foundation conditions and stresses at Shasta and Friant Dams, have resulted in the development of the following approximate formulas for determining the depth of dental treatment:

$$d = 0.002 \, bH + 5 \text{ for } H \geq 150 \text{ feet;}$$
$$d = 0.3b + 5 \qquad \text{for } H < 150 \text{ feet,}$$

where H = height of dam above general foundation level in feet; b = width of weak zone in feet, and d = depth of excavation of weak zone below surface of adjoining sound rock in feet (in clay gouge seams, d should not be less than $0.1H$).

These rules provide a means of approach to the question of how much should be excavated but final judgement must be exercised in the field during actual excavation operations.

Although the preceding rules are suitable for application to foundations with a relatively homogeneous rock foundation with nominal faulting, some dam sites may have several distinct rock types and shears. The effect of rock-type anomalies complicated by large zones of faulting on the overall strength and stability of the foundation requires definitive analysis. Such a study can be performed by the finite element method. This method provides a way to combine the physical properties of various rock types and geologic discontinuities such as faults shears, and joint sets into a value representative of the stress and deformation in a given segment of the foundation. The method also permits substitution of backfill concrete in faults, shears and zones of weak rock, and thus evaluates the degree of beneficiation contributed by the 'dental concrete'.

The data required for the finite element method of analysis includes dimensions and composition of the lithologic bodies and geologic discontinuities, deformation moduli for each of the elements incorporated into the study, and the loading pattern imposed on the foundation by the dam and reservoir.

'Dental treatment' may also be required to improve the stability of rock masses. By inputting data related to the shearing strength of faults, shears, joints, intact rock, pore-water pressures induced by the reservoir and/or groundwater, the weight of the rock mass, and the driving forces induced by the dam and reservoir, a safety factor for a particular rock mass can be calculated.

Protection against Piping

The approximate and analytical methods discussed above will satisfy the stress, deformation and stability requirements for a foundation, but they may not provide suitable protection against piping. Faults and seams may contain material conducive to piping and its accompanying dangers, so to mitigate this condition upstream and downstream cutoff shafts should be excavated in each fault of seam and backfilled with concrete. The dimension of the shaft perpendicular to the seam should be equal to the width of the weak zone plus a minimum of 0.3 m on each end to key the concrete backfill into sound rock. The shaft dimension parallel with the seam should be at least one-half the other dimension. In any instance a minimum shaft dimension of 1.5 m each way should be used to provide working space.

The depth of cut-off shafts may be computed by constructing flow nets and computing the cut-off depths required to eliminate piping effects, or by the methods outlined by Khosla. These methods are particularly applicable for medium to high dams. For low head dams the weighted creep method for determining cut-off depths may be used.

Other adverse foundation conditions may be due to horizontally bedded clay and shale seams, caverns or springs. Procedures for treating these conditions will vary and will depend on field studies of the characteristics of the particular condition to be remedied.

Consolidation

The useful method for consolidating sands and gravels *in situ* is by vibration. In extreme cases the same effect can be obtained with explosives. At Karnafuli Dam, Bangladesh, monsoon floods scoured the diversion channel to depths of 20 m below the general river-bed level in an area that would later be covered by the downstream shell of the rockfill dam. To dewater the hole would probably have endangered the side slopes of the first-stage embankment. It was therefore decided to dump cohesionless material into the water in the hole; after so filling and dewatering the loose material was compacted by detonating buried explosive charges.

Consolidation by grouting: In general there will be cracks, joints and seams within the foundation. Although the specifications should prohibit blasting within 0.5 m of the final excavation limits, there will always remain some loosening of the rock. Openings should be injected with cement both for consolidation and to prevent erosion of infilling material in this zone of maximum hydraulic gradient. Under a gravity dam it is usual to drill primary holes to a depth of about 10 m in a general pattern wherein holes will be at approximately 8 m centres. Each hole will be located and oriented to the best advantage. If there is considerable consumption of grout, intermediate secondary holes will be drilled and grouted, followed by intermediate tertiary holes if required. It is important that grouting pressures be kept low to avoid heaving of the rock and propagation of cracks too fine for the entry of cement but sufficient to accept water.

Where there may later be relatively high stress concentrations the grouting pattern should be spread to consolidate a bulb under the dam. The holes should be located and oriented to intersect as many seams as possible. In sedimentary series it may be advisable to wash soft material from seams and replace it with cement. To be effective for general consolidation this would be slow and expensive, necessitating considerable checking by core drilling. Where the modulus of deformation of the rock mass is low, or variable along the axis of the dam, consolidation grouting can provide considerable improvement. At Repulse Dam, Australia, grouting was done to depths of 5 m over the foundation area. The deformation modulus was measured by loading cables anchored in the rock. The increase in modulus of deformation of the jointed dolerite was of the order of 50%; for the poorest rock it was improved from 2000 to 3000 MPa.

It is understood that the very comprehensive consolidation grouting at Vaiont Dam, Italy resulted in a 100% increase in the modulus of deformation.

Much has been written concerning consumption of grout. It will obviously depend on the porosity of gravels or the width of joints and cracks in rock formations. More importantly, it will depend on the experience of the operators and the attention given to work.

Impermeability by Grouting

Grout Curtain: To impede the flow of water under or around a dam a grout curtain is usually provided. The depth to which this should extend will depend on the type and condition of the rock mass. The suggested depth of grout curtain is

$$D = 1/3H + C,$$

where D = depth in metres; H = height of dam in metres; and C = is a constant varying with the foundation, size of the dam and significance of leakage, say 8 to 25 m.

The grout curtain should extend well into the abutments on the line of the dam. The location of the abutment curtain will be determined relative to rock formations and the provision that will be made for abutment drainage.

When the reservoir is filled and a concrete dam comes under load there is a tendency for cracks to form in the rock upstream from the dam. Their probable extent and location can be determined by finite element studies. It is wise, therefore, to provide a gallery in the dam following and not far distant from the foundation contact. Horizontal extensions of this gallery should also be driven into the abutments. These will permit regrouting, should the grout curtain be fractured or prove inadequate.

One outstanding grout curtain was that provided at the 117-m high Dokan Arch Dam, Iraq. The main rocks in the area from the bottom upwards are: dolomite, overlain by thin-bedded limestone and finally marl. The grout curtain extends into both abutments, its total length being 24 km and its area 450,000 m with holes as deep as 200 m. Apart from this curtain some 580,000 m of rock on the left abutment required grouting. In all, rock sealing and consolidation required 300,000 m of drilling and the injection of 116,000 tons of cement and sand.

It is difficult to prove the effectiveness of a grout curtain, since this would involve many pressurised holes on one side and an equal number of piezometer holes on the other side of the curtain. For this reason considerable attention has been given to the development of geophysical methods.

Resistivity traversing can be used to control the efficiency of grouting in the area of water leakage. The resistivity of the cement grout is lower than that of the rock to be grouted. The drop of resistivity after grouting would indicate the percentage of voids filled with the grout. This method has been used in Yugoslavia and Austria.

Grouting Pressure: The permissible grouting pressure (usually measured at the collar of the hole) will depend on the type of rock and the rock formations, whether bedded sedimentary or massive igneous. The following maximum grouting pressure values for sedimentary rocks are recommended:

i) for solid rocks with steeply inclined fissures

$$P = 30D + 2D^2;$$

ii) for solid rocks with horizontal fissures

$$P = 24D + 0.5D^2,$$

wherein P is the pressure in kPa and D is depth in metres.

FOUNDATION AND ABUTMENT TREATMENT FOR HIGH EMBANKMENT DAMS ON ROCK

General

With the passage of time, the availability of good dam sites with respect to geologic and topographic characteristics has decreased markedly, particularly within the last decade. Formerly, when high dams were considered, the emphasis was on utilisation of concrete and masonry structures, and the existence of a site with prime qualifications as to geologic soundness and topographic configuration was a governing factor in the feasibility of the project. As a result of the great advancements which have been made in the science of soil mechanics and foundation engineering, high embankment dams can now be designed and constructed with a degree of confidence comparable to that associated with concrete structures. However, the criteria of quality for embankment dam sites may differ from those for concrete dams and encompass a much broader range of acceptable conditions.

Excavation of Abutments and Foundation Area

The entire foundation area is usually stripped to rock when the overburden is less than 3 m thick. The engineering advantages of stripping the entire area to rock where the overburden is shallow is the improvement of strength characteristics of the dam, elimination of the need for filters between the embankment and foundation, simplification of drainage, and exposure of foundation rock for inspection and treatment. Where the foundation is stripped to rock under the core contact only, the material left under the shells is invariably described as being equivalent to overlying embankment fills. The choice of excavating to rock under core contact or using a cut-off trench may be influenced by such practical considerations as the width of the core and the character and depth of overburden.

Criteria with Regard to Contact Areas of Core

Maximum permissible slopes at abutments: All experts appear to agree that overhangs in the core contact area should be eliminated. In a few cases vertical faces of limited height and width may be tolerated; however, at most sites a sloping core contact area is provided by rock excavation if necessary. Maximum permissible slopes range from 0.1 horizontal:1.0 vertical to 1.0 horizontal:1.0 vertical. Within this range the slope selected is influenced by practical topographic considerations related to the volume of excavation required, the strength characteristics of the formation, and the bedding and joint system of bedrock at the dam site. In some cases the height and length of the cliffs in relation to the dimensions of the core contact area have also been taken into account.

Upstream-downstream contact slopes: It would appear that the trimming of abutments to provide convergence toward the downstream side is not a commonly accepted practice. As a class, high embankment dams on rock include many rockfill structures with narrow impervious cores. In these designs the length of the path of percolation at the core contact is a matter of concern and, from the response, it appears that a minimum hydraulic gradient of approximately 0.5 horizontal to 1 vertical is usually desired. At three of the structures reported the acceptable hydraulic gradient was obtained by flaring out the core at the abutment contact.

Blasting and trimming criteria: It is generally agreed that rock excavation methods in core contact areas should be controlled by responsible engineers to prevent damage to underlying rock. Controlled blasting with prior approval of patterns and charges seems to be the normal procedure. Presplitting is mentioned in several cases. In one case line drilling was required and at another dam site trimming in the core area was accomplished by hand with pneumatic tools. Where required, the core contact surface has often been smoothened by the combined operations of trimming projections and filling depressions with concrete.

Quality of acceptable rock surface: The adjectives sound, firm and fresh are most frequently used to describe the overall quality of an acceptable rock surface in core contact areas. However, it appears that practical, economic considerations often force the designer to relax these criteria in the presence of local foundation defects. In several cases cut-off trenches in rock were excavated into the general rock surface of the core contact area to reach relatively sound rock in faulted and jointed areas. In another case it was noted that the requirement for a fresh rock surface was relaxed to weathered rock on the upper abutments where the head was low. Defects in the foundation that were impracticable to remove by excavation have been corrected by special surface treatment and grouting.

Criteria with regard to contact area of shell: In principle the overburden and rock in shell contact areas should be equivalent in strength and density to that of the overlying fills. Overburdens most frequently meeting these criteria are sand and gravel deposits in a river bed. Most of the respondents did not indicate any special trimming or grading requirement for contact areas of the shell.

Foundation and Abutment Surface Treatment

Following completion of the necessary excavations in the foundation and abutment areas, additional treatment is generally required to these surfaces prior to the placement of embankment materials. The major purposes of this additional surface treatment are:

   i) To develop a uniform base on which to place and compact embankment material.

   ii) To fill surface and near surface voids and imperfections in the foundation seepage at or adjacent to the contact between the embankment and its foundation.

   iii) To improve the bond between embankment and foundation.
Where rock is exposed to form the foundation in the abutments and valley bottom, substantial treatment is normally required.

   It is general practice to provide a greater degree of treatment beneath the impervious core section than beneath the upstream and downstream shells. In addition, the foundation beneath the embankment filter zones frequently receives the same treatment as in the area beneath the core. The amount of treatment required is primarily dependent on the condition of the exposed rock and the amount of defects present at the surface.

   In the foundation area beneath the core and filter all soil and loose rock remaining from the excavation operations are removed, either by hand or by utilising loading equipment, as conditions permit. The area is then cleaned by brooming and washing with air, water or air-water jets to completely remove all loose and foreign materials. Defects in the surface resulting from seams, joints, faults, shear zones etc. are thoroughly cleaned and backfilled with slush grout, pneumatically applied mortar (such as gunite or shotcrete) or concrete. Such seams are cleaned to the maximum practical depth or to a depth in the order of three times the width of the defect. In zones of badly fractured rock it is common practice to cover the entire area with pneumatically applied mortar or broomed slush grout after a thorough cleaning. Larger depressions in the surface, such as potholes, cavities, solution caverns or channels are thoroughly cleaned by washing and backfilled with concrete, grout or mortar, or if of sufficient size to allow proper compaction, with the contiguous embankment material. Sometimes, the entire core contact foundation area is covered with a thin layer of broomed slush grout and then placing and compacting the first layer of embankment material while the grout is still plastic. Such treatment is of particular value when used on rock which deteriorates rapidly on exposure to the elements. It is common practice to cover exposed friable rock types which tend to slake, spall or degenerate when exposed to air or water, most notably exemplified by some shales, with a coating of mortar, asphalt or asphalt emulsion shortly after final exposure to prevent deterioration in the interval between clean-up and embankment placement.

In other foundation areas outside the core and filter boundaries, special treatment is not normally required. Following the foundation excavation operations, these areas are generally prepared for receiving embankment material by the removal of loose and broken material and levelling of the resultant surface to the extent that fill can be placed and compacted. Levelling may be achieved by placement of hand-tamped backfill or concrete.

Common treatment on the valley abutments calls for reshaping of steps, ledges and overhangs by further excavation or filling with concrete or pneumatically applied mortar. Dental concrete is commonly used to fill undercuts and cavities to develop a suitable slope against which embankment material can be placed and compacted. Where steep abutment slopes are encountered, it is sometimes necessary to use forms to retain the concrete in place. As with the foundation treatment, the degree of improvement is greater beneath the impervious core than beneath the outer shells, and the rock surfaces in the abutments receive similar treatment with respect to cleaning and filling defects as in the foundation area.

Grouting

The grouting programme is developed by analysing prior foundation explorations and geologic reports and from this data a programme is planned to accomplish the desired objectives. The specifications are so prepared that the grouting programme can be modified as grouting is progressing and field conditions dictate.

There are three main objectives in the grouting programme: (1) to reduce the seepage flow through the dam foundation, (2) to prevent possible piping or washing of fines from the core into cracks and fissures in the foundations; and (3) to reduce the hydrostatic pressure in the downstream foundation of the dam. The latter is generally not a problem on most rockfill dams, but for dams on fairly weak foundations and critical abutment configurations this does become a problem and is usually accomplished in conjunction with an abutment drainage system.

The methods of accomplishing the objectives are fairly consistent. Usually a single line of grout holes is used to reduce the leakage through the foundation of the dam. Additional lines of holes are used when the foundation is fairly tight and the rock is closely fractured or jointed. This type of foundation does not allow grout to travel any appreciable distance and the additional lines are necessary to assure adequate reduction in leakage.

To prevent possible piping of the fine core material through the foundation, blanket or consolidation grouting is accomplished as determined by the rock condition. If the core foundation of the dam consists of closely fractured and jointed rock, a blanket grout pattern is used with holes spaced at 3 to 5 m with depths of 7 to 10 m. If the foundation rock is massive, no blanket grouting is done. Localised areas consisting of faults, fissures, or cracks are generally grouted upstream of the cut-off and sometimes downstream. Some agencies have increased the number of curtain grout lines to accomplish the same objective.

The sequence of grouting by most of the reporting agencies is to complete blanket grouting prior to curtain grouting. This is done at relative low pressures to seal the cracks and fissures near the surface which will prevent surface grout leakage while grouting at a greater depth at higher pressures. The curtain is then completed generally by packer grouting starting at the lower depth and progressing upwards. The initial curtain holes are usually 12 m to 30 m apart depending on the spacing necessary to prevent communication between grout holes. The split-spacing method is then used to close out the curtain. A minimum spacing of 3 m is generally required for closure with holes as close as 1 m being used if required. The grout holes are frequently angled to cross cracks, fissures and bedding planes in the foundation to give a more effective grout coverage. Angled holes crossing the normal pattern of grouting are used to check closure in critical areas.

Grout pressure and maximum depths vary appreciably between the reporting agencies but the differences are believed to be mainly in the type of foundation being grouted. The criterion in general is to grout to the maximum pressure that will not cause movement of fracturing of the rock. The maximum depth of holes is generally about half the dam height, but this is varied depending on the tightness of the foundation at depth, as determined by exploration and testing.

The initial mix is usually about 5 to 1 with the mix being thickened to 1 to 1 for closing out the hole. Cement is the primary grouting agent with chemical, bentonite or sand being used for unusual conditions.

Grout caps are used in areas of weak and fractured rock to prevent surface leakage through the foundation and to provide anchorage for grout connections. The cap usually consists of a trench in the foundation filled with concrete. Nominal dimensions are used, usually less than 5 ft deep and a similar dimension in breadth.

At some installations, the grout cap is constructed large enough to allow containment of a grout gallery. This permits construction of the dam during the grouting operation. The gallery also provides access for inspection, future additional grouting and drainage.

With increase in dam height, post-construction grouting from the top of a dam to correct excessive seepage becomes more difficult, especially if the core is inclined. The need for a grouting gallery under the core of high dams should be given careful consideration.

Drainage

The downstream zones of rockfill, pervious fill or filter material serve as drains for seepage that flows through overburden or through unsealed joints in rock of the foundation and abutments.

Galleries or tunnels are sometimes excavated into abutments and foundations to provide access for drilling drain holes and grout holes into the foundation or into the embankment-foundation contact. At the Oroville Dam, a gallery was provided in an excavated trench in the foundation of the core for both purposes. Grout drilling from these tunnels has been questioned because of the danger of uncontrolled flow in/under high heads; consequently defensive design measures must be relied upon. This condition underscores the need for development of a safe technique for such grout drilling, should future corrective measures be needed.

Relief wells and drainage trenches are often used at the downstream toe of dams on pervious overburden to provide relief of seepage pressure and to control seepage discharge without permitting piping. These provisions were not used on the dams reviewed either because there was no pervious overburden, or because there was a cut-off trench and seepage if any was expected to be collected by downstream rockfill zones or filter layers at the base of the embankment. Drainage wells are occasionally used in jointed or stratified rock foundations.

Some caverns found in the foundation at Terminus were filled with compacted impervious material. Where caverns exist, they are customarily filled with grout or concrete. Where necessary and feasible, large caverns are usually blocked with bulkheads to limit the amount of concrete fill. At one dam site, a large-diameter vertical hole was drilled near the axis of the dam to provide convenient access to a cavern whose only known opening was about 1/2 mile downstream.

CHAPTER 19

WORK ORDERS FOR DRILLING OF EXPLORATORY BOREHOLES

Subsurface Exploration for Preparation of DPR for Tunnels of Srisailam left Bank Canal Project
UHL Hydroelectric Project, Stage III, Himachal Pradesh.
Bar Chart
Recording
Pressure and Permeability Test

Consultancy firms/organisations exist which offer services for conducting surveys, determining geological acceptability of sites for dams-tunnels-hydropower plants, hydrology, hydrogeology, flood control and drainage system.

Core drilling provides core samples which provide important supplemental information.

The two major fields of application for drilling are the civil engineering construction industry and the mining industry. In the foundation test-boring business, engineers are primarily interested in the physical character of the material recovered. They want to know the rock classifications, whether the formation is shattered, whether voids exist, permeability and other physical properties. Samples obtained by a drilling operation are studied to ascertain uniaxial compressive strength, triaxial strength, angle of internal friction, modulus of elasticity, poisson ratio, gravity absorption percentage etc.

In-house facilities for exploratory drilling, however, need a large outlay and continuous expenditure for payment to staff deployed, whereas exploratory drilling is needed only occasionally at selected sites. As there are several agencies in the country who own large fleets of drills and the necessary equipment, their services can be conveniently utilised for doing the required exploratory drilling.

Quotations can be conveniently invited from such agencies through newspapers, giving scope of work, period of completion, general terms and conditions and performance guarantee.

After comparing the rates and credentials of contractors, work should be allotted to one with an established reputation and a dedicated, committed and professionally sound workforce.

The allotment of work should be in the formal work order, stipulating special terms and conditions, specifications, work-size of holes, percentage of core recovery, permeability tests, deformation tests, survey by TV camera etc. Schedule of quantities with schedule of rates should also be mentioned.

The period of completion should be specifically mentioned in the work order with a penalty clause in case of delay in execution of the work so that project reports/design criteria of the related work are prepared in time.

Payment for mobilisation etc. of drilling machines and operating equipment should be assessed as per number of drill machines to be deployed and the distances involved from the contractor's headquarters to the site for any particular job.

Copies of tenders floated, quotations received from successful bidders and such orders placed by Thapar Hydroconsult (a division of Karam Chand Thapar and Bros. Ltd., New Delhi) for certain drilling operations are adduced below as a general sample of procedure.

1) Drilling exploratory boreholes for Tunnels I and II of Srisailam Project.
   a) Copy of tender notice
   b) Copy of quotation received
   c) Copy of work order placed with Mining Associates, Asansol India.
2) UHL Hydroelectric Project, Stage III, Himachal Pradesh.
   a) Copy of tender notice
   b) Copy of quotation received
   c) Copy of work order placed with Rawel Singh & Co, Ropar India.

SUBSURFACE EXPLORATION FOR PREPARATION OF DPR FOR TUNNELS OF SRISAILAM LEFT BANK CANAL PROJECT

The project involves preparation of DPR for construction of two tunnels, one of 43.5 km and the other 7.5 km. The consultancy for preparation of the DPR involves carrying out the necessary field investigation including exploration of the subsurface by drilling. The area of the location of the tunnels lies near Amrabad in Nalgonda District of Andhra Pradesh and is about 178 km from Hyderabad. The area is covered by the Amrabad Reserved Forest Area which is also a Wildlife Game Sanctuary. The subsurface investigation will have to be carried out by drilling an adequate number of boreholes located in non-forest areas and in forest areas where cart tracks are available, collecting core samples, conducting water-loss tests etc. The drilling work proposed to be done would be as under and has to be completed in 6 months.

1) 4 boreholes of Nx size, 60 m deep (with core recovery).
2) One borehole 300 m deep with recovery of core for full depth (Nx size).
3) 3 boreholes of about 260 m deep with recovery of cores for the last 50 m only. Rest of the borehole could be a percussion hole of adequate diameter without recovery of any core.
4) 3 inclined boreholes with inclination of 15–45° with depths varying up to 100 m with core recovery for the full depth.

The items of work involved would be as below:
1) Drilling in the overburden.
2) Drilling by percussion drill holes up to 250 m depth without core recovery.
3) Drilling with diamond bits with core recovery for depths up to 300 m.
4) Drilling boreholes with core recovery for holes up to 60 m.
5) Drilling inclined boreholes with inclination of 10 to 45° along with core recovery for depths up to 100 m.
6) Conducting water-loss test by the double-packer method in stages of 5 m each for depths up to 300 m.

The drilling records are to be kept as per ISI.

Note: The cores recovered will have to be stored in proper boxes.

MINING ASSOCIATES

Mining & Drilling Engineers, Atwal Nagar,
S.B. Gorai Road, Asansol-713301 (W.B.)

M/s. Thapar Hydroconsult,
B-6/9, Community Centre,
Safdarjung Enclave,
NEW DELHI-110 029

(Attention: Shri C.P. Chugh)

Sub: Drilling of Exploratory Holes at Srisailam Left Bank Canal Project.

Dear Sirs,

We have received your message conveyed by Mr. Chugh to our Delhi Office regarding your having obtained the order for the above and other associated work from Andhra Pradesh Government and we congratulate you on the same. We have subsequently talked to Mr. Chugh on the phone and he has requested us to send our firm offer and also visit your office on 30th May, 1994 to finalise the agreement for drilling work and in this connection we give our final offer as under:

As per our past correspondence, the quantum of work involved consists of 1 No. 300 metre deep vertical borehole; 4 Nos. 60 metre deep vertical boreholes with coring and 4 Nos. 50 metre core drilling from the depth of 210 metre downwards and non-core drilling in 4 (four) boreholes from the surface to 210 metre depth. All the holes shall be vertical. We will have to supply core boxes and carry out permeability tests in the holes... Now as desired by you, we are giving below our rates and terms for this work subject to a maximum of ± 10% variation in the quantities of core drilling and non-core drilling work as mentioned above.

Rates

1)	1 No. 300 metre vertical borehole: Rate of core drilling shall be	...	Rs. 2,950/- per metre.
2)	4 Nos. 60 metre deep vertical hole: Rate of core drilling shall be	...	Rs. 2,800/- per metre.
3)	4 Nos. 50 metre core drilling from the depth of 210 m downwards: Rate of core drilling shall be	...	Rs. 3,200/- per metre.
4)	4 Nos. Non-core drilling up to 210 metre depth: Rate of drilling shall be	...	Rs. 1,100/- per metre.
5)	Cost of wooden/steel core box (our choice).	...	Rs. 675/- per box.

6) Water percolation tests in the
   holes.   . . .   Rs. 1000/- per test in
                                                            holes under items 1, 3 &
                                                            4 and Rs. 800/- per test
                                                            in hole under item No. 2.

7) Mobilisation & de-mobilisation
   charges.   . . .   Rs. 2,50,000/- L.S.

Terms and Conditions

a) We shall deploy 3 to 4 Longyear-38 and Joy-12B or Joy-7 Diamond Drills simultneously and shall complete the entire work within 8 to 9 months from the date of receipt of commercially acceptable order.

b) The holes will be drilled in Nx size and in case of any mishap or difficulty in the hole which prevents further drilling in Nx size, the hole shall be reduced to Bx size.

c) We shall use double-tube core barrels and shall take all precautions for best core recovery, which very much depends on the type of strata being drilled.

d) Our quoted rates are for drilling through igneous, metamorphic rocks such as granite, quartzite etc. but not through quartz. If we encounter quartz in any hole and total meterage of drilling through quartz in any hole exceeds 5% of total core drilling in each hole, then extra charges of Rs. 2,000/- per metre shall be charged for such excess drilling over 5% over and above the relevant drilling rates for such extra meterage drilling in quartz.

e) We shall try our best to drill the holes to their targeted depths and in case of any serious difficulty due to strata conditions such as encountering any broken or faulty strata etc., which makes further drilling impossible in spite of making all reasonable attempts, the holes shall be closed at that depth and paid for the depth drilled.

f) You will have to provide the borehole sites and the existing roads free of all encumbrances and in time so that our machines do not remain idle on this account or any other reason on your part.

g) If any forest clearance is required, the same shall be arranged by you.

h) We shall submit the daily progress report of the work done at each borehole site to your authorised representative at site in quadruplicate, 2 copies of which shall be retained by him and the other 2 copies shall be returned to us after his signature.

i) You will have to pay us a sum of Rs. $1\frac{1}{2}$ lakhs against Bank Guarantee along with the work order as initial advance of mobilisation. Balance of mobilisation shall be paid immediately on reaching the machines in the area.

j) We shall submit fortnightly bills for the work done in each fortnight and the same shall be payable within the next 15 (fifteen) days, failing which interest will be payable.

k) Our offer is subject to standard force majeure clauses.

l) In case of any dispute of any nature whatsoever between us relating to this work, the same shall be referred for arbitration by either the Chief Drilling Engineer of Central Mine Planning & Design Institute Ltd., Kanke Road, Ranchi or the Director/Professor of Drilling, Indian School of Mines, Dhanbad under the Arbitration Act and his decision shall be final and binding on both the parties.

m) Our offer is valid for a period of 45 days from the date hereof.

We hope you will kindly find our above offer very reasonable.

Thanking you,

Yours faithfully,
FOR MINING ASSOCIATES

(R.B. BANSAL)

MINING ASSOCIATES

ASANSOL (WEST BENGAL)

WORK ORDER
FOR
DRILLING OF EXPLORATORY BOREHOLES FOR
TUNNELS I & II
OF
SRISAILAM PROJECT

THAPAR HYDROCONSULT
(A Division of Karam Chand Thapar & Bros. Ltd.)
NEW DELHI

AUGUST—1994

THAPAR HYDROCONSULT

20 Community Centre, Basantlok Vasant Vihar, New Delhi 110052

M/s Mining Associates
Mining and Drilling Engineer
Atwal Nagar,
SB Gorai Road, Asansol 713 301 (WB)

Sub: Drilling of Exploratory Boreholes (Coring and Non-coring) for Tunnels I & II
  of Srisailam Left Bank Canal Project
Ref: 1. Your Quotation No. MA/TH/CPC/2016, dated 23rd May, 1994
  2. Your letter No. MA/TH/CPC/2216 of 04.07.1994

Dear Sirs,

Your offer for the above work submitted along with your letter referred to as
S1. No. 1 and subsequent discussions held in this regard have been considered by
us and we are pleased to award the above-mentioned work to you subject to the
following terms and conditions:

1. *Scope of Work:* The work consists of drilling of 8 to 12 Nx or larger sized
exploratory boreholes, both coring and non-coring, at locations to be indicated
by THC, for subsurface investigation pertaining to tunnels I & II and related
works of Srisailam Left Bank Canal Project, District Nalgonda, Andhra Pradesh,
and collecting and storing core samples or washed sludge samples in appropriate core boxes as per standard practice, inclusive of all operations, conducting
water-loss/permeability tests in the boreholes and also making available the drilling
equipment and accessories for examining boreholes by borehole camera, and conducting the same as detailed in the General and Special Terms and Conditions and
Specifications, Schedule of Quantities and Schedule of Rates enclosed as Annexures 1, 2 and 3.

2. *Period of Completion:* The work as defined shall be completed within a
period of 9 months inclusive of the mobilisation period, from the date of acceptance
of award of work by you.

Mobilisation of the requisite plant and equipment, men and material shall be
completed within a period of 30 days.

3. *General Terms and Conditions:* The General Terms and Conditions for the
work shall be as detailed in Annexure 1 and shall form part of this work order.

4. *Special Terms and Conditions and Specifications:* The approximate quantum
of work to be carried out and the special terms and conditions and specifications
shall be as detailed in Annexure 2 and shall form part of this work order.

5. *Terms of Payment*

i) Payment for the work will be for the quantum of work carried out as per items of work and as per schedule of rates indicated in Annexure 3, which shall form part of this work order.

ii) 90% of the value of work done fortnightly will be paid to you against submission of the bill in triplicate by you, duly certified by the site representative of THC. Payment will be released within 2 weeks of submission of the bill.

iii) The balance 10% of the value of work done for each borehole will be payable to you after satisfactory completion of the work on each borehole and production of acceptance certificate to be obtained by you from the authorised site representative of THC, for satisfactory completion of the work on the borehole as per the requirements and your submission of a performance bond for Rs. 1.7 lakhs for the entire work, before completion of any one of the boreholes.

iv) A non-refundable mobilisation charge of Rs. 2,50,000/- (Rupees two lakhs fifty thousand only) shall be payable to you for this work. Out of this a sum of Rs. 1,50,000 (Rupees one lakh fifty thousand only) will be payable to you on acceptance of the work order by you and production of a Bank Guarantee for equivalent value in favour of Thapar Hydroconsult from a nationalised bank in an acceptable form and valid for the entire period of the contract. This Bank Guarantee will also be towards performance of the work. (The format for the Bank Guarantee will be supplied on acceptance of the work order.) The balance of Rs. 1,00,000 (one lakh only) will be released immediately after 75% of the equipment as per your programme reaches the site and is certified as such by the site representative of THC. No demobilisation charge will be payable to you on completion of the work and removal of your plant and machinery from the site. The plant, machinery and equipment deployed on this work shall be moved away from the project site after completion of the work.

v) You shall deploy all the necessary plant, machinery and equipment, men and materials from time to time at your cost, for completion of the work as per schedule. You shall improve the existing approaches and make new approaches to the borehole locations, and maintain them during the contract period, make all water-supply arrangements for drinking purposes, drilling and water-loss/permeability tests, lighting arrangements, as per the need of the work, at your own cost. No extra payments on any account will be made.

vi) No extra payments will be made for shifting of the plant machinery and equipment, men and materials from site to site and for transport of men and materials to work sites from places of camp or storage.

6. You shall submit a detailed programme of your mobilisation of plant, machinery and equipment, men and materials, and detailed programme for completion of the work entrusted to you in the form of Detailed Bar Charts, indicating as well the planned physical programme for each borehole, within a period of two weeks from the date of acceptance of the work order by you, for our review.

7. *Performance Guarantee:* You will take all necessary actions to adhere to the programme of work as agreed to between us. You shall give us a performance guarantee in the following manner. The Bank Guarantee for Rs. 1.5 lakhs to be furnished by you for drawing the 1st installment of mobilisation charges will be retained as a part performance guarantee until the end of work and the Bank Guarantee to be submitted will cover this aspect. Upon starting the work and before completion of any one borehole, you shall arrange to submit a further performance guarantee in the form of a Bank Guarantee for a sum of Rs. 1.70 lakhs in the format acceptable to us. The performance guarantee bonds as above will be returned to you on receipt of the final acceptance certificate.

8. Time is of the essence in this work order and you shall be bound by the period of completion as per condition at (2) above and as per the programme to be submitted by you as indicated in condition (6) above. In case you fail to complete the above work by the stipulated date of completion, as stipulated in condition (2) above, you shall be liable to pay liquidated damages at the rate of 1% of the value of work remaining incomplete, per week of delay, subject to a maximum of 5% of the value of the work remaining incomplete. THC will have the right to recover the liquidated damages, if any, as per this condition from the Bank Guarantees submitted by you for mobilisation charges and performance of work and/or the 10% amounts withheld from the running account bills.

We hope this work order as above is acceptable to you. Please sign the duplicate copy of this work order including all Annexures (on all pages) as confirmation of acceptance of this work order and arrange to forward your detailed programme of mobilisation and completion of the above work for acceptance.

We also request you to arrange to start work immediately to ensure completion of work within the stipulated time.

Thanking you,

Yours faithfully,
for THAPAR HYDROCONSULT

O.P. MEHTA   C.S. HEBLI
EXECUTIVE CONSULTANT   SR. EXECUTIVE CONSULTANT
& AUTHORISED SIGNATORY   & TEAM LEADER FOR PROJECT

Encl: 1. General Terms and Conditions   —Annexure 1
      2. Special Terms & Conditions & Specifications —Annexure 2
      3. Schedule of Quantities and Rates   —Annexure 3
      4. Drawing showing location and drill holes —Annexure 4

I hereby accept the above Work Order

                                        FOR MINING ASSOCIATES

ANNEXURE 1: GENERAL TERMS AND CONDITIONS

1.0 Definitions

The following words and expressions shall have the meaning herein assigned to them, unless there is anything in the subject or context inconsistent with the contents of the work order.

1.1 'THC' shall mean Thapar Hydroconsult, a division of Karam Chand Thapar & Bros Ltd., No. 26, Basant Lok Community Centre, Vasant Vihar, New Delhi-110 057, India and shall include their heirs, successors, executors, permitted assigns and legal representatives.

1.2 'Contractor' shall mean M/s. Mining Associates, Mining and Drilling Engineers, having its Registered Office at Atwal Nagar, SB Gorai Road, Asansol-713 301 (West Bengal) and include their heirs, successors, executors, permitted assigns and legal representatives.

1.3 'Works' shall mean and include all permanent and temporary works to be executed, all items and things to be supplied/done and services and activities to be performed by the contractor pursuant to and in accordance with the work order and shall not include the contractor's construction equipment, its related spares and wearing parts.

1.4 The 'Work Order' shall mean the work order issued by THC to the contractor and duly accepted by the contractor by appending his signature on the duplicate copy for the execution of works stated therein, together with all the documents annexed/attached thereto.

1.5 'Site' shall mean the land and other places on, under, in or through which the permanent works or temporary works of the contract are to be executed and any other land and places provided by THC for work space or for any other purpose, for the performance of the contract.

1.6 'Sub-contractor' shall mean any person or firm or company (other than the contractor and the legal representatives, successors and permitted assigns of such person or firm or company) engaged by the contractor with prior written consent of THC/Engineer-in-Charge.

1.7 'Approval' shall mean approval in writing including subsequent written confirmation of previous verbal approval.

1.8 'Final Acceptance Certificate' shall mean the certificate(s) to be issued by the authorised site representative of THC to the contractor.

1.9 'Effective Date' shall mean the date on which the work order is accepted by the contractor.

1.10 'Project' shall mean the work of drilling of exploratory holes, for Tunnels I & II of Srisailam Left Bank Canal Project, as indicated briefly in the Scope of Work and defined in details elsewhere in the work order and its annexures.

1.11 'Contractor's "Bar Chart"' shall mean the programme of mobilisation and execution of work showing the order of sequence and periods of completion of various activities according to which the contractor intends to carry out the works within the period of completion stipulated in the work order for completion of all the works included in the scope of this order.

2.0 Interpretation of the Work Order

2.1 Special Terms and Conditions and Specifications detailed in Annexure 2 shall be read in conjunction with the General Terms and Conditions and specifications of work, drawings and other documents forming part of this order wherever applicable and permissible or the context so requires.

2.2 Wherever it is mentioned that the contractor shall perform certain works or provide certain facilities, it shall be understood to mean that the contractor shall do so at his cost and the work order price shall be deemed to have included cost of such performances and provisions so mentioned.

2.3 The materials, design and workmanship shall satisfy the applicable standards and specifications contained herein and in the ISI codes. Where the Technical Documentation stipulates requirements in addition to those contained in the standard codes and specifications, those additional requirements shall also be complied with.

3.0 Assignment and Subletting

3.1 Except as provided hereinafter, no part of the order or any share or interest therein shall, in any manner or degree, be transferred, assigned, or sublet by the contractor directly or indirectly to any person, firm or company whatsoever without the prior consent in writing of THC, for which the contractor shall give a written request to THC at least 30 days in advance of the proposed date of transfer/assignment/subletting.

3.2 Notwithstanding any transfer, assignment or subletting with the approval of THC as aforesaid, the contractor shall be and shall remain solely responsible and liable to THC for the quality, proper and expeditious execution and performance of the work and for due performance and observance of all the conditions of the contract in all respects, as if such transfer, assignment or subletting had not taken place and as if such work had been done directly by the contractor.

3.3 No action taken by THC under Articles 3.1 and 3.2 above shall absolve or relieve in any manner whatsoever the Contractor of any of his liabilities and obligations including 'period for completion' or give rise to any right to compensation/extension of time or otherwise.

4.0 Law and Jurisdiction

4.1 The order shall be construed and interpreted in accordance with and governed by the Laws of Union of India.

4.2 In respect of all matters or actions arising out of the contract and which may arise at any time, the courts at Delhi shall have exclusive jurisdiction.

5.0 Contractor's Employees

5.1 The contractor shall provide and employ on the site in connection with the execution of the work:
   a) Only such technical personnel as are skilled and experienced in their respective callings and such subagents, foremen and leading hands as are competent to do or give proper supervision to the work they are required to perform or supervise, and
   b) Such skilled, semi-skilled and unskilled labour as is necessary for the proper and timely execution of the work.

6.0 Liability, Indemnity and Insurance

6.1 The contractor shall be liable to THC for the performance of the services in accordance with the provisions of the contract and any loss suffered by the THC as a result of default of the contractor in such performance will be borne by the contractor. The liability will be limited to the performance guarantee, and not directed towards any other losses to THC.

6.2 The contractor shall keep THC fully and effectively indemnified against all losses, damage, injuries, deaths, expenses, actions, proceedings, demands, costs and claims, including but not limited to, legal fees and expenses suffered by THC or any Third Party where such loss, damage, injury or death is the result of a wrongful action, negligence or breach by the contractor or his associate subcontractors, or the personnel or agents of either of them, including the use or violation of any copyright work or literary property or patented invention, article or appliance.

6.3 The following types of insurance shall constitute the minimum insurance in regard to the requirements of providing indemnity. The contractor shall arrange for the insurance at his cost and arrange to pay the premiums from time to time to keep the insurance policies valid during the period of this contract.
   a) *Contractor's Plant and Machinery Policy*
      The contractor's plant and machinery policy shall cover the full replacement value of the plant and machinery. The policy shall also be endorsed to cover plant and machinery working underground.
   b) *Motor Vehicle Insurance Policy*
      The cars and commercial vehicles policy shall cover direct physical damage to the vehicle as well as liability to third parties.
      It shall apply to all authorised drivers who possess, a valid driving licence in India.
   c) *Insurance of Workmen and Personnel, and Work as a Whole*
      The contractor shall arrange for insurance of workmen against mishaps, accidents and losses due to any eventuality on the work at his own cost.
   d) *Workman's Compensation*
      The contractor shall be responsible for extending workman's compensation to his employees, as per the prevalent laws.

7.0 If the contractor or his subcontractor or their employees shall break, deface or destroy any property belonging to THC and others during the execution of the work, the same shall be made good by the contractor at his own expense.

8.0 The contractor shall be responsible for all arrangements necessary for executing the work, including pumping of all water requirements for all work sites and work purposes, the contractor's colony/camp etc.

9.0 Obligations of THC

9.1 THC shall be responsible for making available the site as required for the execution of the works on an 'as is where is' basis, free of rights of third parties.

9.2 THC shall provide the topographical, geotechnical, hydrological and other technical information as is available with them, excluding restricted data, for the performance of the contractor's obligations under this work order.

9.3 All fossils, coins, articles of value or antiquity and structures and other remains or things of geological or archeological interest discovered on the site of the work shall as between THC and the contractor be deemed the absolute property of THC. The contractor shall take reasonable precautions to prevent his workmen or any other persons from removing or damaging any such article or thing and shall immediately upon recovery thereof and, before removal, inform THC of such discovery and carry out, at the expense of THC, disposal of the same as shall be directed by THC.

9.4 THC will arrange for any authorisation/permission and clearance as may be required from proper authorities, well in time for the conduct of the work by the contractor. THC shall arrange to mark the borehole points on the ground well in advance.

10.0 Clearance of Site on Completion

On completion of the work, the contractor shall hand over on an 'as is where is basis' roads and masonry constructions excluding any removable fixtures or fittings established by the contractor at site, as required by THC at no extra cost to THC. The contractor shall clear away and remove from the site all remaining construction equipment, surplus materials, rubbish and temporary work of every kind, and leave the whole of the site of work clean and in a workman-like condition to the satisfaction of THC. In case the contractor fails to remove/clear the above, THC shall remove/clear the same at the cost of the contractor.

11.0 Labour and Compliance with Labour/Industrial and Other Laws

11.1 The contractor shall make his own arrangements for the engagement of all labour, local or otherwise and provide for the transport, housing, feeding and payment thereof at his own cost and under his own arrangements

11.2 The contractor shall at his expense ensure due compliance with all applicable and governing Indian laws including industrial and labour laws, rules and regulations and bylaws both of the Central and State Governments and all other local authorities and shall keep THC harmless and indemnified in respect thereof.

11.3 The contractor shall ensure due compliance with the provisions of the relevant Minimum Wages Act, Payment of Wages Act, Contract Labour (Regulation and Abolition) Act, Workmen's Compensation Act, E.P.F. Act and other labour/industrial laws in force.

12.0 Acceptance Certificate

12.1 A Final Acceptance Certificate shall be issued to the contractor at his request by THC within thrity days after completion of the work or the date of rectification of outstanding deficiencies/damages/ defects, whichever is later.

12.2 The contractor shall be responsible for payment of all Central and State Govt. taxes as applicable, including Income Tax and/or Surcharge. Sales Tax, if any, shall not be the responsibility of the Contractor. THC will deduct tax at source against payments as applicable under the law.

13.0 Secrecy

All information, data and drawings furnished/disclosed by THC to the contractor and all drawings, calculations, models, technical information and the like supplied by the contractor to THC or THC's representative, shall be treated by the contractor and his agents, subcontractors and employees as confidential, and shall not be used by any of them without the previous written consent of THC, except in connection with the execution, operation and maintenance of the work.

14.0 Force Majeure

14.1 *Definition*

a) For the purpose of this contract, 'Force Majeure' means an event which is beyond the reasonable control of a party, and which makes a party's performance of its obligations hereunder impossible or so impractical as reasonably to be considered impossible in the circumstances, and includes, but is not limited to war, riots, civil disorder, earthquake, fire, explosion, storm, flood, or other adverse weather conditions, strikes, lockouts, or other industrial action (except where such strikes, lockouts or other industrial actions are within the power of the party invoking Force Majeure to prevent) confiscation, or any other action by governmental agencies.

b) Force Majeure shall not include (i) any event which is caused by the negligence or intentional action of a party or such party's subconsultants or agents or employees, nor (ii) any event which a diligent party could reasonably have been expected to both (a) take into account at the time of the conclusion of this contract and (b) avoid or overcome in the carrying out of its obligations hereunder.

c) Force Majeure shall not include insufficiency of funds or failure to make any payment required hereunder.

14.2 *No Breach of Contract*

The failure of a party to fulfil any of its obligations hereunder shall not be considered to be a breach of, or default under, this contract in so far as such inability arises from the event of Force Majeure, provided that the party affected by such an event has taken all reasonable precautions, due care and reasonable alternative measures, all with the objective of carrying out the terms and conditions of this contract.

14.3 *Measures to be Taken*

a) A party affected by an event of Force Majeure shall take all reasonable measures to remove such party's inability to fulfil its obligations hereunder with a minimum of delay.

b) A party affected by an event of Force Majeure shall notify the other party of such event as soon as possible and in any event not later than fourteen (14) days following the occurrence of such event, providing evidence of the nature and cause of such event, and shall similarly give notice of the restoration of normal conditions as soon as possible.

c) The parties shall take all reasonable measures to minimise the consequences of any event of Force Majeure.

14.4 *Extension of Time*

Time shall be considered of the essence in the contract. If, however failure of the contractor to complete the work as per the stipulated dates referred to above is due to increase in the quantity of work to be done under the contract, or Force Majeure, an appropriate extension of time shall be given. The consultant shall request such extension within one month of the cause of such delay and in any case before expiry of the contract period.

14.5 *Payments*

During the period of their inability to perform the services as a result of an event of Force Majeure, the contractor shall be entitled to continue to be paid under the terms of this contract.

14.6 *Consultation*

Not later than thirty (30) days after the contractor, as the result of an event of Force Majeure, has become unable to perform a material portion of the services, the parties shall consult with each other with a view to agreeing on appropriate measures to be taken under the circumstances.

15.0 Disputes in regard to interpretation of the contract and any matter related to implementation of the contract shall be settled by mutual discussions. Decisions arrived at between the General Manager, THC and Managing Director of Mining Associates, shall be final.

ANNEXURE 2: SPECIAL TERMS AND CONDITIONS AND SPECIFICATIONS FOR THE WORK

1. The scope of the work constitutes drilling 8 to 12 boreholes with minimum 80% core recovery in sound rock and a minimum as can be practically achieved by following laid-down drilling practices in weaker rocks, the depth of the boreholes ranging from 50 to 300 m as per drawing showing location of drill holes enclosed as Annexure 4.

   The boreholes and tests envisaged are as below:
   i) 4 Nx size coring boreholes of approximate depths of 60 m each, with collection of cores and/or sludge samples for the entire depth.
   ii) 3 non-coring boreholes of 114 mm size, by DTH machines, up to a depth of around 210 m, with collection of sludge samples every 3 to 5 m depth.
   iii) Drilling the depths of the above 3 non-coring bore holes beyond 210 m up to about 260 m, with Nx size coring boreholes and recovering cores and/or sludge samples for these depths beyond 210 m.
   iv) Drilling one Nx size coring borehole for a depth of about 300 m and recovering core and/or sludge samples for the entire length.
   v) Conducting water-loss/permeability tests in the boreholes at 3 m or 5 m intervals with single- or double-packer method.
   vi) Drilling inclined Nx size coring boreholes up to 45° inclination, if necessary up to 100 m depth, and recovering cores and/or sludge samples for the entire depth.

2. The contractor shall deploy 3 to 4 drill rigs of adequate capacity with all accessories, tools and tackles for the execution of coring drill holes. He shall also deploy adequate numbers of DTH drill rigs to drill the non-coring boreholes. If need be, the contractor shall deploy additional numbers of drill rigs, both non-coring and coring type, if at any stage it is noticed that their deployment is a necessity for completion of the total work within the stipulated time frame.

3. The drilling machine to be deployed shall be of suitable capacity so as to finish the holes in NX size to entire depth. Double-tube core barrels shall be used for recovery of cores from coring holes. Normally 1.5 m and 3-m long double-tube core barrels shall be used. In case of encountering hard strata yielding 100% core recovery, core barrels of 6 m length could also be used. Reduction of boreholes up to Bx size coring where necessary is permissible with due prior permission to be obtained in writing from THC.

4. The drill machines shall be suitably anchored in concrete/sleeper foundation so that the vertical/inclined alignment of the borehole is not disturbed due to incidental vibrations of the drilling operation.

5. Arrangement of water for performance of drilling operations and water-percolation tests will have to be made by the contractor at drill sites from existing nallahs, wells and water springs by deployment of suitable pressure pump/pipelines/ transportation by water tankers as may be necessary, at his own cost.

6. The cores shall be stored and preserved in wooden/galvanised steel sheet core boxes to be supplied by the contractor. The sample of the core box shall be approved by the Site Engineer In-charge of THC. Sludge will be collected and preserved at intervals of 3 m to 5 m from zones of the boreholes which do not yield cores. Even in zones where the drilling machine has started penetrating rapidly, sludge shall be collected and preserved in the core box with an indication by wooden gutka of the probable depth of such a zone. The sludge collected shall be approximately one kg for each sample.

7. Arrangement for wash boring, i.e., drive hammer, drive pipe, drive shoe, wash-boring pipe, shall be the responsibility of the contractor at no extra cost to THC.

8. Shifting of the drill machines and other plant, equipment, tools and tackles from one site to another shall be done by the contractor at his own cost.

9. In case there is jamming of drill rods due to loose strata or encountering of pebbles etc., the contractor may resort to cement or bentonite grouting of the hole for stabilising the sides and redrill the hole through the grouted portions for continuing the borehole. Grouting and redrilling of holes if need be shall be carried out by the contractor at his own cost.

10. In case of jamming difficulties and if the hole cannot proceed further, the contractor shall, after informing the site representative of THC, drill another hole at a location close to the abandoned hole by preferably the non-coring method to the depth already reached and by non-coring and/or coring technique for the remaining depth to reach the targeted depth. Payment for the new hole shall be made at half the schedule rates, up to the depth already drilled in the abandoned hole. An unfinished/abandoned hole shall be paid at the rates indicated in the Schedule of Rates. Payment for the new hole for depths beyond the depth already reached in the abandoned hole and up to the targeted depth shall be made at the scheduled rates.

    If the contractor does not take up drilling of the new hole or does not complete the new hole to the designated depth due to the fault of the contractor, THC shall have the right to recover the cost of the abandoned hole already paid to the contractor, from the bills payable to the contractor for the work carried out on other holes of this work order.

11. Any additional work on this Project, if required by THC, shall be carried out by the contractor at the same applicable rates and conditions as already agreed to as per this order.

12. For non-core drilling, DTH drills with 14 kg/cm^2 compressor will be deployed to reach the 210 m depth. In case DTH drilling is not possible to reach the targeted depth due to encountering of groundwater or caving in of the strata being drilled, requisite size rotary/diamond core bits shall be used for successful destructive drilling. In case rotary drilling with roller bits is not possible due to the mismatched drillability factor of metamorphosed rocks, only diamond core drilling shall be adopted and core samples shall be collected in suitable

intervals for which instructions will be given on site. Such core drilling will be paid at scheduled rates.

Samples of cuttings for non-core drilling shall be collected at suitable intervals at the behest of the THC representative in polythene bags suitably marked for identification. The sample of cutting for non-coring drilling shall be at least one kg for each sample to be collected at every 3 m to 5 m of borehole depth.

13. Water-loss tests/permeability tests in the boreholes shall be conducted using standard equipment, adequate capacity reciprocating pump set, and mechanical single packers or mechanical double packers as the case may demand. Water-loss tests done in stages of 3 m or 5 m depths, as directed by the Engineer-in-Charge. They shall be conducted after initially soaking the hole with water under suitable pressure for at least one hour, in rising stages of pressure from 1 kg/sq. cm up to a maximum of 10 kg/sq. cm. The range of pressure to be adopted for each depth to be tested will be as per direction of the Engineer-in-Charge. If need be, in the case of large losses, tests at lower or smaller intervals of pressure should be taken at no extra cost. Three readings shall be taken for each pressure stage.

14. When called upon by THC the drilling machine shall be placed at its disposal or that of other agencies for TV camera photographing the inhole strata and conduction of other tests inside the borehole. The duration of such periods will be 4–8 hours at selected sites. If the machine remains idle for more than 2 days after completion of the borehole at the instance of THC, idling charges of Rs. 2500/- per mechine per day shall be payable from the third day.

15. The contractor shall abide by the rules regarding entry into forest areas, carrying out the work without disturbing the environment and abide by the environmental safeguards stipulated by State and Central Govts. for this project. Drilling work in locations inside the Reserve Forest areas will be permitted only from dawn to dusk and the contractor shall follow the same strictly. In locations with no environmental restrictions, the contractor will be permitted to work in all 3 shifts if he so desires.

16. The contractor shall submit a daily progress report on drilling of each borehole with a weekly report for the abstract of the depths of holes drilled during the week.

17. The contractor shall submit a daily progress report of the work done at each borehole site to the authorised representative of THC at the site in quadruplicate and two copies of the same will be returned to him duly acknowledging receipt.

ANNEXURE 3: SCHEDULE OF QUANTITIES AND SCHEDULE OF RATES OF WORK

Sl. No.	Item	Unit	Qty in m	Rate per Unit	Amount
1.	Mobilisation charges for plant and machinery, men and materials	L/S			2,50,000
2.	Drilling non-coring bore holes of 100–114 mm size using downhole drills of 200 psi capacity, including all operations, collection of sludge samples at intervals of 3–5 m and storing them in core boxes as per specifications.				
	0–210 m	m	630	1100	6,93,000
3.	Drilling of vertical coring Nx size boreholes and recovering of core by using suitable plant and equipment and double-tube core barrels, including all operations, collection and storing of cores in core boxes as per specifications.				
	a) 0–100 m	m	340	2800	9,52,000
	b) 100–200 m	m	110	2950	3,24,500
	c) 200–300 m	m	240	3200	7,68,000
4.	Conducting water-loss/permeability tests in boreholes in stages using single or double packers placed at 3–5 m apart and at water pressures ranging from 1 kg/sq. cm up to 10 kg/sq. cm in increasing sequence depending upon the hydrogeology, transmissibility and permeability of the formations encountered.				
	a) From 0 to 100 m	Nos.	70	800	56,000
	b) From 100 to 300 m	Nos.	70	1000	70,000
5.	Supply of wooden or GI sheet core boxes in suitable sizes, each to store cores from 4 to 5 m depths of the boreholes or sludge samples with				

(Contd.)

Annexure 3. Continued

Sl. No.	Item	Unit	Qty in m	Rate per Unit	Amount
	suitable partitions, marker blocks, handle painting etc.	Box	160	675	1,08,000
6.	Drilling of inclined up to 45°, coring, Nx size boreholes and recovering cores by using suitable plant and equipment and double-tube core barrels, including all operations, collection and storing of cores and sludge samples in core boxes as per specifications. For 0–100 m	m	–	3,220	–
7.	Providing use of drilling and other equipment at borehole locations, including running of plant and machinery, all other operations for conducting tests inside the boreholes and logging or examination of boreholes by camera etc. for a shift of eight hours.	per Shift	–	4,000	–
	Total				32,21,500

Notes: a) The above noted quantities of items of work to be executed are approximate and may vary up to +25% depending on the need of the work as will be decided by the Site Engineer-in-Charge of THC.

b) The rates indicated shall hold good for all quantities of items of work to be executed and remain firm during the period of this contract.

c) The core boxes in item (4) will be as per sample approved by the Site Engineer-in-Charge of THC.

d) The rates indicated for the inclined coring boreholes will apply to the presently proposed 5 coring boreholes, as indicated in Annexure 2, and also for any additional boreholes that may be necessary during the course of the work.

e) The rates indicated for the water-loss/permeability tests shall also apply to the inclined boreholes.

f) In case of collapse of a borehole and adoption of cement grouting for its stabilisation, the necessary redrilling in set cement and reaming of the borehole will not be paid for separately.

g) Payment for mobilisation as outlined in Item No. 1 and for the work as detailed in Items 2 to 7, shall be released as indicated in the Terms of Payment set forth in this covering letter (THC: 102-1(F): 94).

BALLARPUR INDUSTRIES LIMITED

Thapar House, 124, Janpath, New Delhi-110001

UHL H.E. PROJECT STAGE-III
HIMACHAL PRADESH

DRILLING OF EXPLORATORY HOLES

Kind Attn: S. Gurcharan Singh

Quotations are requested for drilling 2 nos. exploratory holes of 30 to 40 m depth on power-house site near Durg Temple at the confluence of the Beas River and Khaddar Nala about 160 km from Pathankot. The formations to be drilled are conglomerates, sandstone etc. Water is available within 500 m. Stage pump may have to be deployed for executing drilling operations.

A high percentage of core recovery is needed to assess the details of the strata and hence double-tube core barrels with bottom discharge features and diamond bits of 16/30 spc or 20/40 spc with suitable loading will be needed while coring the strata. Casing strings of suitable size, if needed, are to be provided by the contractor.

Holes are to be drilled in Nx size but if reduction is called for, holes up to Bx size will also be acceptable.

It may be noted that any borehole not drilled to the targeted depth will not be liable for payment.

Duration of work: The work is required to be completed within 45 days.

Cores will be stacked in core boxes of standard size. If additional work is called for, it will be executed on quoted rates.

Any other information required can be obtained by contacting us personally.

You are requested to intimate the model/make of drilling machines you propose to deploy on the job. The quotation is required within a week.

for BALLARPUR INDUSTRIES LTD

C.P. CHUGH
CONSULTANT.

M/s Rawel Singh & Co.
Bella Road
Ropar-140 001
Punjab.

RAWEL SINGH & CO.

Bela Road, Near Parmar Hospital, ROPAR-140001 (Pb.)

Shri C.P. Chugh
Consultant
Ballarpur Industries Ltd.,
The Thapar Group, Thapar House,
124, Janpath, New Delhi-110 001.

Subject: Quotation for UHL H.E. Project Stage III Himachal Pradesh - Drilling
             of exploratory holes.
Reference: Your office enquiry No. THC: 102-I(C): 93, July 6, 1993.

Sir,
    In this regard, we take pleasure in submitting our quotation for the work
required in your above-referred letter for UHL H.E. Project Stage III as follows:

Sr. No.	Description	Rate
1.	Drilling of 2 Nos. exploratory holes of 30 to 40 metre depth of Nx/Bx size at Power-house site near Durg temple at confluence of Beas River and Khadder Nala in conglomerate, sandstone etc. with D/T core barrel and casing, wherever necessary.	Rs. 2737/-per metre. (Rs. Two thousand seven hundred & thirty-seven per metre).

Terms and Conditions.

1. The work shall be taken up after the rainy season and 15 days after firm award
   to us.
2. JOY 12 B/LONGYER-34 Bore Drill Machine, diesel driven, shall be deployed
   on work.
3. Full payment shall be taken for each completed borehole within 15 days through
   Demand Draft payable at Ropar.
4. The shifting charges of Rs. 10,000/- shall be taken on reaching our equipment
   at the work site.
       Assuring you of our best services and thanking you,

Yours faithfully,
For Rawel Singh & Co

THAPAR HYDROCONSULT

20 Community Centre, Basantlok Vasant Vihar, New Delhi 110057

M/s Rawel Singh & Co
Bela Road,
Near Parmar Hospital
Ropar-140 001 (Pb)

Sub: Order for Drilling of Exploratory Holes at Project Site for UHL Hydroelectric Project Stage-III in Himachal Pradesh.

Ref: Your Quotation No. 2737/RSC dated 09.08.1993 and 2748/RSC dated 30.10.1993.

Dear Sirs,

With reference to your offer vide your above-mentioned letters for above-quoted work and subsequent discussions we had on date, we are pleased to award the above-mentioned work to you subject to the following conditions:

1 General Terms and Conditions

The order will be governed by the general terms and conditions attached as Annexure I which will be read in conjunction with this letter.

2 Scope of Work

The scope of work will be as defined in Annexure II attached.

3 Period of Completion

The work as defined in Annexure II shall be completed in a period of four months, effective 1st March 1994, i.e., by 30th June 1994. You are requested to *submit a Bar Chart* providing the programme showing the order of sequence and periods of completion of various activities according to which you intend to carry out the work within the stipulated time. The Bar Chart may please be submitted latest by 28th February '94.

Time is of the essence in this work order. In case you fail to complete the above work by the stipulated date of completion, i.e., 30th June '94, you shall pay liquidated damages at the rate of 25% of the value of this work order per week of delay subject to the maximum of 2% of the value of this work order.

4 Schedule of Rates

a) One time 'Mobilisation Charges' of Rs. 10,000/- per machine subject to a maximum amount of Rs. 20,000/-.

b) For shifting the machines from site to site the contractor will be paid Rs. 3,500/- (Rupees Three Thousand Five Hundred only) per shift. However, this charge is not applicable for initial transfer of machines to the work site and removal of the machines from the work site. The cost for these shifts are provided for in (a) above.

c) Rs. 3,000/- per metre (Rupees Two Thousand Seven Hundred Thirty-Seven only per metre) will be payable for the actual drilling done and accepted by our authorised representative (more details available in Annexure II). This rate is applicable for boreholes up to 60 m. For the borehole of 150 m ($\pm$ 20%) to be drilled at Bbabhuri Dhar Tunnel Intake the rate applicable will be Rs. 3,800/- (Rupees Three Thousand Eight Hundred only) per m. In case any other borehole is required to be drilled between 60 and 150 m or more than 150 m besides the boreholes indicated in the scope of work, the rate for the same will be mutually agreed upon.

d) In case permeability tests are required and are expressly ordered by us, then Rs. 1000/- (Rupees One Thousand only) will be payable for each permeability test for double/single packers placed 5 m apart, for borehole of depth up to 60 m and the rate will be Rs. 2000/- for each such test for boreholes beyond 60 m depth.

e) If observation of groundwater tables is required and expressly ordered by THC, then the actual cost of materials (i.e. piezometer tip, PVC pipes, head cap, foundation and protective enclosure) will be reimbursed against documentary proof. The contractor will also be entitled to 15% of the cost of the materials to cover their labour for installing the same.

These rates mentioned above are firm for carrying out the work and in case any additional work beyond what has been specified in Annexure II is ordered, the same unit rates as indicated above will be applicable for the additional quantities of work so ordered.

5 Payment Terms

a) The one-time mobilisation charge as mentioned above will be paid to you after your equipment has reached the site of work and the work has started. Payment will be made against your certified bill.

b) After completion of each borehole, 90% of the amount due to you for that specific borehole will be paid on submission of your bill duly certified by our authorised site representative confirming satisfactory completion of the said bore.

c) After each shift the shifting charge will be payable against submission of your certified bills.

d) The balance 10% amount of the rate will be retained as retention money and will be paid to you within one month after all the work is satisfactorily completed, against submission of your bills along with the final acceptance certificate to be obtained by you from our authorised site representative.

All bills are to be submitted in triplicate.

Please sign and return the duplicate copy of this letter, including all annexures (on all pages) as a confirmation of acceptance of this order. Also please forward your detailed Bar Chart for completion of the above work latest by 28th February '94 and arrange to start work immediately so as to ensure completion by 30th June 1994.

Thanking you,

Yours faithfully,
for THAPAR HYDROCONSULT

COL. B.B. SHARMA
MANAGER ENGINEERING.

Encl: 1. General Terms and conditions   —Annexure I
      2. Scope of Work   —Annexure II
      3. Drawing showing location and drill holes —Annexure III

I hereby accept the above Work Order

For RAWEL SINGH & CO

ANNEXURE I: GENERAL TERMS AND CONDITIONS

1.0 Definition

The following words and expressions shall have the meaning herein assigned to them, unless there is anything in the subject or context inconsistent with the contents of this work order.

1.1 'THC' shall mean Thapar Hydroconsult Head Office at Thapar House, 124 Janpath, New Delhi-110 001, India and shall include its successors and assigns.

1.2 'Contractor' shall mean M/s. Rawel Singh & Co, a Company having its Registered Office at Bela Road, Near Parmar Hospital, Ropar-140 001 (Pb) and include its legal representatives, successors and permitted assigns.

1.3 'Works' shall mean and include all permanent and temporary works to be executed, all items and things to be supplied/done and services and activities to be performed by the contractor pursuant to and in accordance with the order and shall not include the contractor's construction equipment, its related spares and wearing parts.

1.4 The 'Order' shall mean the order placed by THC on the contractor and duly accepted by the contractor by appending his signature on the duplicate copy for the execution of works together with all the documents annexed/attached thereto.

1.5 'Site' shall mean the land and other places on, under, in or through which the permanent works or temporary works of the contract are to be executed and any other land and places provided by THC for work space or for any other purpose, for the performance of the contract.

1.6 'Sub-contractor' shall mean any person or firm or company (other than the contractor) (and the legal representatives, successors and permitted assigns of such person or firm or company) engaged with prior written consent of THC/Engineer-in-Charge.

1.7 'Approved' shall mean approval in writing including subsequent written confirmation of previous verbal approval.

1.8 'Final Acceptance Certificate' shall mean the certificate(s) to be issued by the authorised representative of THC to the contractor.

1.9 'Effective Date' shall mean the date on which the order is issued.

1.10 'Project' shall mean the drilling of exploratory holes in line with the scope of work defined elsewhere in this order.

1.11 'Contractor's "Bar Chart"' shall mean the programme showing the order of sequence and periods of completion of various activities according to which the contractor intends to carry out the works within the period of completion stipulated in the order for completion of all the works included in the scope of this order.

2.0 Interpretation of the Order

2.1 Special order Conditions, if any, shall be read in conjunction with the General Terms and Conditions and specifications of work, drawings and other documents forming part of this order wherever applicable and permissible or the context so requires.

2.2 Wherever it is mentioned that the contractor shall perform certain work or provide certain facilities, this shall be understood to mean that the contractor shall do so at his own cost and the order price shall be deemed to have included cost of such performances and provisions so mentioned.

2.3 The materials, design and workmanship shall satisfy the applicable standards, specifications contained herein and codes referred to. Where the technical documentation stipulates requirements in addition to those contained in the standard codes and specifications, those additional requirements shall also be complied with.

3.0 Assignment and Subletting

3.1 Except as provided hereinafter, no part of the order or any share or interest therein shall, in any manner or degree, be transferred, assigned, or sublet by the contractor directly or indirectly to any person, firm or company whatsoever without the prior consent in writing of THC, for which the contractor shall give a written request to THC at least 30 days in advance of the proposed date of transfer/assignment/subletting.

3.2 Notwithstanding any transfer, assignment or subletting with the approval of THC as aforesaid, the contractor shall be and shall remain solely responsible and liable to THC for the quality, proper and expeditious execution and performance of the work and for due performance and observance of all the conditions of the contract in all respects, as if such transfer, assignment or subletting had not taken place and as if such work had been done directly by the contractor.

3.3 No action taken by THC under this article shall absolve or relieve in any manner whatsoever the contractor of any of his liabilities and obligations including 'period for completion' or give rise to any right to compensation/extension of time or otherwise.

4.0 Law and Jurisdiction

4.1 The order shall be construed and interpreted in accordance with and governed by the Laws of Union of India.

4.2 In respect of all matters or actions arising out of the contract and which may arise at any time, the courts at Delhi shall have exclusive jurisdiction.

5.0 Contractor's Employees

5.1 The Contractor shall provide and employ on the site in connection with the execution of the Work:
   a) Only such technical personnel as are skilled and experienced in their respective callings and such subagents, foremen and leading hands as are

competent to do or give proper supervision to the work they are required
to perform or supervise, and

b) Such skilled, semi-skilled and unskilled labour as is necessary for the proper
and timely execution of the work.

6.0 Liability & Insurance

6.1 The contractor shall be liable to THC for the performance of the services in
accordance with the provisions of the contract and any loss suffered by THC
as a result of default of the contractor in such performance will be borne by
the contractor.

6.2 The contractor shall keep THC fully and effectively indemnified against all
losses, damage, injuries, deaths, expenses, actions, proceedings, demands, costs
and claims, including but not limited to legal fees and expenses suffered by
THC or any Third Party when such loss, damage, injury or death is the result
of a wrongful action, negligence or breach by the contractor or his associate
subcontractors, or the personnel or agents of either of them, including the use
or violation of any copyright work or literary property or patented invention,
article or appliance.

6.3 The following types of insurance shall constitute the minimum insurance in
regard to the above requirements:

a) **Contractor's Plant and Machinery Policy**
The contractor's plant and machinery policy shall cover the full replacement
value of the plant and machinery. The policy shall also be endorsed to cover
plant and machinery working underground.

b) **Motor Vehicle Insurance Policy**
The cars and commercial vehicles policy will cover direct physical damage
to the vehicle as well as liability to third parties.

c) It shall apply to all authorised drivers who possess a valid driving licence
in India.

d) **Workman's Compensation Policy**
The contractor will be responsible to extend Workman's Compensation for
his employees.

7.0 If the contractor or his subcontractor or their employees shall break, deface or
destroy any property belonging to THC and others during the execution of the
work, the same shall be made good by the contractor at his own expense.

7.1 The contractor shall be responsible for all arrangements including pumping
of all water requirements for all work sites and work purposes, contractor's
colony/camp etc.

8.0 Obligations of THC

8.1 THC shall be responsible for making available the site as required for the
execution of the works on an 'as is where is' basis, free of rights of third
parties.

8.2 THC shall provide the topographical, geotechnical, hydrological and other technical information as is available with them, excluding restricted data, for the performance of the contractor's obligations under this order.

8.3 All fossils, coins, articles of value or antiquity and structures and other remains or things of geological or archaeological interest discovered on the site of the work shall as between THC and the contractor be deemed the absolute property of THC. The contractor shall take reasonable precautions to prevent his workmen or any other persons from removing or damaging any such article or thing and shall immediately upon recovery thereof and, before removal, inform THC of such discovery and carry out, at the expense of THC, disposal of the same.

9.0 Clearance of Site on Completion

On completion of the work, the contractor shall hand over on an 'as is where is' basis roads and masonry constructions, excluding any removable fixtures or fittings established by the contractor at the site, as required by THC at no extra cost to THC and the contractor shall clear away and remove from the site all remaining construction equipment, surplus materials, rubbish and temporary work of every kind, and leave the whole of the site and work area clean and in a workman-like condition to the satisfaction of THC. In case the contractor fails to remove/clear the above, THC shall remove/clear the same at the cost of the contractor.

10.0 Labour and Compliance with Labour/Industrial and other Laws:

10.1 The contractor shall make his own arrangements for the engagement of all labour, local or otherwise and provide for the transport, housing, feeding and payment thereof at his own cost and under his own arrangements.

10.2 The contractor shall at his own expense ensure due compliance with all applicable and governing Indian laws including industrial and labour laws, rules and regulations and bylaws both of the Central and State Governments and all other local authorities and shall keep THC harmless and indemnified in respect thereof.

10.3 The contractor shall ensure due compliance with the provisions of the relevant Minimum Wages Act, Payment of Wages Act, Contract Labour (Regulation and Abolition) Act, Workmen's Compensation Act, E.P.F. Act and other labour/industrial Laws in force.

11.0 Acceptance Certificate

11.1 A Final Acceptance Certificate shall be issued to the contractor at his request by THC within thirty days after completion of the work or the date of rectification of outstanding deficiencies/damages/defects, whichever is later.

11.2 The contractor shall be responsible for payment of his income tax and/or surcharge. THC will deduct tax at source against payments as applicable under the law.

12.0 Secrecy

All information, data and drawings furnished/disclosed by THC to the Contractor and all drawings, calculations, models, technical information and the
like supplied by the contractor to THC or THC's representative will be treated
by the contractor and his agents, subcontractors and employees as confidential, and shall not be used by any of them without the previous written consent
of THC, except in connection with the execution, operation and maintenance
of the work.

ANNEXURE II: SCOPE OF WORK

1. The scope of drilling constitutes 9 to 10 bore holes with a minimum 70% core recovery in sound rock and a minimum 60% core recovery in other rocks ranging from 20 to 150 m deep at UHL-III Hydroelectric Project, Distt. Mandi, H.P. as per drawing, showing location of drill holes, and attached Annexure III.

2. The drilling machine to be deployed will be of suitable capacity so as to finish the holes in NWX size to entire depth. Reduction of boreholes up to Bx size where necessary is permissible.

3. As the formations comprise weaker/friable and fragmented rocks, double-tube barrels NWG, NWL and NWM will be utilised to obtain the highest core recovery. In case of non-observance of these precautions and washing away of the core of any borehole, the contractor will be required to drill alternative holes at adjacent sites at his own cost. Wherever necessary uncapping of boreholes will have to be done by the contractor to successfully finish bore inlets to the targeted depths.

4. The drill machines will be suitably anchored in concrete/sleeper foundation so that alignment of the boreholes is not disturbed due to incidental vibrations of the drilling operation.

5. Arrangement of water for performance of drilling operations and water-percolation tests will have to be made by the contractor at drill sites from existing nallahs and water springs by deployment of suitable pressure pump/pipelines/mechanical packers/water meters/piezometers.

6. The cores will be stored and preserved in wooden core boxes to be supplied by the contractor. Sludge as collected for non-coreable zones will be stored and preserved at proper places in the core box to observe their stratification and geological features. Arrangement for wash boring, i.e., drive hammer, drive pipe, drive pipe shoe, wash-boring pipe will be the responsibility of the contractor at no extra cost to THC.

7. Shifting of the drill machines from one site to other will be done by the contractor at his own cost. THC will reimburse the amount specified in the Schedule of Rates.

8. Borehole losses, if any, due to jamming of drill/casing/drive pipe strings and their non-retrievance or due to any other reason will be borne by the contractor.

9. Work done on any unfinished/abandoned hole will not be considered for payment by THC.

10. Any additional work, if required by THC, will be carried out by the contractor at the same applicable rates and conditions as already agreed to as per this order.

11. The contractor shall provide a final report which is to contain borehole logs with details in a format approved by THC. The report will be submitted along with recovered material, such as cores, sludge etc.

12. Details of the boreholes are as under:

   i) The drill hole in the centre of Khaddar nallah along the tailrace tunnel alignment shall extend 2 below the invert level of the tailrace tunnel. If rock is not encountered up to that depth, then the hole shall be taken to a depth of minimum 5 m in rock. Water-percolation tests are to be performed in the hole for every 3-m segment. Expected depth is 50 m.

   ii) The depth of hole in the Khaddar terrace on the left bank of the Khaddar nallah along the tailrace tunnel alignment shall also extend up to 5 m in rock. Water-percolation tests shall be performed for every 3-m segment. Expected depth is 50 m.

   iii) A drill hole at the centre of the power-house will be drilled through the existing exploratory trench. The drill hole shall extend 5 m below the draft tube level. The groundwater table shall be observed in this hole and water-percolation tests shall be performed. Expected depth is 40 m.

   iv) One borehole in Thana nallah along the alignment of Khaddar tunnel shall extend 3 m below the invert level of the tunnel. Expected depth is 50 m.

   v) One borehole on the slope of the hill along the Bhabhuri Dhar tunnel alignment shall extend 2 m below Bhabhuri Dhar tunnel invert level. Expected depth is 30 m.

   vi) Two shallow boreholes located on either contact of Palampur thrust with expected depth of 20 m each shall extend 2 m in rock. It is intended to delineate the contact of the thrust.

   vii) Two holes shall be drilled at Bhabhuri Dhar tunnel intake. One hole shall be 50 m inside from the right bank of Ranakhad and the other 50 m towards Ranakhad from the first outcrop of rock seen about 300 m from the intake face of the tunnel. Expected depth of holes is estimated to be 50 m and 150 m respectively.

Both these drill holes shall be drilled 5 m in rock or 5 metres below the invert level tunnel, whichever is less. The groundwater table will be observed with the help of a piezometer.

   viii) One drill hole is to be bored in the middle of Ranakhad along the axis of the intake alignment. The drill hole will extend 5 m in rock or to a depth of 20 m, whichever is less. Water-percolation tests will be performed in this borehole.

RAWEL SINGH & CO.

Bela Road, Near Parmar Hospital, Ropar-140001 (Pb.)

The General Manager,
26, Community Centre,
Basantlok, Vasant Vihar,
New Delhi - 110057

Sub: Acceptance of order for drilling of exploratory holes at project site for UHL
Hydroelectric Project stage-III in Himachal Pradesh.
Ref: Your order No. 102-1 (F): 94 dated 18, February, 94.

Sir,
We take pleasure in accepting the order referred above and thank you for the same. We further assure you that we shall complete the work within time schedule of order and to your entire satisfaction.
Thanking you and assuring our best services.

DA/Order returned duly signed

BAR CHART

A bar chart shows the time schedule programme for the work of exploratory/production drilling monthwise. This becomes an essential annexure for any contractual documentation so that the contractor feels obliged to get the job executed timely. The bar chart showing the time schedule programme for the work of exploratory holes at the project site for UH1 Hydroelectric Project Stage III in Himachal Pradesh (India) is depicted in Fig. 19.1.

In this chart mobilisation and demobilisation periods are also shown along with those of permeability tests.

RECORDING

Information, samples and cores are of little value to anyone if not properly recorded or marked for positive identification. It is extremely important that all required data be properly recorded as it is obtained. Samples and cores must be promptly labelled, packed and stored in accordance with the job specifications.

Borehole logs, strata sheets, and daily reports should include everything accomplished during the work shift. A strata record is needed whenever a drill is used to drill rock for quarry or mineral exploration. It is very important that the information submitted on the record be accurate and correct in every respect. It must include a brief but complete description of the strata; the core recovered, the angle of the bedding or vein as indicated on the core; whether fault zones, cavities, caving zones, or other unusual features were encountered, and if so, at what depths; the number of surveys taken; the location of wedges and of casing or tools left in the hole; and the signature of the driller (see Fig. 19.2).

PRESSURE AND PERMEABILITY TESTS

It is desirable to know whether a soil or rock formation will permit the passage of water and, if so, how much and where the leakage will occur.

Pressure tests are normally required in investigations of proposed dam sites, reservoir locations and other projects in which fluid pressures above or below normal are to be created. They are usually conducted by the drill crew, often periodically as the hole is advanced. A typical equipment set-up is illustrated in Fig. 19.3.

Equipment for pressure testing consists of a pump capable of delivering a specified volume of water at a specified maximum pressure, in accordance with job specifications; an accurate water meter which provides a flow reading in litres, also normally requested in the specification; hoses and valves to permit the measured water flow to be directed into the hole or bypassed, and to permit retaining the water in the hole under specified pressure; an accurate pressure gauge of adequate capacity for the pressures required; approved packers—single or double—for isolating any specified section of the hole; and an adequate supply of clean water.

Packers of various types are available. When holes are drilled in rock formation retaining their drilled size, leather cup/rubber packers can be conveniently used. Pneumatic packers duly inflated with compressed air can be used in all types of rock.

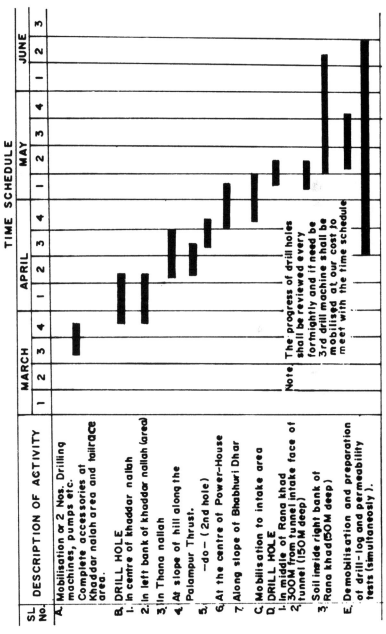

Fig. 19.1: Sample for chart.

GEOLOGICAL AND TECHNICAL DATA LOG OF DRILL BORE HOLE

BORE HOLE NO. BH
SHEET NO.

1 PROJECT   2. LOCATION

3 FEATURE   4. CO-ORDINATES

5 GROUND LEVEL EL.   6 COLLAR EL

7 ANGLE WITH HORIZONTAL   8. TOTAL DEPTH OF HOLE   9. STARTED ON

10 COMPLETED ON   11. TYPE AND MAKE OF MACHINE

12 CLIENT   13. DRILLING AGENCY

LITHOLOGY		SCALE (METRE)	CORE RECOVERY/DRILL /DRIVE RUN (%)	ROCK QUALITY DESIGNATION (%)	SIZE OF CORE PIECES					SIZE OF CASING	SIZE OF HOLE	SIZE OF BIT USED	DEPTH OF WATER LEVEL (M)		DRILL WATER LOSS IN HOLE DURING DRILLING	COLOUR OF RETURN/ WASH WATER	PERMEABILITY (LUGEON)	RATE OF PENETRATION cm/Hr	REMARKS
	LOG				<10 cm	10-25 cm	25-75 cm	75-150cm	>150 cm				BEFORE	AFTER					
DESCRIPTION OF STRATA																			

Fig. 19.2: Bore hole log data

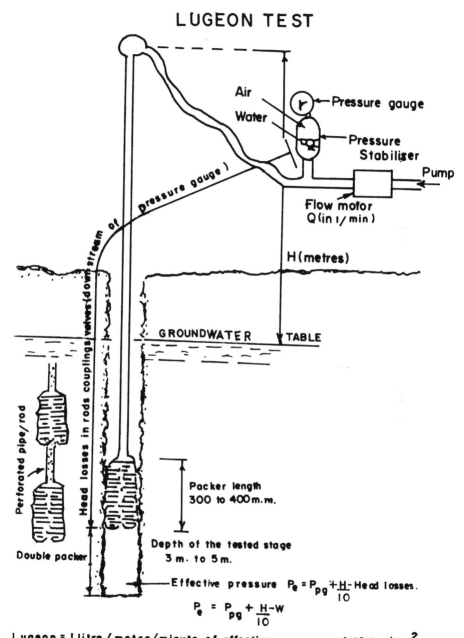

Fig. 19.3: Typical equipment set-up for pressure tests.

IN-SITU PERMEABILITY(WATER PERCOLATION)TESTS DATA   SHEET NO:

1. PROJECT: Hydroelectric project UHL-III2 FEATURES:

3. CO-ORDINATES: 4. DRILL-HOLE NO: 5. TOTAL DEPTH OF DRILL HOLE:

6. GROUND LEVEL EL: 7. COLLAR EL: 8 GROUNDWATER EL :

9. STARTED ON: 10.COMPLETED ON: 11 SIZE OF HOLE

12 TESTING AGENCY

Test Section From	To	Meter Reading of Water intake (Litre) Initial	After 5 Minutes	After 10 Minutes	After 15 Minutes	Water intake Litre/5 Minutes First 5 Minutes	Second 5 Minutes	Third 5 Minutes	Average of Last two	Water intake L/M of hole	Water level below GL(M) Before	After	Water pressure of collar kg/cm² or d.s.l	Height of water Swivel(M)	LUGEONS	Feet/Year	Cm/sec	Size of hole/Size of casing/packer
1	2	3	4	5	6	7	8	9	10	11	12	13	14	15	16	17	18	
1.50	3.00	40	56	71	86	16	15	15	15.0	2.00	—	1.55	0.5	1.90	17.2	87.2	8.43X10⁻⁵	HX CASING/
		95	117	138	160	22	21	22	21.5	2.87			1.0			85.8	8.29X10⁻⁵	HX SINGLE
REBOUND		170	198	225	252	28	27	27	27.0	3.60			1.5			81.5	8.15X10⁻⁵	PACKER
		260	278	295	312	18	17	17	17.0	2.27			1.0			67.9	6.56X10⁻⁵	
		320	332	343	355	12	11	12	11.5	1.53			0.5			66.8	6.45X10⁻⁵	
3.00	4.50	410	425	437	450	13	14	13	13.5	1.80	—	3.60	0.5	2.0	16.2	74.8	7.23X10⁻⁵	— DO—
		455	474	492	510	19	18	19	18.0	2.40			1.0			72.6	7.01X10⁻⁵	
REBOUND		518	541	563	589	25	24	24	24.0	3.20			1.5			78.5	7.38X10⁻⁵	
		606	610	633	650	16	17	17	17.0	2.27			1.0			68.6	6.63X10⁻⁵	
		665	676	686	696	11	10	10	10.0	1.33			0.5			55.3	5.34X10⁻⁵	

Fig. 19.4: Sample record of in-situ permeability (water percolation) tests.

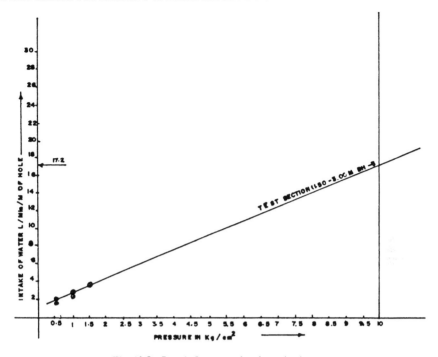

Fig. 19.5: Sample Lugeon value determination.

Placing the packers requires lowering them to the desired location and then expanding them. Double packers require a perforated pipe or rod between the packers to permit water to escape.

Pump, water meter, gauge and valves should be assembled and tested before placing packers. When the packers are in place, the pump discharge should be diverted into the packer rods or pipe gradually, until the specified pressure is indicated on the gauge.

Water-meter reading and exact time should be noted. Pressure should be maintained for the specified time of the test and a water-meter reading again taken to establish the amount of 'take'. All this information should be recorded on the Pressure Test Report.

If 'holding' tests are required, the specified pressure should again be developed and the valve on the line should be closed. Pressure drop and time should be recorded in the appropriate spaces on the test report. Be sure the pressure gauge is placed between the valve and the hole.

After all required tests are completed in one zone, reposition the packers for testing the next section. Repeat the procedure as required until the entire hole is tested.

Occasionally, in 'seamy' or weathered rock, the rate of loss will be so great that the maximum volume of water required will escape without developing the required pressure. Record the volume and pressure attained.

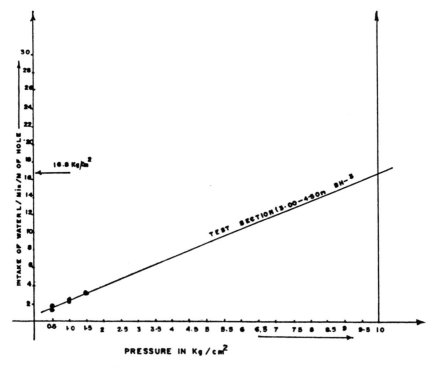

Fig. 19.6: Sample Lugeon value determination.

The data of *in-situ* permeability percolation tests of borehole 3 of UHL Stage-III Hydro-Electric Project, Jogindernagar carried out in June 1994 is given in Fig. 19.4.

Lugeon: This specifies the coefficient of permeability of the material. It is defined as the intake of water in litres per minute and per metre section of hole. During the test the pressure is slowly increased up to limit that upheaval does not occur. As illustrated in Fig. 19.4 pressure has been increased from 0.5 kg/sq. cm with increments of 0.5 kg/sq. cm and up to 1.5 kg/sq. cm. Here the rock type has layers of sandstone and conglomerate. To calculate Lugeon value the graph is plotted between pressure (kg/sq.cm) and intake of water in L/min/m of the hole. A straight line is obtained. When corresponding to the pressure value of 10 kg/sq. cm, the value of intake of water is noted and thus is called the Lugeon value. In Fig. 19.5 and 19.6 Lugeon values are 17.2 and 16.8 to depths of 1.5–3 m and 3–4.5 m respectively. Further, according to the following Lugeon values the types of strata are indicated:

< 1	Totally impermeable (as per design requirement)
1 to 5	Impermeable
5 to 20	Semi-pervious
> 50	Open/free flow.

INDEX

Printed in India

Printed and bound by CPI Group (UK) Ltd, Croydon, CR0 4YY

23/10/2024

01777667-0016